Medicinal Plants and the Rhizosphere

Medicinal Plants and the Rhizosphere

Edited by

Manoj Kumar Mishra
Department of Biotechnology
S.R. Institute of Management and Technology
Lucknow, Uttar Pradesh, India

and

Shalini Tiwari
Division of Natural Sciences and Mathematics
University of the Ozarks, Clarksville
AR 72830, USA

CABI

CABI is a trading name of CAB International

CABI
Nosworthy Way
Wallingford
Oxfordshire OX10 8DE
UK

Tel: +44 (0)1491 832111
E-mail: info@cabi.org
Website: www.cabi.org

CABI
200 Portland Street
Boston
MA 02114
USA

T: +1 (617)682-9015
E-mail: cabi-nao@cabi.org

A catalogue record for this book is available from the British Library, London, UK.

ISBN-13: 9781836990451 (hardback)
9781836990468 (ePDF)
9781836990475 (ePub)

DOI: 10.1079/9781836990475.0000

Commissioning Editor: Rebecca Stubbs
Editorial Assistant: Emma McCann
Production Editor: Rosie Hayden

Typeset by Exeter Premedia Services Pvt Ltd, Chennai, India
Printed in the USA

Contents

Contributors

Abhishek Awasthi: School of Studies in Biotechnology, Shaheed Mahendra Karma Vishwavidyalaya, Jagdalpur, Bastar-494001, Chhattisgarh, India. Email: drabhishekawasthi28@gmail.com

Ahmad Mahmood: School of Interdisciplinary Engineering and Sciences (SINES), National University of Sciences and Technology (NUST), Islamabad, Pakistan. Email: ahmad.mahmood@sines.nust.edu.pk

Ankita Chaubey: iGlobal Research and Publishing Foundation, B2A, B-Block Extension, Peepal Road, Uttam Nagar, New Delhi-110059, India. Email: drankitachaubey@gmail.com. Orcid: https://orcid.org/0000-0003-0922-461X

Ankita Yadav: Dr B R Ambedkar Center for Biomedical Research, University of Delhi, Delhi 110007, India. Email: ankitaofficial1116@gmail.com

Archana Prasad: Grape Physiology and Breeding Lab, Department of Fruit Tree Sciences, Institute of Plant Sciences, Agricultural Research Organization (ARO) – Volcani Institute, Rishon Lezion 7505101, Israel. Email: archanaambt@gmail.com

Ashutosh K. Pandey: Department of Life Sciences, Sharda School of Bio-Science and Technology, Sharda University, Greater Noida, Uttar Pradesh, India. Email: Aashu.p20@gmail.com

Chongtham Allaylay Devi: Amity Institute of Horticultural Studies and Research, Amity University, Noida, Uttar Pradesh, India. Email: chongthamallaylaydevi@gmail.com

Debasis Mitra: Department of Microbiology, Graphic Era (Deemed to be University), 566/6, Clement Town, Dehradun 248002, Uttarakhand, India. Email: debasismitra3@gmail.com

Dharmsheel Shrivastav: Department of Biotechnology and Microbiology, School of Sciences, Noida International University, Gautam Budh

Nagar-201308, Uttar Pradesh, India. Email: Dharmsheel.shrivastav1@gmail.com. Orcid: https://orcid.org/0000-0002-2022-3290

Gauri Saxena: Department of Botany, University of Lucknow, Lucknow-226007, Uttar Pradesh, India. Email: gaurigupta72@yahoo.com

Karishma Kumari: Department of Agronomy, Horticulture and Plant Science, South Dakota State University, Brookings, SD 57007, USA. Email: Karishma.kumari@jacks.sdstate.edu

Kumar Shreshtha: National Centre for Aquatic Animal Health, Cochin University of Science and Technology, Ernakulam, Kerala 682016, India. Email: krshreshtha13@gmail.com. Orcid: https://orcid.org/0009-0000-0591-4686

Lomas Kumar Tomar: School of Biomedical Engineering and Health Sciences, Shobhit University, Meerut, Uttar Pradesh, India. Email: lomas.iitd@gmail.com

Manoj Kumar Mishra: Department of Biotechnology, S.R. Institute of Management and Technology, Lucknow, Uttar Pradesh, India. Email: manoj30biotech@gmail.com. Orcid: https//orcid.org/0000-0001-7836-6254

Md Tareq Bin Salam: Soil, Water and Environmental Discipline, Khulna University, Khulna, Bangladesh; Department of Biochemistry and Molecular Biology, Oklahoma State University, Stillwater, OK 74075, USA. Email: tareqss@swe.ku.ac.bd

Meenu Singh Sengar: Department of Agronomy, Horticulture and Plant Science, South Dakota State University, Brookings, SD 57007, USA. Email: meenu.sengar@sdstate.edu

Muhammad Arshad: Department of Environmental Sciences, University of Yamanashi, Kofu 400-8510, Yamanashi, Japan. Email: marshad@iese.nust.edu.pk

Neha Verma: Department of Biotechnology, Meerut Institute of Engineering and Technology, Meerut 250005, India. Email: neha.emo@gmail.com. Orcid: https://orcid.org/0000-0002-1369-0931

Periyasamy Panneerselvam: Crop Production Division, ICAR–Central Rice Research Institute, Cuttack 753006, India. Email: panneerselvamicar@gmail.com

Pooja Singh: Department of Biotechnology and Microbiology, School of Sciences, Noida International University, Gautam Budh Nagar, Uttar Pradesh2 01308, India. Email: pooja.singh1@niu.edu.in

Pradeep Kumar: Department of Agricultural and Biosystems Engineering, South Dakota State University, Brookings, SD 57007, USA. Email: Pradeep.Kumar5170@jacks.sdstate.edu

Pragati Saxena: Department of Biotechnology and Microbiology, School of Sciences, Noida International University, Gautam Budh Nagar, Uttar Pradesh 201308, India. Email: pragati.saxena@niu.edu.in

Pranesh Prabhu: Department of Biotechnology, Institute of Engineering and Technology, Lucknow 226021, India. Email: praneshprabhuwork@gmail.com. Orcid: https://orcid.org/0000-0001-8963-8307

Puthem Robindro Singh: ICAR–Directorate of Onion and Garlic Research, Pune, Maharashtra 410505, India. Email: puthemrobindrosingh@gmail.com

Rahul Kumar: Department of Microbiology, Graphic Era (Deemed to be University), 566/6, Clement Town, Dehradun 248002, Uttarakhand, India. Email: rahulkumar.phdmb@gmail.com

Rajat Awasthi: Department of Biotechnology, Institute of Engineering and Technology, Lucknow 226021, India. Email: rajatawasthi.lko@gmail.com. Orcid: https://orcid.org/0009-0007-9208-6698

Rajesh Dangal: Department of Agricultural and Biosystems Engineering, South Dakota State University, Brookings, SD 57007, USA. Email: Rajesh.dangal@jacks.sdstate.edu

Renu Verma: Department of Biotechnology, Faculty of Engineering and Technology, Rama University, Mandhana, Kanpur, Uttar Pradesh 209217, India. Email: renuverma.fet@ramauniversity.ac.in

Ryota Kataoka: Department of Environmental Sciences, University of Yamanashi, Kofu 400-8510, Yamanashi, Japan. Email: rkataoka@yamanashi.ac.jp

Sakshi Anand: Department of Biotechnology, Jaypee Institute of Information Technology, Noida, Uttar Pradesh 201309, India. Email: Anand.sakshi64@gmail.com

Samreen Abbas: Department of Biotechnology, Institute of Engineering and Technology, Lucknow 226021, India. Email: abbsam1996@gmail.com. Orcid: https://orcid.org/0009-0002-5985-5356

Saniya Shakir: Department of Biotechnology, Institute of Engineering and Technology, Lucknow 226021, India. Email: er.saniya.shakir4@gmail.com. Orcid: https: orcid.org/0009-0003-2973-215X

Saurabh Gangola: Department of Microbiology, Graphic Era University, Dehradun, Uttarakhand 248001, India. Email: Saindsaurabh@gmail.com

Sergio de los Santos Villalobos: Instituto Tecnológico de Sonora, 5 de febrero 818 sur, Ciudad Obregón 85000, Sonora, Mexico. Email: dlsantosv@gmail.com

Seyit Uguz: Biosystems Engineering, Faculty of Agriculture, Uludag University, Bursa, Turkey. Email: seyit@uludag.edu.tr

Shalini Tiwari: Division of Natural Sciences and Mathematics, University of the Ozarks, Clarksville, AR 72830, USA. Email: stiwari@ozarks.edu

Shilpy Singh: Department of Biotechnology and Microbiology, School of Sciences, Noida International University, Gautam Budh Nagar, Uttar

Pradesh 201308, India. Email: shilpy.singh24@gmail.com. Orcid: https://orcid.org/0000-0003-2274-9090

Shiv Rattan: Department of Ornamental Plants and Agricultural Biotechnology, Agricultural Research Organization (ARO) – Volcani Institute, Rishon Lezion 7505101, Israel. Email: shivrattan.negi@gmail.com

Shraddha Sahu: Department of Biotechnology, Faculty of Engineering and Technology, Rama University, Mandhana, Kanpur, Uttar Pradesh 209217, India. Email: drshraddha.fet@ramauniversity.ac.in

Sourav Chattaraji: Centre for Industrial Biotechnology Research, School of Pharmaceutical Sciences, Siksha 'O' Anusandhan (Deemed to be University), Bhubaneswar, Odisha, India. Email: souravchattaraj@soa.ac.in

Syeda Irsa Mazhar Kazmi: Soil, Water and Environment Discipline, Khulna University, Khulna 9208, Bangladesh. Email: Irsa.mazhar@sines.nust.edu.pk

Udit Yadav: School of Sciences, IFTM University, Moradabad 244102, India. Email: udityadav48@gmail.com

Vandana Anand: Department of Botany, Institute of Science, BHU, Varanasi, India. Email: vandanaanand1987@gmail.com

Varun Kumar Sharma: Department of Biotechnology and Microbiology, School of Sciences, Noida International University, Gautam Budh Nagar, Uttar Pradesh 201308, India. Email: varungenetics@gmail.com. Orcid: https://orcid.org/0000-0001-8575-6939

Ved Vrat Verma: Department of Biotechnology, School of Engineering and Technology, Noida International University, Gautam Budh Nagar, Uttar Pradesh 201308, India. Email: vedvratverma@gmail.com

Waleed Asghar: Department of Biochemistry and Molecular Biology, Oklahoma State University, Stillwater, OK 74075, USA. Email: Waleedasghar978@gmail.com

Xufei Yang: Department of Biosystems and Agricultural Engineering and Animal Science, Michigan State University, East Lansing, MI 48824, USA. Email: yangxuf1@msu.edu

Zain-Ul Abideen Usmani: Department of Agricultural and Biosystems Engineering, South Dakota State University, Brookings, SD 57007, USA. Email: ZainulAbideen.Usmani@sdstate.edu

Zhongli Chen: Key Laboratory of the Three Gorges Reservoir Region's Eco-Environment, Chongqing University, Chongqing 400030, People's Republic of China. Email: zhongli.chen@cqu.edu.cn

Preface: A Comprehensive Understanding of Medicinal Plant Microbiomes

The relationship between plants and their associated microbial communities represents one of the most profound and complex interactions in nature. Over the past two decades, our understanding of these interactions has undergone a dramatic transformation, driven by advances in molecular biology, metagenomics, and next-generation sequencing technologies. The field of plant microbiomics has evolved from a descriptive science documenting microbial presence to a mechanistic discipline revealing how microorganisms fundamentally shape plant physiology, metabolism, and disease resistance.

In this perspective, medicinal plants are especially important. Traditional medical systems around the world have been using these herbs for generations to treat a wide range of illnesses. Their production of secondary metabolites, which are complex chemical molecules with a variety of biological actions, has long been credited with their therapeutic value. The synthesis, accumulation, and bioavailability of these medicinally significant chemicals, however, appear to be significantly influenced by the microbes that inhabit various plant compartments, such as the rhizosphere, phyllosphere, and endophytic habitats. An expanded phenotype that characterizes the medicinal plant as a holobiont rather than an isolated organism is represented by the symbiotic networks formed between plants and their microbiota.

In order to compile the most recent information on microbial diversity in medicinal plants, this book brings together top researchers and practitioners from many fields and geographical areas. From the soil–root interface (the rhizosphere) to internal plant tissues (endophytes) and aerial surfaces (the phyllosphere), the chapters thoroughly cover a variety of research areas. Modern research findings are presented in each chapter while being understandable to people who are unfamiliar with these subjects.

Key themes throughout this volume include the following.

- The spatial distribution and ecological organization of microbial communities across different plant microhabitats, including compartments such as roots, stems, leaves, and internal tissues.
- The basic processes that underlie interactions between microbes and plants, such as the function of root exudates, metabolite communication, and the generation of stress responses and plant immunity.
- The importance of microbial variety for secondary metabolism in medicinal plants, including the production of bioactive chemicals, their bioaccumulation, and their potential for therapeutic use.
- The microbiomes of medicinal plants are different from those of non-medicinal species, and these differences are driven by ecological and evolutionary causes.
- The functional roles of specialized microbes, especially endophytic bacteria and fungi, on plant health, stress tolerance, and the synthesis of compounds that are crucial for pharmaceuticals.
- Modern molecular approaches including high-throughput sequencing, metagenomic analysis, metabolomic profiling, and network pharmacology for investigating plant microbiome structure and function.

We believe this book will serve as an essential resource for students, researchers, clinicians, and industry professionals working at the intersection of plant biotechnology, microbiology, phytochemistry, and agricultural science. The information presented here provides a solid foundation for future research into optimizing medicinal plant production through microbiome-informed agricultural practices, developing disease-resistant cultivars, and discovering novel bioactive compounds with therapeutic applications.

We acknowledge the contributions of all chapter authors, whose expertise and dedication have made this comprehensive volume possible. We are particularly grateful to the scientific community for the collective advancement of knowledge in plant–microbe interactions and medicinal plant research.

This volume represents a pivotal moment in plant science, a time when we recognize that plants are never truly singular organisms but rather complex ecological communities where microbial partners are essential for defining phenotype, metabolism, and medicinal potential. We hope this book will inspire continued innovation in understanding and harnessing the power of plant microbiomes for human health and agricultural sustainability.

Dr Manoj Kumar Mishra
Department of Biotechnology
S.R. Institute of Management and Technology
Lucknow, Uttar Pradesh, India

Dr Shalini Tiwari
Division of Natural Sciences and Mathematics
University of the Ozarks, Clarksville
AR 72830, USA

1 Microbial Communities in Different Plant Compartments

Kumar Shreshtha[1], Rajat Awasthi[2], Manoj Kumar Mishra[3] and Neha Verma[4*]

[1]National Centre for Aquatic Animal Health, Cochin University of Science and Technology, Ernakulam, India
[2]Department of Biotechnology, Institute of Engineering and Technology, Lucknow, India
[3]Department of Biotechnology, S.R. Institute of Management and Technology, Lucknow, India
[4]Department of Biotechnology, Meerut Institute of Engineering and Technology, Meerut, India

1.1 Introduction

Plants exist not as solitary organisms but as dynamic metaorganisms composed of the host and an extensive community of associated microbes. The microbiome can consist of bacteria, fungi, archaea, and viruses. Soil microbiomes exist in every plant compartment, from the root–soil interface to internal plant tissues, and to aerial surfaces of the plant (Fig. 1.1).

The overall association between plants and microbes is not incidental – it has evolved over millions of years of coadaptation and cohabitation, thus allowing plants to thrive in many stressful and diverse environments. Plant-associated microbiome consciousness in agriculture, ecology, and biotechnology has created a surge of research in this topic. These microbiomes are the basis of plant growth, health, and productivity, either directly or indirectly. Microbiomes help plants acquire nutrient elements that are otherwise unavailable (e.g. nitrogen and phosphorus), help regulate key plant hormones, suppress pathogens, and enhance plant tolerance to abiotic stressors, including drought, salinity, and temperature extremes. In return, plants provide carbon-rich root exudates, exploitable structural niches within the plant, and metabolic byproducts that assist with the survival and activity of the microbes (Bulgarelli *et al.*, 2013; Turner *et al.*, 2013).

*Corresponding author: neha.emo@gmail.com

(eds M.K. Mishra and S. Tiwari)
DOI: 10.1079/9781836990475.0001

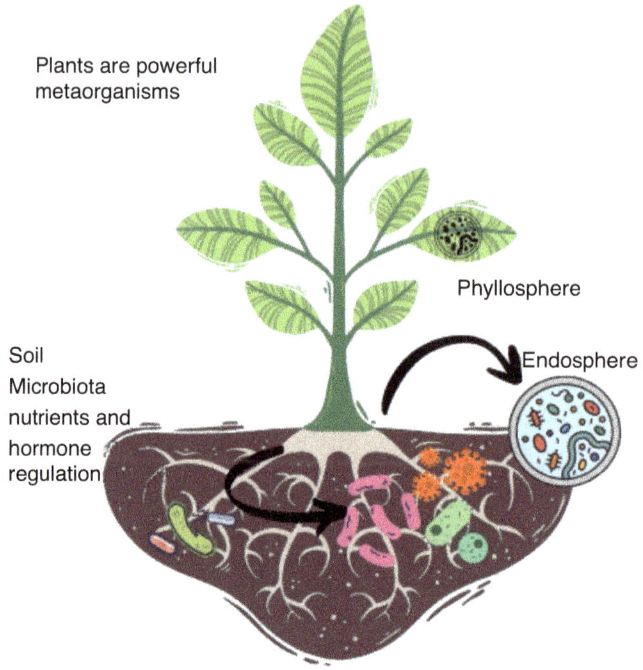

Fig. 1.1. Plant microbiota interactions. The endosphere, microbiota in soil, and phyllosphere increase plant nutrients, stress resistance, defense against pathogens, and sustainability.

Historically, our understanding of plant–microbe interactions was constrained by culture-based techniques that could only account for a small portion of microbial diversity. However, with advancements in high-throughput sequencing technologies and a host of omics techniques such as metagenomics, metatranscriptomics, metaproteomics, and metabolomics, we can look into the intricate multilayered relationships between plants and microbes. These techniques offer researchers an unprecedented opportunity to characterize microbial communities regarding microbial diversity, function, and biological response to environmental and developmental signals (Amann *et al.*, 1995; Fierer and Jackson, 2006).

The plant holobiont, i.e. a plant along with its microbes and environment, has been considered as one system that guides new advances in sustainable agriculture systems. Rather than exclusively working towards strategies to improve the host plant's genetics, there is a growing understanding that enhancing and manipulating the plant microbiome could yield similar, if not better, improvements in crop yield, stress resilience, and ecological sustainability, due to the availability of unique environmental niches that are created differently in each plant compartment, influenced by the plant's physical structure, biochemical gradients within each compartment, and the host plant through signals the plant produces. The rhizosphere is highly populated because the nutrients exuded become available to microorganisms. Conversely, in the endosphere, microbes must overcome host immune systems to achieve stable internal colonization. The phyllosphere is nutrient limited and under environmental stress, yet specialized microbes inhabit plant surfaces under extreme conditions.

Understanding spatiotemporal dynamics is essential to facilitate the development of site-specific microbial applications (Berendsen *et al.*, 2012; Vandenkoornhuyse *et al.*, 2015). The temporal dynamics of the microbiome are also a significant aspect of microbiome biology. Microbial communities do not exist in a vacuum; they fluctuate with plant developmental stages, environmental disturbance, and agricultural practices. Fluctuating plant–microbe–environment interactions provide a nuanced challenge but also an opportunity. Developing universal microbial solutions under fluctuating conditions is more difficult, but available knowledge on specific plant stages can help inform intervention strategies designed to achieve maximum impact. Soil type, pH, moisture, and organic matter are the most well characterized environmental variables that strongly influence microbiome structure, while climate change can present another layer of complexity by introducing new stressors and creating or modifying phenotypic responses of the plant (Pugesgaard *et al.*, 2020; Velásquez *et al.*, 2022). Given these variations, a higher-order systems understanding of plant, microbe, and environment is needed.

Recent research has also begun to identify the roles of microbial secondary metabolite compounds that serve as signaling molecules, antimicrobials or growth regulators. These molecules mediate complicated interspecies interactions and affect plant responses. With metabolomics and synthetic biology advances, we can now utilize these compounds for specific purposes. Additionally, microbiome transplantation, based on principles learned from medical sciences, opens new avenues for exploration. While fecal microbiota transplants are used in human health, soil and root microbiota transplants hold promise for providing disease resistance or improving soil health in agroecosystems. The prospect of designing synthetic microbial consortia for specific crops and in specified environments is equally enticing. These consortia can be assembled with well-characterized strains of bacteria with complementary functions. Delivery systems designed to enhance colonization and persistence will be delivered via seed coatings, root dips, or foliar sprays (Mendes *et al.*, 2013; D'Hondt *et al.*, 2021).

The regulatory framework for microbial products is still not consistent. Regulations and high-quality standards should be created to enhance the success of the product. Farmers and related workers must be educated and informed about effectiveness and economic benefits. Additionally, ethical considerations about bioengineering, biodiversity conservation, and ecological safety must be considered when using and marketing these products. Unintended consequences may arise from interventions made to the microbial community. Because of this, a precautionary strategy founded on rigorous scientific validation will be needed (de Lorenzo *et al.*, 2018; Geisen *et al.*, 2019).

1.2 Plant Compartments and Associated Microbial Niches

Plants house highly specialized microbial communities in different anatomical compartments, subject to host selection pressures and special environmental conditions. The important microbiota niches connected to plants are the rhizosphere, endosphere, and phyllosphere (Fig. 1.2). Knowledge of the structure, dynamics, and functional capacity of microbial communities, all of which support the ecological interactions between plants and microbes, as well as the ability to use these processes in sustainable agriculture, is necessary for an early understanding of these processes (Compant *et al.*, 2019; Berg *et al.*, 2020).

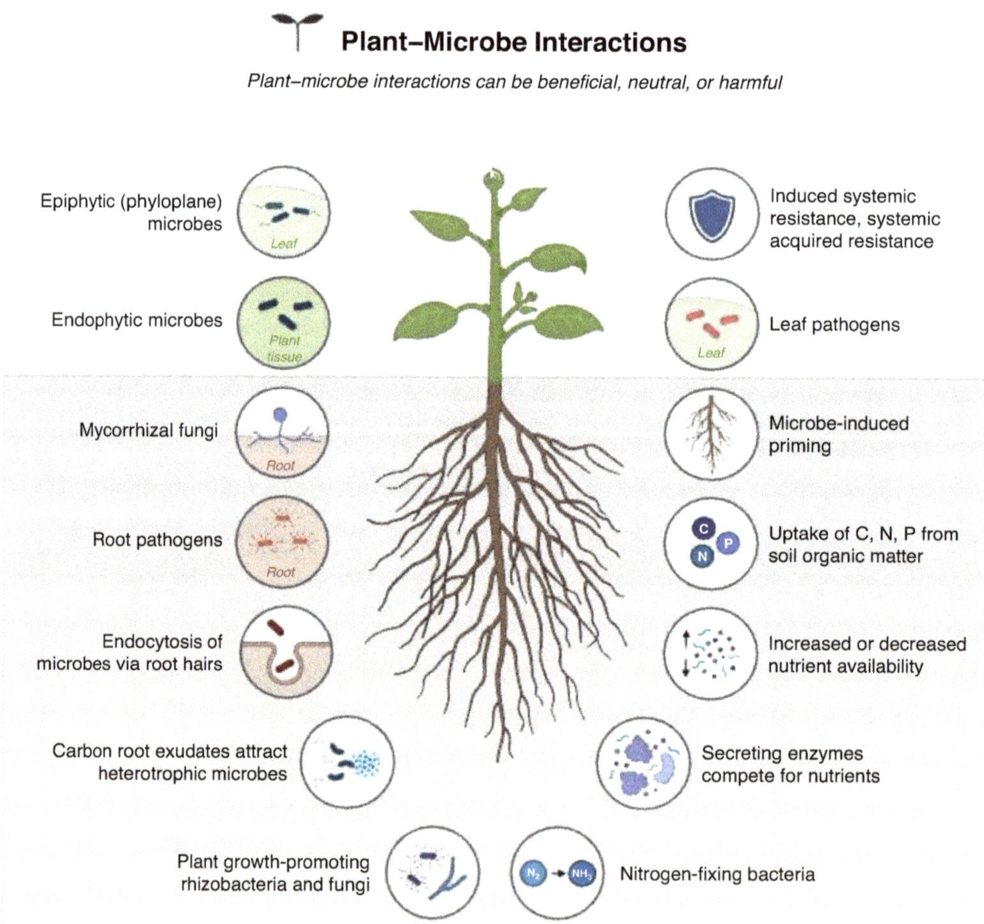

Fig. 1.2. Different regions of the plant compartment.

1.2.1 Rhizosphere

The rhizosphere is the soil region existing between the microbes and the plant roots, with factors including root exudates, mucilage, senescence, and microbial activity, so it is also called the secondary genome of plants. It serves as the interface of dynamic biochemical activities involving nutrient transfer, competition by microbes, signaling, and defense mechanisms (Mendes *et al.*, 2013; Philippot *et al.*, 2013; Shreshtha *et al.*, 2024b).

Exudates from roots act as chemical stimuli for microorganisms and therefore govern the activity and structure of microbial societies. The exudates are rich in low molecular weight organic compounds, including sugars, amino acids, phenolics, and secondary metabolites as potential nitrogen and carbon sources. Correspondingly, plant growth-promoting rhizobacteria (PGPR) and arbuscular mycorrhizal fungi (AMF) induce nutrient solubilization, nitrogen fixation, hormonal regulation, and increased water absorption (Vessey, 2003; Bais *et al.*, 2006).

Prominent bacterial families of the rhizosphere include *Pseudomonas*, *Bacillus*, *Azospirillum*, *Rhizobium*, and *Streptomyces* which have plant growth promotion and biocontrol potential. Mycorrhizal fungi, e.g. *Glomus* and *Rhizophagus*, form mutualistic relationships with roots, increasing the absorption surface and phosphorus acquisition (Smith and Read, 2008; Alori *et al.*, 2017). Plant genotype, growth stage, soil type, and external inputs such as fertilizers and pesticides govern the assembly of the rhizosphere microbiome. Recent evidence has shown that crop rotation, organic amendment, and conservation tillage hold promise for augmenting the rhizosphere's functional robustness and microbial diversity (Bender *et al.*, 2016; Toju *et al.*, 2018). Novel tools emphasizing metagenomics have aided scientists in moving away from taxonomy and exploring microbial metabolic networks and other networks, including resistance and keystone. Multi-omes integration outlines establishing and maintaining microbial networks in the rhizosphere and details quorum sensing, nutritional cycles, and plant pathogen suppression (Fitzpatrick *et al.*, 2018; Jacoby and Kopriva, 2019; Shreshtha *et al.*, 2025).

1.2.2 Endosphere

The endosphere includes internal plant cells with endophytic microbes that live without inducing disease. They can infect the reproductive organs, leaves, stems, and roots through natural openings and wounds and settle inside the internal plant milieu (Rosenblueth and Martínez-Romero, 2006; Hardoim *et al.*, 2008).

The endosphere is a highly selective microenvironment, fueled by plant immunity and biochemical cues. Endophytic fungi and bacteria typically establish mutualistic connections, offering nutrient mobilization, protection against stress, and secondary metabolite assembly benefits. The genera *Herbaspirillum*, *Enterobacter*, *Serratia*, and *Burkholderia* are representative bacterial endophytes. Fungal endophytes such as *Colletotrichum*, *Fusarium*, and

Piriformospora indica have been linked with heightened drought tolerance and protection against pathogens (Ryan *et al.*, 2008; Rodriguez *et al.*, 2009). The endosphere harbors less dense yet highly functionally specialized microbes relative to the rhizosphere. These microbial assemblages can modulate host gene expression and concentrations of hormones and, in selected cases, offer transgenerational benefits. Endophytes are additionally deployed as biological vessels of new compounds or transgenic characteristics, providing future candidates for next-generation biofertilizers and biopesticides (Hardoim *et al.*, 2008; White *et al.*, 2019).

Fluorescence *in situ* hybridization (FISH), single-cell genomics-based experiments, and confocal microscopy improve knowledge about endophytic colonization and host–microbe interactions. However, maintaining constant field conditions challenges endophyte stability and productivity (Stone *et al.*, 2000; Sessitsch *et al.*, 2012).

1.2.3 Phyllosphere

The phyllosphere is the total of all plant parts in the air: stems, fruits, flowers, and primary foliage. This compartment is unfavorable for microbial colonization due to minimal nutrients, drying out, irradiation with UV light, and changing temperatures (Lindow and Brandl, 2003; Vorholt, 2012). It shelters different microbial taxa crucial for plant fitness and ecological interactions. Typical inhabitants of the phyllosphere include bacteria such as *Sphingomonas*, *Methylobacterium*, *Pseudomonas*, and *Micrococcus*. Other fungi like *Cladosporium* and *Alternaria* are ubiquitous (Delmotte *et al.*, 2009; Bringel and Couée, 2015; Africa *et al.*, 2025).

These microorganisms are epiphytes (surface colonizers) or may infect inner plant parts and grow as endophytes. Phyllosphere microorganisms regulate plant physiology through the alteration of stomata activity, the release of volatile organic compounds (VOCs), and the secretion of antimicrobial compounds, reducing foliar pathogens (Ritpitakphong *et al.*, 2016; Remus-Emsermann and Schlechter, 2018.

The focus is increasingly on the atmospheric interaction of phyllosphere microbiomes. Certain phyllosphere bacteria can metabolize atmospheric methane and fix nitrogen and are thus thought to play a wider ecological role than was realized (Jumpponen and Jones, 2010; Sapkota *et al.*, 2015). Also, microbial biofilms on the leaf surface are involved in the plant's water balance and light reflection. The structure of the phyllophere microbial communities is highly influenced by the surrounding environment, leaf shapes, and plant maturity, which means that farming practices like foliar application of biocontrol compounds or excess fertilizers can support or disrupt these communities (Müller *et al.*, 2016; Liu *et al.*, 2020; Shreshtha *et al.*, 2024b).

New technology for precision agriculture has started to deliver helpful microbes to the phyllosphere, enhancing crop resistance and sustainable growth. Combining all regions, i.e. the rhizosphere, endosphere, and phyllosphere, plays a decisive role in plant nutrition and development. Analyzing these regions, interactions, and responses to internal and external stimuli is fundamental to developing sustainable agriculture.

1.2.4 Spermosphere

The spermosphere is a unique and significant microhabitat around a germinating seed. This transient zone appears during early seed germination and serves as a recruitment hotspot for microbes, with recruitment primarily influenced by seed exudates that supply carbon and nitrogenous compounds (Nelson, 2004; Johnston-Monje and Raizada, 2011). Microbial populations of the spermosphere decide seedling vigor, nutrient uptake, and disease resistance at initial stages. Seed exudates such as amino acids, sugars, and organic acids preselectively activate symbiotic microbes, including PGPRs such as *Bacillus*, *Paenibacillus*, and *Azospirillum* (Hardoim *et al.*, 2008; Compant *et al.*, 2019). At the same time, these exudates will also attract seedborne and soilborne pathogens, so the spermosphere becomes a combat zone between pathogenic and beneficial microbes. More recent research has put forward the theory of a "seed microbiome," that is, microorganisms carried within or on the surface of the seed (Truyens *et al.*, 2015; Shade *et al.*, 2017).

These microbes can colonize the spermosphere and seedling tissues early, laying the basis for future rhizosphere and endosphere interactions. Microbial inoculants and biopriming are being researched as seed treatments to modulate the spermosphere microbiome and enhance germination and crop establishment (Johnston-Monje *et al.*, 2016; Nelson, 2018).

1.2.5 Anthosphere and carposphere

The anthosphere, i.e. the flower environment, and carposphere, i.e. the fruit environment, contain specialized microbiomes. Flowers perform a specific function to attract pollinators and emit different volatile compounds, nectar, and lipids, which influence the microbial communities. These microbiomes can alter reproductive growth by influencing pollen viability, stigma receptivity, and seed set (Fridman *et al.*, 2012; Aleklett *et al.*, 2014).

Pseudomonas, *Erwinia*, *Bacillus*, and yeasts are the microbial genera frequently found in the anthosphere. Some of these microorganisms can offer defense against herbivores or flower diseases. Others change pollinator behavior and plant fitness by affecting nectar chemistry and floral scent (Jacquemyn *et al.*, 2013; Aleklett *et al.*, 2014). Additionally, fruits serve as nutrient-dense niches for bacteria and fungi that can break down sugars and organic acids. The carposphere contains many members of the genera *Lactobacillus*, *Acetobacter*, and *Penicillium* (Droby and Wisniewski, 2018). Some microbes are linked to postharvest fermentation processes and the mechanism by which plant seeds spread, and are even accountable for postharvest spoilage. It therefore becomes essential to understand this interaction where pollinator and microbial spoilage affect profits (Janisiewicz and Korsten, 2002; Berg *et al.*, 2014).

1.2.6 Caulosphere

The microbial community associated with the plant's stem, mainly around the surface tissue, is called the caulosphere. This provides the connecting link between the rhizosphere and phyllosphere and the above-ground parts via microbial pathways (Sasse *et al.*, 2018; Compant *et al.*, 2019). It is also essential for disease spread and upward microbial movement, but this is less studied. Aerosol deposition, direct colonization through lenticels or wounds, and upward migration from the rhizosphere are the three ways in which microorganisms can enter the caulosphere (Müller *et al.*, 2016; Ritpitakphong *et al.*, 2016).

Xylella, *Clavibacter*, and *Ralstonia* are among the bacterial species that migrate along the xylem and use the plant's internal transport system to colonize the entire system. However, *Streptomyces* and endophytic *Bacillus* species can also increase plant defense mechanisms and migrate systemically (Jacoby and Kopriva, 2019; Liu *et al.*, 2020), as suggested by new techniques such as sophisticated imaging and DNA stable isotope probing, which enabled the capture of the movement and activity of microorganisms in the stem region. These studies revealed a novel surface of long-distance signaling, microbial translocation, and niche connectivity (Sessitsch *et al.*, 2012; Sasse *et al.*, 2018).

1.3 Interaction and Crosstalk Between Compartments

Plant sections all harbor growth-promoting microbes. First, rhizosphere microbes colonize and then aid in the establishment of endospore and caulosphere bacteria in their areas (Compant *et al.*, 2010; Berg *et al.*, 2014). More importantly, plants' systemic signaling pathways may be manipulated by an incursion of microbial volatiles into the rhizosphere or phyllosphere. Further communication between the compartments can be fostered by reactive oxygen species, secondary metabolites, and phytohormones (Berendsen *et al.*, 2012; Raaijmakers and Mazzola, 2012).

Microbes enter tissues and interact through competitive exclusion, secretion of effector molecules, and quorum sensing. System-level understanding is needed to engineer microbiomes. New developments in spatial omics, single-cell genomics, and synthetic biology shed light on how microbes in different plant regions affect plant performance.

The new spatial microbiome ecology offers instruments to investigate microbial population dynamics along gradients of plant tissue and environmental boundaries (Toju *et al.*, 2018; Wippel *et al.*, 2021).

1.4 Temporal Dynamics Across Compartments

Microbiomes change throughout a plant's life as influenced by the physiology of the plant, human processes, and the surrounding climate. Early microbes prepare conditions for later ones, by modifying or enhancing existing conditions (Chaparro *et al.*, 2013; Fitzpatrick *et al.*, 2018).

While core microbiomes that remain throughout development will probably have more direct effects on plant health, transient microbial consortia can afford some temporary relief (Turner *et al.*, 2013; Bulgarelli *et al.*, 2015).

Time series metagenomics analyses and longitudinal field trials illuminate the temporal dynamics of plant-associated microbial communities in the identification of correct temporal windows within which a particular microbial treatment can be administered to harvest the benefits of enhanced crop resilience, for example bio-inoculation during flowering or fostering drought resistance during early vegetative growth (Massalha *et al.*, 2017; Liu *et al.*, 2020; Shreshtha *et al.*, 2024a).

1.5 Functions of Microbial Communities in Plant Health

Microorganisms affect growth in different parts of the plant. These microbes supply nutrients and biocontrol services to reduce harsh environmental conditions like heavy metals, drought, and salty conditions (Mendes *et al.*, 2013; Shreshtha *et al.*, 2024b). Nutrient acquisition is one of the principal foundations of microbial life associated with plants. Certain microbes improve the bioavailability and absorption of vital elements such as sulfur, potassium, iron, phosphorus, and nitrogen in low-nutrient environments. This involves several microbial processes, such as siderophore synthesis, phosphate solubilization, nitrogen fixation, and the enzymatic breakdown of organic matter (Fierer and Jackson, 2006; Alori *et al.*, 2017).

Microbial activity also directly influences plant development. Microbes promote plant growth by synthesizing phytohormones, including auxins, cytokinins, and gibberellins. Auxins such as indole-3-acetic acid promote lateral root development, which allows the plant to take up more water and nutrients. Certain microbes possess ACC deaminase (an enzyme that decreases the levels of the ethylene precursor ACC) so that stress-induced ethylene in the plant is reduced and root growth is promoted (Vessey, 2003; Bashan *et al.*, 2014).

Microbial defense functions are also vital. Beneficial microbes defend plants against pathogens by competing, inhibiting growth with antibiotics, parasitizing, and activating the immune system. *Bacillus subtilis*, for example, secretes antibiotics and lipopeptides that kill pathogens directly while fungi like *Trichoderma* spp. demonstrate mycoparasitism by secreting enzymes that degrade the cell wall of pathogenic fungi. These beneficial microbes can further trigger induced systemic resistance (ISR) in the plant (Compant *et al.*, 2010; Raaijmakers and Mazzola, 2012).

Microbial communities also play a significant role in plant resistance to abiotic stress. Mycorrhizal fungi project hyphal networks, enhancing the surface area where water may be absorbed. Some microbes synthesize osmoprotectants such as proline and glycine betaine to enable plants to preserve osmotic balance under salinity. Others detoxify heavy metals or synthesize antifreeze compounds under stress (Geisen *et al.*, 2019; Liu *et al.*, 2020).

Microorganisms influence larger ecological processes beyond the single-plant scale. These effects include the formation of soil structure and controlling greenhouse gas flows. The exudates of microorganisms and the network of fungal hyphae are essential for soil aggregation, and bacterial taxa, such as methanotrophs and denitrifying bacteria, participate in greenhouse gas cycling (Van der Heijden *et al.*, 2008; Delgado-Baquerizo *et al.*, 2018).

To summarize, microbial communities are complex and essential to the health of plants. They are a living interface of the plant and the environment, which confers plant adaptation, productivity, and resilience (Hacquard and Schadt, 2015; Berg *et al.*, 2020).

1.6 Microbiome Engineering and Biostimulants

The scientific community and agribusiness are leveraging the increasing understanding of plant microbiomes into possible strategies for designing microbiomes and biostimulant technologies to promote crop health and yield. Reductions in populations of any bacteria or fungi in the microbiome would be brought about by applications of microorganisms or stimulation of native microbiomes (Compant *et al.*, 2019; D'Hondt *et al.*, 2021). Microbiome engineering refers to the adjustment of microbial communities toward pragmatically specific agricultural goals. To offer some resilience to crops, synthetic microbial communities (SynComs) are being developed with traits such as nitrogen fixation, phosphate solubilization, and pathogen suppression (Toju *et al.*, 2018; Wippel *et al.*, 2021).

Another microbiome engineering-level step is to enhance the traits of microbes by genetic techniques and synthetic biology. These methods would use tools like CRISPR-Cas9 to engineer microbes with better colonization and stress tolerance (de Lorenzo *et al.*, 2018; Velásquez *et al.*, 2022). Biostimulants, natural products or microorganisms that promote plant growth and wellbeing, are principles associated with microbiome engineering. Microbial biostimulants include *Azospirillum*, *Bacillus*, *Pseudomonas*, and *Trichoderma* spp. Nonmicrobial biostimulants include humic acids, protein hydrolyzates, and seaweed extracts (Bhattacharyya and Jha, 2012; Bashan *et al.*, 2014). Opportunities in microbiome engineering and biostimulants depend highly on soil type or environment, genotype, and climate. As with all promising technologies, its absolute acceptance should be based on field experiments, surveys, and input from expert agronomists (Geisen *et al.*, 2019; Berg *et al.*, 2020).

1.7 Challenges in Applying Plant Microbiomes

Although the potential of plant microbiomes to transform agriculture is well accepted, their field application is still beset by complications (Table 1.1). These range across biological, technical, ecological, economic, and regulatory domains (Compant *et al.*, 2019; D'Hondt *et al.*, 2021).

One of the significant challenges is the intrinsic variability and complexity of plant–microbe interactions. What is effective under controlled greenhouse conditions tends to fall short in field situations because of soil heterogeneity, climatic variability, and microbial competition (Philippot *et al.*, 2013; Berg *et al.*, 2014).

A second barrier is the host specificity and ecological compatibility of microbial inoculants. Plants preferentially attract microbes with root exudates; even positive microbes can lose this niche if they are not well suited to local conditions (Sessitsch *et al.*, 2012; Wippel *et al.*, 2021). Technical barriers involve producing, stabilizing, and delivering live microbial products. The issues of storage, transportation, and shelf-life create hurdles for wide implementation. Formulation techniques, such as encapsulation, are being

Table 1.1. Key challenges in plant microbiota application.

Challenging area	Description	Implication	References
Biological complications	Variability in plant microbiota integration across habitat	Inconsistency in biological efficacy between greenhouse and field trials	(Chen *et al.*, 2013; Philippot *et al.*, 2013; Paul *et al.*, 2021)
Host selectivity and suitability	Microbial inoculants may not succeed if not adapted to the plant or soil	Inhibit effectiveness across the broad agronomy biosphere	(Berendsen *et al.*, 2012)
Technical hurdles	Problems with storage, distribution, and production	Scalability difficulties in marketable use	(Xu and Shanklin, 2016)
Regulatory and biosecurity issues	Fragmented framework, biosecurity danger of engineered microbiota	Developing safe microbial products, validating them in the field, and regulating them to prevent ecological and biosecurity risks	(Sow *et al.*, 2022)
Financial, ethical, and uptake barriers	Agronomist hesitation, sustainable concerns, education, and information	Developing microbial products, testing them on crops, and showing that they are effective, affordable, and safe	(Xiang and Gao, 2023)

considered but these have limitations when it comes to scale (Bashan *et al.*, 2014; Bender *et al.*, 2016).

In general, regulatory regimes are fragmented and lack uniformity at the regional level. Biosafety issues concerning genetically engineered organisms themselves or unforeseen ecological consequences compound matters regarding regulation (de Lorenzo *et al.*, 2018; Velásquez *et al.*, 2022). Gaining adoption and furthering the economy is still a barrier. Farmers may hesitate to purchase microbiome products due to their irregular behavior and uncertain return on investment (Compant *et al.*, 2019; Geisen *et al.*, 2019).

In addition, ecological and ethical concerns raise the risk of horizontal gene flow or displacement of native microbial populations. Farmers need education and training to trust scientists in bridging the field for application (Hacquard and Schadt, 2015; Berg *et al.*, 2020).

1.8 Multi-omics Approaches in Microbiome Studies

Multi-omics approaches have changed the perspective about understanding and managing microbiomes associated with plants, including metabolomics, proteomics, transcriptomics, and genomics (Turner *et al.*, 2013; Berg *et al.*, 2020). Genomic studies and analyses demonstrate microbial diversity and functional gene content. Through metagenomic sequencing, the unknown taxa and metabolic pathways were discovered (Fierer and Jackson, 2006; Delgado-Baquerizo *et al.*, 2018).

As indicated by Massalha *et al.* (2017) and Fitzpatrick *et al.* (2018), transcriptomics depicts dynamic gene expression, thus showing the responses of microbes to stresses or developmental stages. Proteomics connects genetic potential and function by identifying structural proteins and enzymes involved in plant–microbe interactions (Knief *et al.*, 2012; Müller *et al.*, 2016). This discipline investigates signaling compounds, antimicrobial metabolites, and hormones affecting plant–microbe interactions (Bringel and Couée, 2015; Jacoby and Kopriva, 2019). One would not actualize the full potential of metabolites when integrating this with genome, transcriptome, and proteome data to advocate for systems-based predictions and precision microbiome management (D'Hondt *et al.*, 2021; Wippel *et al.*, 2021). Even if overcoming technical barriers relating to the integration and standardisation of data seems next to impossible, multi-omics opens a broad vista for predictive microbiome science and agriculture (Geisen *et al.*, 2019; Liu *et al.*, 2020).

1.9 Synthetic Biology and Future Innovations

The combined approach of agricultural systems to microbiome research and synthetic biology is innovative. With synthetic biology, microbial genomes can be deliberately manipulated to enhance traits such as antimicrobial activity, nitrogen fixation, and stress resistance (de Lorenzo *et al.*, 2018; Velásquez *et al.*, 2022). Tools such as CRISPR-Cas9 and TALENs are allowing microbial

Table 1.2. Synthetic biology and microbiota are merging to transform sustainable agronomy.

Application area	Key points	Reference
Genome engineering	CRISPR-Cas9 and TALENs allow accurate editing for stress resistance, biogeochemical cycling, and antimicrobial characteristics	(Sow *et al.*, 2022)
SynComs	Designed consortia enhance the use of nutrients, microbial resistance, and stress modification	(Shayanthan *et al.*, 2022)
Climate-smart application	Engineered microbes increase carbon sequestration and lower greenhouse gases, along with purifying soils	(Hacquard *et al.*, 2015)
Biosafety and public concerns	GMO perception, eco-friendly risk, and regulatory concerns impede assumption	(Linquist *et al.*, 2015)
Collective study and biofoundries	Worldwide partnerships are accelerating synthetic biology and microbiome research and supporting sustainable agronomy	(Andreozzi *et al.*, 2019)

GMO, genetically modified organism.

gene editing to create beneficial actions in response to specific plant signals (Compant *et al.*, 2019; D'Hondt *et al.*, 2021). Synthetic microbial consortia, or SynComs, are aimed at replicating or improving natural microbiomes. These consortia could program microbes for pathogen suppression, stress tolerance, or nutrient cycling (Toju *et al.*, 2018; Wippel *et al.*, 2021).

Genome-modified microbes can be involved in climate-smart agriculture for detoxifying contaminated soils, decreasing greenhouse gases, and enhancing carbon sequestration (Van der Heijden *et al.*, 2008; Delgado-Baquerizo *et al.*, 2018). Biosafety and regulation still pose a threat to the development of climate-smart, genetic engineering and there are still public concerns about genetically modified organisms around the world (Hacquard and Schadt, 2015; Berg *et al.*, 2020).

The cooperative research described here, along with practices established in biofoundries, pushes toward improvement. Synthetic biology for microbiomes can lead to an agricultural system that is more sustainable and resilient (Table 1.2) (Compant *et al.*, 2019; D'Hondt *et al.*, 2021).

1.10 Conclusion

Plant health, growth, and resistance depend on microorganisms and their communities. These microorganisms act between the rhizosphere, endosphere, and phyllosphere to mediate nutrient uptake, stress management, and disease protection (Turner *et al.*, 2013; Berg *et al.*, 2020).

If sustainable agriculture is to be actualized, the microcommunity needs to be activated via improvements in multi-omics, microbiome engineering, and synthetic biology (Toju *et al.*, 2018; D'Hondt *et al.*, 2021). However, the real difficulties of ecology, complexity, scalability, and regulatory frameworks impede these factors (Geisen *et al.*, 2019; Velásquez *et al.*, 2022). Therefore, introducing prospects in AI, biosensors, and synthetic biology alliances could revolutionize agriculture and render it resilient to climate change (de Lorenzo *et al.*, 2018; Wippel *et al.*, 2021).

References

Africa, A.J., Setati, M.E., Hitzeroth, A.C. and Blancquaert, E.H. (2025) Exploring the evolution of microbial communities from the phyllosphere and carposphere to the grape must of *Vitis vinifera* L. cv's Chardonnay and Pinot noir. *Food Microbiology* 130, 104780. DOI: 10.1016/j.fm.2025.104780.

Aleklett, K., Hart, M. and Shade, A. (2014) The microbial ecology of flowers: An emerging frontier in phyllosphere research. *Botany* 92(4), 253–266. DOI: 10.1139/cjb-2013-0166.

Alori, E.T., Glick, B.R. and Babalola, O.O. (2017) Microbial phosphorus solubilization and its potential for use in sustainable agriculture. *Frontiers in Microbiology* 8, 971. DOI: 10.3389/fmicb.2017.00971.

Amann, R.I., Ludwig, W. and Schleifer, K.H. (1995) Phylogenetic identification and in situ detection of individual microbial cells without cultivation. *Microbiological Reviews* 59(1), 143–169. DOI: 10.1128/mr.59.1.143-169.1995.

Andreozzi, A., Prieto, P., Mercado-Blanco, J., Monaco, S., Zampieri, E. *et al.* (2019) Efficient colonization of the endophytes *Herbaspirillum huttiense* RCA24 and *Enterobacter cloacae* RCA25 influences the physiological parameters of *Oryza sativa* L. cv. Baldo Rice. *Environmental Microbiology* 21(9), 3489–3504. DOI: 10.1111/1462-2920.14688.

Bais, H.P., Weir, T.L., Perry, L.G., Gilroy, S. and Vivanco, J.M. (2006) The role of root exudates in rhizosphere interactions with plants and other organisms. *Annual Review of Plant Biology* 57, 233–266. DOI: 10.1146/annurev.arplant.57.032905.105159.

Bashan, Y., de-Bashan, L.E., Prabhu, S.R. and Hernandez, J.P. (2014) Advances in plant growth-promoting bacterial inoculant technology: Formulations and practical perspectives (1998–2013). *Plant and Soil* 378(1–2), 1–33. DOI: 10.1007/s11104-013-1956-x.

Bender, S.F., Wagg, C. and van der Heijden, M.G.A. (2016) An underground revolution: Biodiversity and soil ecological engineering for agricultural sustainability. *Trends in Ecology & Evolution* 31(6), 440–452. DOI: 10.1016/j.tree.2016.02.016.

Berendsen, R.L., Pieterse, C.M.J. and Bakker, P.A.H.M. (2012) The rhizosphere microbiome and plant health. *Trends in Plant Science* 17(8), 478–486. DOI: 10.1016/j.tplants.2012.04.001.

Berg, G., Grube, M., Schloter, M. and Smalla, K. (2014) The plant microbiome and its importance for plant and human health. *Frontiers in Microbiology* 5, 491. DOI: 10.3389/fmicb.2014.00491.

Berg, G., Rybakova, D., Grube, M. and Köberl, M. (2020) The plant microbiome explored: Implications for experimental botany. *Journal of Experimental Botany* 71, 6455–6461. DOI: 10.1093/jxb/eraa407.

Bhattacharyya, P.N. and Jha, D.K. (2012) Plant growth-promoting rhizobacteria (PGPR): Emergence in agriculture. *World Journal of Microbiology and Biotechnology* 28(4), 1327–1350. DOI: 10.1007/s11274-011-0979-9.

Bringel, F. and Couée, I. (2015) Pivotal roles of phyllosphere microorganisms at the interface between plant functioning and atmospheric trace gas dynamics. *Frontiers in Microbiology* 6, 486. DOI: 10.3389/fmicb.2015.00486.

Bulgarelli, D., Schlaeppi, K., Spaepen, S., Van Themaat, E.V.L and Schulze-Lefert, P. (2013) Structure and functions of the bacterial microbiota of plants. *Annual Review of Plant Biology* 64, 807–838. DOI: 10.1146/annurev-arplant-050312-120106.

Bulgarelli, D., Garrido-Oter, R., Münch, P.C., Weiman, A., Dröge, J. *et al.* (2015) Structure and function of the bacterial root microbiota in wild and domesticated barley. *Cell Host and Microbe* 17(3), 392–403. DOI: 10.1016/j.chom.2015.01.011.

Chaparro, J.M., Badri, D.V. and Vivanco, J.M. (2013) Rhizosphere microbiome assemblage is affected by plant development. *ISME Journal* 8(4), 790–803. DOI: 10.1038/ismej.2013.196.

Chen, W., Wellings, C., Chen, X., Kang, Z. and Liu, T. (2013) Wheat stripe (yellow) rust caused by *Puccinia striiformis* f. sp. *tritici*. *Molecular Plant Pathology* 15, 433–446. DOI: 10.1111/mpp.12116.

Compant, S., Duffy, B., Nowak, J., Clément, C. and Barka, E.A. (2010) Use of plant growth-promoting bacteria for biocontrol of plant diseases: Principles, mechanisms of action, and future prospects. *Applied and Environmental Microbiology* 71, 4951–4959. DOI: 10.1128/AEM.01452-10.

Compant, S., Samad, A., Faist, H. and Sessitsch, A. (2019) A review on the plant microbiome: Ecology, functions, and emerging trends in microbial application. *Journal of Advanced Research* 19, 29–37. DOI: 10.1016/j.jare.2019.03.004.

de Lorenzo, V., Prather, K., Chen, G.Q., O'Day, E., Kameke, C. *et al.* (2018) The power of synthetic biology for bioproduction, remediation and pollution control. *EMBO Reports* 19, e45658. DOI: 10.15252/embr.201845658.

Delgado-Baquerizo, M., Oliverio, A.M., Brewer, T.E., Benavent-González, A., Eldridge, D.J. *et al.* (2018) A global atlas of the dominant bacteria found in soil. *Science* 359(6373), 320–325. DOI: 10.1126/science.aap9516.

Delmotte, N., Knief, C., Chaffron, S., Innerebner, G., Roschitzki, B. *et al.* (2009) Community proteogenomics reveals insights into the physiology of phyllosphere bacteria. *Proceedings of the National Academy of Sciences USA* 106(38), 16428–16433. DOI: 10.1073/pnas.0905240106.

D'Hondt, K., Kostic, T., McDowell, R., Eudes, F., Singh, B.K. *et al.* (2021) Microbiome innovations for a sustainable future. *Nature Microbiology* 6(2), 138–142. DOI: 10.1038/s41564-020-00857-w.

Droby, S. and Wisniewski, M. (2018) The fruit microbiome: A new frontier for postharvest biocontrol and postharvest biology. *Postharvest Biology and Technology* 140, 107–112. DOI: 10.1016/j.postharvbio.2018.03.004.

Fierer, N. and Jackson, R.B. (2006) The diversity and biogeography of soil bacterial communities. *Proceedings of the National Academy of Sciences USA* 103(3), 626–631. DOI: 10.1073/pnas.0507535103.

Fitzpatrick, C.R., Copeland, J., Wang, P.W., Guttman, D.S., Kotanen, P.M. *et al.* (2018) Assembly and ecological function of the root microbiome across angiosperm plant species. *Proceedings of the National Academy of Sciences USA* 115(6), E1157–E1165. DOI: 10.1073/pnas.1717617115.

Fridman, S., Izhaki, I., Gerchman, Y. and Halpern, M. (2012) Bacterial communities in floral nectar. *Environmental Microbiology Reports* 4(1), 97–104. DOI: 10.1111/j.1758-2229.2011.00309.x.

Geisen, S., Briones, M.J., Gan, H., Behan-Pelletier, V.M., Friman, V.P. *et al.* (2019) A methodological framework to embrace soil biodiversity. *Soil Biology and Biochemistry* 136, 107536. DOI: 10.1016/j.soilbio.2019.107536.

Hacquard, S. and Schadt, C.W. (2015) Towards a holistic understanding of the beneficial interactions across the populus microbiome. *New Phytologist* 205(4), 1424–1430. DOI: 10.1111/nph.13133.

Hacquard, S., Garrido-Oter, R., González, A., Spaepen, S., Ackermann, G. *et al.* (2015) Microbiota and host nutrition across plant and animal kingdoms. *Cell Host and Microbe* 17(5), 603–616. DOI: 10.1016/j.chom.2015.04.009.

Hardoim, P.R., van Overbeek, L.S. and van Elsas, J.D. (2008) Properties of bacterial endophytes and their proposed role in plant growth. *Trends in Microbiology* 16(10), 463–471.

Jacoby, R. and Kopriva, S. (2019) Metabolic niches in the rhizosphere microbiome: An ecological perspective. *New Phytologist* 223, 105–112. DOI: 10.1111/nph.15765.

Jacquemyn, H., Lenaerts, M., Tyteca, D. and Lievens, B. (2013) Microbial diversity in the floral nectar of seven *Epipactis* (Orchidaceae) species. *MicrobiologyOpen* 2(4), 644–658. DOI: 10.1002/mbo3.103.

Janisiewicz, W.J. and Korsten, L. (2002) Biological control of postharvest diseases of fruits. *Annual Review of Phytopathology* 40(1), 411–441. DOI: 10.1146/annurev.phyto.40.120401.130158.

Johnston-Monje, D. and Raizada, M.N. (2011) Conservation and diversity of seed associated endophytes in Zea across boundaries of evolution, ethnography and ecology. *PLOS ONE* 6(6), e20396. DOI: 10.1371/journal.pone.0020396.

Johnston-Monje, D., Lundberg, D.S., Lazarovits, G., Reis, V.M. and Raizada, M.N. (2016) Bacterial populations in juvenile maize rhizospheres originate from both seed and soil. *Plant and Soil* 405, 337–355. DOI: 10.1007/s11104-016-2826-0.

Jumpponen, A. and Jones, K.L. (2010) Seasonally dynamic fungal communities in the Quercus macrocarpa phyllosphere differ between urban and nonurban environments. *New Phytologist* 186(2), 496–513. DOI: 10.1111/j.1469-8137.2010.03197.x.

Knief, C., Delmotte, N., Chaffron, S., Stark, M., Innerbner, G. *et al.* (2012) Metaproteogenomic analysis of microbial communities in the phyllosphere and rhizosphere. *Frontiers in Microbiology* 3, 483. DOI: 10.3389/fmicb.2012.00483.

Lindow, S.E. and Brandl, M.T. (2003) Microbiology of the phyllosphere. *Applied and Environmental Microbiology* 69(4), 1875–1883. DOI: 10.1128/AEM.69.4.1875-1883.2003.

Linquist, S., Cottenie, K., Elliott, T.A., Saylor, B., Kremer, S.C. *et al.* (2015) Applying ecological models to communities of genetic elements: The case of neutral theory. *Molecular Ecology* 24(13), 3232–3242. DOI: 10.1111/mec.13219.

Liu, H., Brettell, L.E., Qiu, Z. and Singh, B.K. (2020) Microbiome-mediated stress resistance in plants. *Trends in Plant Science* 25(8), 733–743. DOI: 10.1016/j.tplants.2020.03.014.

Massalha, H., Korenblum, E., Malitsky, S., Shapiro, O.H. and Aharoni, A. (2017) Live imaging of root–bacteria interactions in a microfluidics setup. *Proceedings of the National Academy of Sciences USA* 114(17), 4549–4554. DOI: 10.1073/pnas.1618584114.

Mendes, R., Garbeva, P. and Raaijmakers, J.M. (2013) The rhizosphere microbiome: Significance of plant beneficial, plant pathogenic, and human pathogenic microorganisms. *FEMS Microbiology Reviews* 37(5), 634–663. DOI: 10.1111/1574-6976.12028.

Müller, D.B., Vogel, C., Bai, Y. and Vorholt, J.A. (2016) The plant microbiota: Systems-level insights and perspectives. *Annual Review of Genetics* 50, 211–234. DOI: 10.1146/annurev-genet-120215-034952.

Nelson, E.B. (2004) Microbial dynamics and interactions in the spermosphere. *Annual Review of Phytopathology* 42(1), 271–309. DOI: 10.1146/annurev.phyto.42.121603.131041.

Nelson, E.B. (2018) The seed microbiome: Origins, interactions, and impacts. *Plant and Soil* 422(1), 7–34. DOI: 10.1007/s11104-017-3289-7.

Paul, C.E., Eggerichs, D., Westphal, A.H., Tischler, D. and van Berkel, W.J.H. (2021) Flavoprotein monooxygenases: Versatile biocatalysts. *Biotechnology Advances* 51, 107712. DOI: 10.1016/j.biotechadv.2021.107712.

Philippot, L., Raaijmakers, J.M., Lemanceau, P. and Putten, W.H. (2013) Going back to the roots: The microbial ecology of the rhizosphere. *Nature Reviews Microbiology* 11(11), 789–799. DOI: 10.1038/nrmicro3109.

Pugesgaard, S., Petersen, S.O., Chirinda, N., Olesen, J.E. and Jørgensen, U. (2020) Crop management effects on nitrous oxide emissions from temperate agricultural soils. *Soil Biology and Biochemistry* 146, 107805. DOI: 10.1016/j.soilbio.2020.107805.

Raaijmakers, J.M. and Mazzola, M. (2012) Diversity and natural functions of antibiotics produced by beneficial and plant pathogenic bacteria. *Annual Review of Phytopathology* 50, 403–424. DOI: 10.1146/annurev-phyto-081211-172908.

Remus-Emsermann, M.N. and Schlechter, R.O. (2018) Phyllosphere microbiology: At the interface between microbial individuals and the plant host. *New Phytologist* 218(4), 1327–1333. DOI: 10.1111/nph.15054.

Ritpitakphong, U., Falquet, L., Vimoltust, A., Berger, A., Métraux, J.-P. *et al.* (2016) The microbiome of the leaf surface of *Arabidopsis* protects against a fungal pathogen. *New Phytologist* 210(3), 1033–1043. DOI: 10.1111/nph.13808.

Rodriguez, R.J., White Jr, J.F., Arnold, A.E. and Redman, A.R.A. (2009) Fungal endophytes: Diversity and functional roles. *New Phytologist* 182(2), 314–330. DOI: 10.1111/j.1469-8137.2009.02773.x.

Rosenblueth, M. and Martínez-Romero, E. (2006) Bacterial endophytes and their interactions with hosts. *Molecular Plant-Microbe Interactions* 19(8), 827–837. DOI: 10.1094/MPMI-19-0827.

Ryan, R.P., Germaine, K., Franks, A., Ryan, D.J. and Dowling, D.N. (2008) Bacterial endophytes: Recent developments and applications. *FEMS Microbiology Letters* 278(1), 1–9. DOI: 10.1111/j.1574-6968.2007.00918.x.

Sapkota, R., Knorr, K., Jørgensen, L.N., O'Hanlon, K. and Nicolaisen, M. (2015) Host genotype is an important determinant of the cereal phyllosphere mycobiome. *New Phytologist* 207(4), 1134–1144. DOI: 10.1111/nph.13418.

Sasse, J., Martinoia, E. and Northen, T. (2018) Feed your friends: Do plant exudates shape the root microbiome? *Trends in Plant Science* 23(1), 25–41. DOI: 10.1016/j.tplants.2017.09.003.

Sessitsch, A., Hardoim, P., Döring, J., Weilharter, A., Krause, A. *et al.* (2012) Functional characteristics of an endophyte community colonizing rice roots as revealed by metagenomic analysis. *Molecular Plant-Microbe Interactions* 25(1), 28–36. DOI: 10.1094/MPMI-08-11-0204.

Shade, A., Jacques, M.A. and Barret, M. (2017) Ecological patterns of seed microbiome diversity, transmission, and assembly. *Current Opinion in Microbiology* 37, 15–22. DOI: 10.1016/j.mib.2017.03.010.

Shayanthan, A., Ordoñez, P.A.C. and Oresnik, I.J. (2022) The role of synthetic microbial communities (SYNCOM) in sustainable agriculture. *Frontiers in Agronomy* 4, 896307. DOI: 10.3389/fagro.2022.896307.

Shreshtha, K., Prakash, A., Pandey, P.K., Pal, A.K., Singh, J. *et al.* (2024a) Isolation and characterization of plant growth promoting rhizobacteria from cacti root under drought condition. *Current Research in Microbial Sciences* 8, 100319. DOI: 10.1016/j.crmicr.2024.100319.

Shreshtha, K., Raj, S., Pal, A.K., Tripathi, P., Choudhary, K.K. *et al.* (2024b) Isolation and identification of rhizospheric and endophytic bacteria from cucumber plants irrigated with wastewater: Exploring their roles in plant growth promotion and disease suppression. *Current Research in Microbial Sciences* 7, 100256. DOI: 10.1016/j.crmicr.2024.100256.

Shreshtha, K., Tripathi, P., Lal, J.A. and Tripathi, V. (2025) Metagenomics: A powerful tool for screening of beneficial biocontrol agents. In: Tripathi, V. (ed.) *Bio-Control Agents for Sustainable Agriculture: Diversity, Mechanisms and Applications*. Springer, Singapore, pp. 23–40. DOI: 10.1007/978-981-96-3232-9_2.

Smith, S.E. and Read, D. (2008) Colonization of roots and anatomy of arbuscular mycorrhizas. In: *Mycorrhizal Symbiosis*, 3rd edn. pp. 42–90. DOI: 10.1016/B978-012370526-6.50004-0.

Sow, S.L.S., Brown, M.V., Clarke, L.J., Bissett, A., van de Kamp, J. *et al.* (2022) Biogeography of Southern Ocean prokaryotes: A comparison of the Indian and Pacific sectors. *Environmental Microbiology* 24(5), 2449–2466. DOI: 10.1111/1462-2920.15906.

Stone, J.K., Bacon, C.W. and White Jr, J.F. (2000) An overview of endophytic microbes: Endophytism defined. In: *Microbial Endophytes*. Marcel Dekker, New York, USA, pp. 17–44. DOI: 10.1201/9781482277302.

Toju, H., Peay, K.G., Yamamichi, M., Narisawa, K., Hiruma, K. *et al.* (2018) Core microbiomes for sustainable agroecosystems. *Nature Plants* 4(5), 247–257. DOI: 10.1038/s41477-018-0139-4.

Truyens, S., Weyens, N., Cuypers, A. and Vangronsveld, J. (2015) Bacterial seed endophytes: Genera, vertical transmission and interaction with plants. *Environmental Microbiology Reports* 7(1), 40–50.

Turner, T.R., James, E.K. and Poole, P.S. (2013) The plant microbiome. *Genome Biology* 14(6), 209. DOI: 10.1186/gb-2013-14-6-209.

Van der Heijden, M.G.A., Bardgett, R.D. and Van Straalen, N.M. (2008) The unseen majority: Soil microbes as drivers of plant diversity and productivity in terrestrial ecosystems. *Ecology Letters* 11(3), 296–310. DOI: 10.1111/j.1461-0248.2007.01139.x.

Vandenkoornhuyse, P., Quaiser, A., Duhamel, M., Le Van, A. and Dufresne, A. (2015) The importance of the microbiome of the plant holobiont. *New Phytologist* 206(4), 1196–1206. DOI: 10.1111/nph.13312.

Velásquez, A.C., Castroverde, C.D.M. and He, S.Y. (2022) Plant–pathogen warfare under changing climate. *Nature Reviews Microbiology* 20, 391–405. DOI: 10.1038/s41579-021-00657-1.

Vessey, J.K. (2003) Plant growth promoting rhizobacteria as biofertilizers. *Plant and Soil* 255, 571–586. DOI: 10.1023/A:1026037216893.

Vorholt, J.A. (2012) Microbial life in the phyllosphere. *Nature Reviews Microbiology* 10(12), 828–840. DOI: 10.1038/nrmicro2910.

White, J.F., Kingsley, K.L., Zhang, Q., Verma, R., Obi, N. *et al.* (2019) Review: Endophytic microbes and their potential applications in crop management. *Pest Management Science* 75(10), 2558–2565. DOI: 10.1002/ps.5527.

Wippel, K., Tao, K., Niu, Y., Zgadzaj, R., Kiel, N. *et al.* (2021) Host preference and invasiveness of commensal bacteria in the Lotus and Arabidopsis root microbiota. *Nature Microbiology* 6(9), 1150–1162. DOI: 10.1038/s41564-021-00941-9.

Xiang, W. and Gao, J. (2023) Do not be anticlimactic: Farmers' behavior in the sustainable application of green agricultural technology—a perceived value and government support perspective. *Agriculture* 13(2), 247. DOI: 10.3390/agriculture13020247.

Xu, C. and Shanklin, J. (2016) Triacylglycerol metabolism, function, and accumulation in plant vegetative tissues. *Annual Review of Plant Biology* 67(1), 179–206. DOI: 10.1146/annurev-arplant-043015-111641.

2 The Rhizosphere: Microbial Diversity and Its Impact on Plants

Puthem Robindro Singh[1], Chongtham Allaylay Devi[2*] and Ashutosh K. Pandey[3*]

[1]ICAR-Directorate of Onion and Garlic Research, Pune, India
[2]Amity Institute of Horticultural Studies and Research, Amity University, Noida, India
[3]Department of Life Sciences, Sharda School of Bio-Science and Technology, Sharda University, Greater Noida, India

2.1 Introduction

The rhizosphere is a dynamic contact between plant roots and the surrounding soil that is essential for medicinal plants' health and growth. It is inhabited by numerous microorganisms such as bacteria, fungi, actinomycetes, archaea, protozoa, invertebrates, and other species that interact with each other (Pathan *et al.*, 2020; Ling *et al.*, 2022). These microbial communities establish symbiotic, mutualistic, and antagonistic relationships with plant roots, profoundly affecting the growth, health, and secondary metabolite production of medicinal plants. Furthermore, the manufacture of secondary metabolites essential to the therapeutic properties of plants, such terpenoids, flavonoids, and alkaloids, is influenced by the microbial community (Berg and Smalla, 2009). The growth, productivity, and therapeutic efficacy of medicinal plants are directly impacted by the rhizosphere, which mediates nutrient availability and stress resilience (Chaparro *et al.*, 2012). It is a dynamic environment where a variety of biological, chemical, and physical interactions occur between the roots and the soil. These interactions can significantly affect plant growth and health (Mendes *et al.*, 2013).

Continuous cropping is an increasingly common method of agricultural as well as medicinal plant production due to limited arable land resources worldwide. Constant cropping may result in declining soil fertility and

*Corresponding author: chongthamallaylaydevi@gmail.com and Aashu.p20@gmail.com

 Medicinal Plants and the Rhizosphere (eds M.K. Mishra and S. Tiwari)
DOI: 10.1079/9781836990475.0002

reduced crop production and quality, posing significant challenges for agriculture and allied sectors. Several investigations have revealed that continious cultivation causes a two-fold reduction in plant quality. The first is an increase in the abundance of soilborne pathogenic microorganisms, which makes crops more susceptible to diseases and pests; the second is the accumulation of root exudates or decomposition products, known as autotoxins, which lead to oxidative stress responses and a decrease in plant immunity, affecting plant growth and development (Zhang *et al.*, 2020). Autotoxins, which contain organic acids, phenolic acids, quinones, coumarins, flavonoids, and steroids as secondary metabolites, perform defense and signaling functions throughout plant development (Zhang *et al.*, 2020).

According to Bhattacharyya and Jha (2012), the microbial realm is the world's biggest untapped store of biodiversity, with just a small percentage (<10%) of the microscopic landscape known to date. Microbes make up the vast majority of life on Earth. Microorganisms, while being the most miniature forms of life, play an important part in every aspects of life on Earth. However, there is very little research into rhizosphere bacteria elucidating the role of these microorganisms with special reference to medicinal plants. Therefore, based on advancements in medicinal plant microbiome research, the current study highlights the diversity of microorganisms, their functions in relation to medicinal plants, influences of microbiome research and its future prospects.

2.2 Importants Aspects of the Rhizosphere

2.2.1 Root exudates

Plant roots discharge an assortment of substances into the rhizosphere, including sugars, amino acids, organic acids, and enzymes. These root exudates serve several purposes, including fostering beneficial microorganisms, boosting nutrient availability, and controlling soil pH (Vessey, 2003).

2.2.2 Microbial activity

In the rhizosphere, bacteria, fungi, and actinomycetes flourish (Fig. 2.1). These bacteria are crucial for the development of soil structure, disease prevention, and nutrient cycling. Some beneficial microorganisms can help plants to combact adverse effects, while others, like nitrogen-fixing bacteria, can enhance plant nutrition.

2.2.3 Soil structure and dynamics

The rhizosphere's activity at plant roots affects the texture, structure, and porosity of the soil. The soil may be broken up by the roots, which would

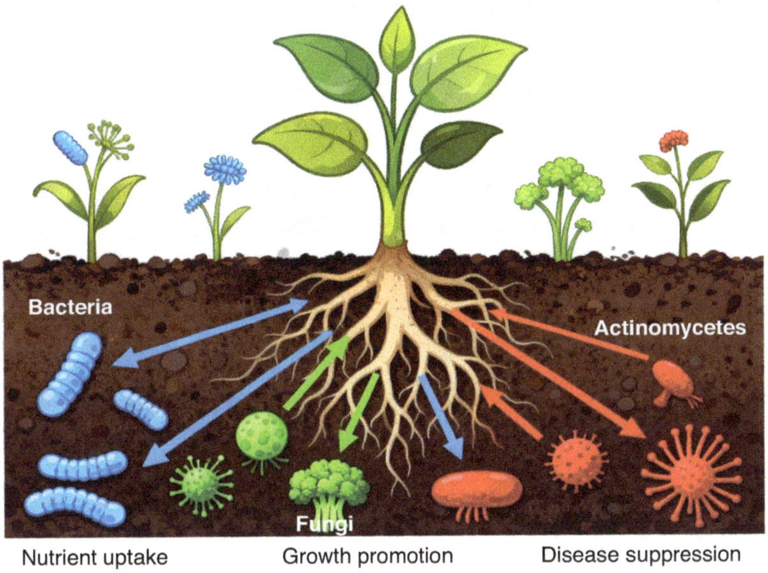

Fig. 2.1. Rhizosphere microbial diversity and plant benefits. Diverse bacteria, fungi, and actinomycetes in the root zone enhance nutrient uptake, growth promotion, and pathogen suppression.

increase water penetration and aeration. Moreover, soil aggregate formation is encouraged by the organic compounds found in root exudates (Vessey, 2003).

2.2.4 Nutrient cycling

The rhizosphere, a critical site for nutrient exchange, is where roots receive essential nutrients including potassium, phosphorus, and nitrogen. Through the fixation and degradation of nitrogen, microorganisms can aid in the availability of nutrients.

2.2.5 Plant–microbe interactions

In the rhizosphere, interactions between microbes and plants may be both beneficial and detrimental. While symbiotic connections, such as those between legumes and nitrogen-fixing bacteria, assist plants to obtain nutrients, pathogenic microbes can cause adverse effects to plants.

2.3 Diversity of Microbes and Their Roles in the Rhizosphere

2.3.1 Fungi

Fungi have a considerable impact on the development, health, and production of secondary metabolites in medicinal plants. Their diverse interactions with plants—whether saprophytic, endophytic, or symbiotic—have a significant influence on the overall pharmacological properties and quality of medicinal crops.

Important functions of fungi in medicinal plants include the following.

2.3.1.1 Symbiotic associations

The symbiotic relationships between medicinal plants and other species, such as microorganisms (fungi and bacteria), animals, and other plants, are mutually beneficial. These interactions can aid plant development, promote health, and provide medicinal advantages (Berendsen *et al.*, 2012). Below is an overview of key types of symbiotic associations and their examples.

MYCORRHIZAL FUNGI. Mycorrhizal fungi create symbiotic associations with the roots of many medicinal plants, enhancing water and nutrient absorption, particularly phosphorus, while also getting carbohydrates from the plant (Smith and Read, 2008).

Medicinal plant examples include:

- *Withania somnifera (ashwagandha)*: Arbuscular mycorrhizal fungi (AMF) increase root biomass and concentration of withanolides, the plant's active chemical substances.
- *Curcuma longa (turmeric)*: Mycorrhizal associations improve curcumin production and plant growth.

Their roles in medicinal plants include:

- boosting secondary metabolite levels, such as flavonoids, alkaloids, and terpenoids, which are significant in plant treatments
- promoting plant resilience to environmental stress, resulting in consistent production and quality of therapeutic compounds.

CYANOBACTERIA–PLANT ASSOCIATIONS. Cyanobacteria fix atmospheric nitrogen, which benefits medicinal plants in aquatic and wetland habitats (Khan and Lee, 2013). For example, in azolla (traditional medicine plant), symbiotic cyanobacteria deliver nitrogen to the plant, increasing its medicinal potential.

POLLINATION AND SEED DISPERSAL MUTUALISM. Interactions with insects, birds, and mammals aid medicinal plants' reproduction and dissemination. These creatures benefit from nectar, pollen, and fruits. For example, in *Rauwolfia serpentina* (Indian snakeroot), insect pollinators boost seed set, ensuring the plant's viability for therapeutic harvest.

2.3.1.2 *Endophytic fungi*

Endophytes are fungi that dwell within plant tissues without causing damage. Functions in medicinal plants include the following.

- *Secondary metabolite enhancement*: Endophytic fungi frequently create bioactive substances including alkaloids, tannins, and terpenoids that imitate or supplement the medicinal plant's natural components. For example, *Taxomyces andreanae*, an endophytic fungus in yew trees, generates taxol, a powerful anticancer chemical (Wink, 2010).
- *Disease resistance*: Endophytes strengthen the plant's immune response and create antimicrobial compounds to fight against infections.
- *Stress tolerance*: Endophytes assist medicinal plants to tolerate abiotic stress, such as drought and salt stress, by modifying stress-responsive pathways. Endophytic fungi, such as *Tinospora cordifolia* (guduchi), promote the formation of alkaloids and phenolic chemicals. In *Azadirachta indica* (neem), endophytic interactions increase bioactive chemical levels, including azadirachtin (Khan and Lee, 2013).
- *Decomposers and soil health improvers*: Saprophytic fungi break down organic debris in soil, recycling nutrients needed for medicinal plant development. They produce enzymes (e.g. cellulases and ligninases) that degrade complex organic compounds into bioavailable forms, increasing soil fertility (Smith and Read, 2008).
- *Biocontrol agents*: Certain fungi protect medicinal plants by inhibiting pathogens. *Trichoderma* species are well known for their ability to inhibit soilborne diseases while encouraging plant development and health. These fungi also promote systemic resistance in plants, which strengthens them against infections. (Lugtenberg and Kamilova, 2009).
- *Secondary metabolite modulation*: Fungi interact with medicinal plants to stimulate or upregulate the production of bioactive compounds, including the following.
  - Alkaloids: Fungi regulate pathways that improve alkaloid content in plants such as *Catharanthus roseus* (source of vincristine and vinblastine).
  - Phenolics and flavonoids: Fungal interactions enhance antioxidant and antiinflammatory compounds.
  - Terpenoids: Mycorrhizal fungi and endophytes can enhance terpenoid concentrations, critical for plants like *Artemisia annua* (source of artemisinin).
- *Pathogen defense.*
  - Fungi generate antimicrobial compounds and compete with harmful species for resources in the rhizosphere. This reduces the threat of diseases like root rot or wilt in medicinal plants.

- *Abiotic stress tolerance.*
  - Fungi like AMF improve water uptake and osmotic adjustment under drought conditions.
  - They help plants cope with heavy metal toxicity by immobilizing metals in the soil or sequestering them in fungal hyphae.

2.3.1.3 Production of bioactive compounds for medicinal use

Some fungi associated with medicinal plants create pharmacologically significant chemicals. Examples include *Penicillium* sp. that produce penicillin and *Fusarium* sp. linked with medicinal plants that produce fumonisins.

2.3.1.4 Impact of fungi on medicinal plant quality

Fungi influence the yield and quality of medicinal plants by enhancing:

- nutrient uptake and stress resilience
- the concentration and diversity of therapeutic phytochemicals
- resistance to pathogens, ensuring healthier plants with fewer agrochemical inputs.

2.3.1.5 Applications in cultivation of medicinal plants

- *Biofertilizers*: Mycorrhizal fungi are used to enhance nutrient availability and plant growth.
- *Biocontrol agents: Trichoderma* and similar fungi are employed to manage pests and diseases.
- *Endophyte biotechnology*: Researchers are isolating endophytic fungi to increase the production of valuable secondary metabolites in plants.

2.3.2 Bacteria

Bacteria are essential to the growth, health, and pharmacological properties of medicinal plants (Table 2.1). Their interactions with plants, which can be mutualistic or antagonistic, have a substantial impact on plant physiology, stress resistance, and bioactive chemical synthesis. These helpful microorganisms are known as plant growth-promoting rhizobacteria (PGPR), endophytes, and other functional bacteria in the rhizosphere.

Table 2.1. Bacteria and their roles in medicinal plants (Vessey, 2003; Tshikhudo *et al.*, 2023).

Bacteria	Role	Examples
Rhizobium	Nitrogen fixation, growth promotion	Leguminous medicinal plants
Pseudomonas fluorescens	Disease suppression, secondary metabolite enhancement	*Catharanthus roseus*
Bacillus subtilis	Induced systemic resistance, stress tolerance	*Withania somnifera*
Azospirillum	Auxin production, root growth stimulation	*Artemisia annua*
Streptomyces spp.	Antibiotic production, pathogen inhibition	Various medicinal plants
Burkholderia spp.	Phosphate solubilization, heavy metal detoxification	*Curcuma longa*

Key roles of bacteria in medicinal plants include the following.

2.3.2.1 *Plant growth promotion*

NUTRIENT ACQUISITION

- *Nitrogen fixation*: According to Tshikhudo *et al.* (2023), nitrogen-fixing bacteria such as *Rhizobium, Azospirillum,* and *Bradyrhizobium* transform atmospheric nitrogen into ammonia that plants can absorb.
- *Phosphate solubilization: Pseudomonas, Bacillus,* and *Burkholderia* are among the bacteria that produce organic acids and enzymes that solubilize insoluble phosphate so that plants can use it (Wu *et al.*, 2021a).
- *Potassium and zinc mobilization*: Some bacteria increase the bioavailability of zinc and potassium, which are necessary for plant development and enzymatic activity (Wu *et al.*, 2021b; Tshikhudo *et al.*, 2023).

PHYTOHORMONE PRODUCTION. Numerous PGPR generate hormones that promote root development and general plant growth, such as auxins, cytokinins, and gibberellins (Tshikhudo *et al.*, 2023). For instance, *Azospirillum* encourages the development of lateral roots by producing indole-3-acetic acid (IAA).

2.3.2.2 *Disease resistance*

- *Biocontrol agents*: Bacteria protect medicinal plants by competing with pathogens for resources, generating antibiotics, and creating systemic resistance. Examples include *Pseudomonas fluorescens* which generates siderophores and antifungal compounds that inhibit diseases such as *Fusarium* and *Phytophthora. Bacillus subtilis* generates lipopeptides that suppress fungal and bacterial infections.

- *Induced systemic resistance (ISR)*: Certain bacteria activate ISR, allowing the plant's defensive systems to respond more quickly and efficiently to pathogens.

2.3.2.3 Stress tolerance

- *Abiotic stress mitigation*: Bacteria enhance the plant's ability to tolerate environmental stresses such as drought, salinity, and heavy metal toxicity. Mechanisms include the following.
  - Osmolyte production: Bacteria like *Bacillus* and *Enterobacter* produce osmoprotectants that help plants to maintain water balance under drought or salinity stress.
  - iACC deaminase production: Enzymes produced by bacteria reduce ethylene levels in stressed plants, minimizing growth inhibition.
- *Bioremediation of soil contaminants*: Bacteria degrade toxic compounds in the soil or immobilize heavy metals, protecting medicinal plants from toxicity.

2.3.2.4 Enhanced production of secondary metabolites

- *Stimulation of bioactive compound production*: Bacteria have a role in producing secondary metabolites of therapeutic value, such as alkaloids, flavonoids, terpenoids, and phenolics (Wink, 2010). *Pseudomonas putida* increases alkaloid synthesis in plants such as *Catharanthus roseus* (a source of vinblastine and vincristine) while *Bacillus* spp. are known to enhance plant phenolic and flavonoid synthesis, which improves antioxidant qualities.
- *Synthesis of bioactive compounds*: Certain bacterial endophytes can create antibiotics or anticancer drugs on their own, which adds to the plant's medicinal potential.

2.3.2.5 Rhizosphere engineering

- *Soil fertility improvement*: Decomposer bacteria degrade organic materials and recycle nutrients in the soil. Their metabolic activities improve soil structure and fertility, resulting in an ideal environment for medicinal plant growth.
- *Siderophore production*: Bacteria such as *Pseudomonas* make siderophores that bind iron, increasing its availability to plants while depriving pathogens of this critical mineral.

2.3.2.6 Biocontrol of pests and pathogens

- *Antimicrobial production*: Some bacteria generate antibiotics and lytic enzymes that kill pathogenic fungi, bacteria, and nematodes. *Streptomyces* spp. generate streptomycin and other antibiotics that are useful against plant diseases. *Bacillus thuringiensis* generates insecticidal toxins, which protect plants from pests.
- *Competition and quorum sensing interference*: Beneficial bacteria compete with diseases for resources and space in the rhizosphere. They also impede pathogen communication (quorum sensing) to avoid coordinated assaults.

2.3.2.7 Entophytic properties

Endophytic bacteria live inside plant tissues without causing harm and provide various benefits.

- Enhance nutrient uptake and modulate stress responses.
- Influence secondary metabolite pathways to increase medicinal compound production.
- For example, endophytic bacteria in *Azadirachta indica* (neem) are known to enhance the production of bioactive compounds like azadirachtin.

Applications in cultivation of medicinal plants include the following.

- *Biofertilizers*: Bacteria like *Azospirillum*, *Rhizobium*, and phosphate solubilizing bacteria are used to enhance soil fertility and plant growth.
- *Biocontrol agents*: *Bacillus* and *Pseudomonas* are commercially used to control pathogens.
- *Biotechnological approaches*: Bacteria are employed in cocultures or inoculations to enhance the yield and quality of medicinally important phytochemicals.

2.3.3 Actinomycetes

Actinomycetes are a distinct category of filamentous, Gram-positive bacteria renowned for their capacity to create a diverse variety of bioactive substances as well as their importance in the health and growth of medicinal plants (Table 2.2). These soil-dwelling bacteria are well known for their antimicrobial activities, as well as their contributions to nutrient cycling, plant growth promotion, and secondary metabolite enhancement.

Roles of actinomycetes in medicinal plants include the following.

2.3.3.1 Antimicrobial activity

Actinomycetes, notably *Streptomyces* species, generate more than 70% of therapeutically effective antibiotics, including streptomycin, tetracycline, and erythromycin. They inhibit soilborne plant diseases by generating

Table 2.2. Actinomycetes beneficial to medicinal plants.

Actinomycete species	Function	Medicinal plant	References
Streptomyces spp.	Antibiotic production, pathogen suppression	*Withania somnifera*	(Sreevidya *et al.*, 2016; Chaurasia *et al.*, 2018)
Micromonospora spp.	Phytohormone production, growth promotion	*Curcuma longa*	(Singh *et al.*, 2018)
Nocardia spp.	Antimicrobial compound production	*Azadirachta indica*	(El-Tarabily and Sivasithamparam, 2006)

antibiotics, antifungal metabolites, and lytic enzymes. *Streptomyces griseoviridis* is a biocontrol agent that protects plants from fungal infections such as *Fusarium*.

2.3.3.2 Plant growth promotion

- *Production of plant growth hormones*: Actinomycetes generate phytohormones such as IAA, gibberellins, and cytokinins, which promote root and shoot growth. Enhanced root development promotes nutrient and water intake, resulting in improved plant growth (Nalini and Prakash, 2017).
- *Nitrogen fixation*: Some actinomycetes, like *Frankia*, fix atmospheric nitrogen in symbiotic associations with actinorhizal plants. This process enriches soil nitrogen levels, benefiting medicinal plants (Nalini and Prakash, 2017).

2.3.3.3 Biocontrol of pathogens and pests

Actinomycetes act as effective biocontrol agents due to their ability to:

- produce antifungal, antibacterial, and antiviral compounds that suppress a wide range of pathogens
- induce systemic resistance in plants, fortifying them against diseases
- compete with pathogens for nutrients and space in the rhizosphere, reducing their harmful effects (Nalini and Prakash, 2017).

2.3.3.4 Soil fertility and nutrient cycling

Actinomycetes break down organic matter by creating extracellular enzymes such as cellulases, ligninases, and chitinases, which return nutrients to the soil. They improve soil structure and fertility, allowing medicinal plants to flourish more effectively (Nalini and Prakash, 2017).

2.3.3.5 *Stress tolerance in plants*

Actinomycetes assist medicinal plants to withstand abiotic conditions such as drought, salt, and heavy metal toxicity by producing osmoprotectants and antioxidant enzymes that reduce oxidative damage in plants. Heavy metals are sequestered in plant cell walls, limiting toxicity exposure (Selim *et al.*, 2021).

2.3.3.6 *Enhancement of secondary metabolite production*

Actinomycetes have a role in the manufacture of medicinally significant secondary metabolites in plants, including alkaloids, flavonoids, and terpenoids. *Catharanthus roseus*, which is connected with actinomycetes, has a higher alkaloid content, which improves its medicinal benefits. Certain actinomycetes also create bioactive chemicals directly, which enhance the therapeutic characteristics of plants (Nalini and Prakash, 2017).

2.3.3.7 *Applications in medicinal plant cultivation*

- *Biocontrol agents*: Actinomycetes are used to develop biopesticides and biofungicides to protect medicinal plants from pathogens.
- *Biofertilizers*: They are incorporated into biofertilizers to improve soil fertility, nutrient availability, and plant growth.
- *Secondary metabolite enhancement*: Actinomycete inoculants are used to stimulate the production of medicinal compounds, increasing the pharmacological quality of plants.

2.3.4 Other microorganisms: Protozoa and archaea

2.3.4.1 *Protozoa in medicinal plants*

Protozoa, which are single-celled eukaryotic microorganisms, have an indirect but important impact on medicinal plant development and health. Protozoa, unlike bacteria or fungi, are best recognized for their ecological roles in the soil and rhizosphere, where they interact with other microorganisms and impact plant growth (Table 2.3).

Roles of protozoa in medicinal plants include the following.

Table 2.3. Protozoa-mediated benefits in specific medicinal plants.

Protozoa	Function	Medicinal plant	References
Acanthamoeba spp.	Nutrient cycling, bacterial grazing	*Ocimum sanctum* (Tulsi)	(Thauer *et al.*, 2008)
Paramecium spp.	Promotion of beneficial rhizobacteria	*Withania somnifera*	(Foissner, 1999)
Tetrahymena spp.	Enhancement of rhizosphere nutrient cycling	*Aloe vera*	(Clarholm, 1985)

- *Nutrient cycling*: Protozoa control microbial populations in the soil by feeding on bacteria and fungi. This predation causes the release of nutrients such as nitrogen, phosphorus, and other micronutrients, which are then accessible to plants.
- *Impact on medicinal plants*: Protozoa-mediated nutrient release promotes medicinal plant development and output by guaranteeing an adequate supply of critical nutrients. For instance, greater nitrogen availability enhances secondary metabolite formation, which improves plant medicinal qualities. As an example, *Paramecium* species increase root-associated bacterial activity in medicinal plants such as *Withania somnifera* (ashwagandha).
- *Disease suppression*: Protozoa assist in preventing disease by feeding on harmful fungi and bacteria. They regulate microbial populations, which reduces the potential of soilborne diseases in medicinal plant production.
- *Soil structure and rhizosphere interactions*: Protozoa improve soil structure by increasing particle aggregation through mobility and contact with other microbes. This produces an ideal environment for medicinal plant roots to develop and absorb nutrients efficiently.
- *Influence on secondary metabolite production*: Protozoa subsequently boost secondary metabolite synthesis by controlling microbial populations that interact with plants. Plants with well-balanced rhizosphere communities frequently have greater quantities of alkaloids, flavonoids, and terpenoids, which are important components in medicinal plants.

Applications in medicinal plant cultivation include the following.

- *Sustainable soil management*: Protozoa play a vital role in organic agricultural systems because they improve nutrient cycling and microbial diversity, minimizing the demand for artificial fertilizers.
- *Integrated disease management*: Protozoa assist in biological control measures by lowering the number of plant diseases in the rhizosphere.
- *Rhizosphere engineering*: Understanding and regulating protozoa–plant interactions have the potential to increase medicinal plant yield and phytochemical content.

2.3.4.2 Archaea in medicinal plants

Archaea, a unique class of single-celled microbes, are well known for their ability to survive in adverse settings. While archaea have received less attention in plant ecosystems than bacteria and fungi, they are becoming more recognized for their function in the rhizosphere and contributions to plant development and health. Their metabolic plasticity and interactions with other soil microbes have an impact on nutrient cycle, soil fertility, and medicinal plants' stress tolerance (Table 2.4).

Table 2.4. Archaea and their roles in medicinal plants.

Archaea	Function	Medicinal plant	Reference
Nitrososphaera viennensis	Ammonia oxidation, nitrogen cycling	*Withania somnifera, Azadirachta indica* (Neem)	(Pester *et al*., 2012; Prosser and Nicol, 2012)
Halobacterium spp.	Salinity stress alleviation	*Aloe vera*, *Salicornia*, *Withania somnifera*	(Oren, 2010, 2011)
Methanobrevibacter spp.	Methanogenesis and nutrient release, Methane production, organic matter cycling	*Ocimum sanctum* (Tulsi), *Curcuma longa* (Turmeric)	(Thauer *et al*., 2008, 2010)
Sulfolobus spp.	Sulfur cycling and heavy metal detoxification	*Curcuma longa*	(Baker *et al*., 2024)
Archaeoglobus spp.	Sulfur metabolism, soil fertility	*Ocimum sanctum* (Tulsi)	(Stetter *et al*., 1993)

Roles of archaea in medicinal plants include the following.

NUTRIENT CYCLING AND SOIL FERTILITY

- *Methanogens*: Methanogenic archaea produce carbon dioxide and methane from organic materials, which contributes to the carbon cycle. Nutrients including phosphorus and nitrogen are released during this breakdown process, allowing plants to absorb them.
- *Ammonia oxidizers*: *Nitrososphaera* and *Nitrosopumilus* are examples of ammonia-oxidizing archaea (AOA), which are essential to nitrification because they transform ammonia into nitrite, a type of nitrogen that plants can use. In settings where bacteria are less active, such as low-nitrogen soils, their activity is particularly crucial.

IMPACT ON MEDICINAL PLANTS

- In therapeutic plants such as *Ocimum sanctum* (holy basil) and *Withania somnifera* (ashwagandha), increased nitrogen availability stimulates growth and raises the synthesis of secondary metabolites.
- In nitrogen-deficient soils, archaea-mediated nitrification is especially important because it promotes the growth of nitrogen-dependent medicinal plants like *Ocimum sanctum* (holy basil) and *Azadirachta indica* (neem).

STRESS TOLERANCE IN PLANTS

- *Abiotic stress alleviation*: By enhancing soil structure and water retention, archaea assist plants to adapt to harsh environmental factors including heat, drought, and excessive salt.
- Halophilic archaea: They are found in saline soils and, by preserving osmotic pressure and ionic equilibrium, increase plant adaptation in

salt-affected locations. For instance, *Halobacterium* species help medicinal plants such as *Withania somnifera* and *Aloe vera* by improving soil structure in salty situations.

RHIZOSPHERE INTERACTIONS. Archaea in the rhizosphere interact with bacteria and fungi to maintain microbial populations and contribute to a healthy ecosystem. They regulate nutrition availability, pathogen control, and plant defense responses.

BIOREMEDIATION AND DETOXIFICATION. Archaea may metabolize contaminants and heavy metals, which reduces soil toxicity. For example, sulfur-metabolizing archaea cleanse polluted soils, making it possible to grow medicinal plants such as *Curcuma longa* (turmeric) in polluted locations.

POTENTIAL ROLE IN SECONDARY METABOLITE PRODUCTION. Emerging research demonstrates that archaea may impact secondary metabolite production by modifying the rhizosphere microbiota. Examples include increased phenolic and flavonoid synthesis in medicinal plants as a result of food availability and archaeal stress signaling.

Applications of archaea in medicinal plant cultivation include the following.

- *Biofertilizers*: AOA may be used as biofertilizers to improve nitrogen availability in medicinal plant production.
- *Stress management*: Halophilic archaea might help cultivation of therapeutic plants in salty or drought-prone areas.
- *Biotechnological approaches*: Archaea's unique metabolic capabilities can be used to promote soil health and maximize bioactive chemical synthesis in medicinal plants.

2.4 Roles of Rhizospheric Microbes in Medicinal Plant Health

- *Nutrient cycling and uptake*: Microbes increase the bioavailability of minerals including nitrogen, phosphate, and potassium, ensuring that medicinal plants receive enough nourishment to create active phytochemicals (Lugtenberg and Kamilova, 2009).
- *Production of bioactive compounds*: Numerous rhizospheric microorganisms produce secondary metabolites, including siderophores, antibiotics, and antifungal substances, which shield plants from disease and affect the synthesis of secondary metabolites in plants that have therapeutic value (Wu *et al.*, 2021b).
- *Disease suppression*: By creating antimicrobial chemicals, competing for resources and space, or causing systemic resistance in plants, beneficial bacteria inhibit soilborne diseases.

- *Stress tolerance*: Microbes increase plant tolerance to abiotic stimuli (such as drought, salt, and heavy metals) by altering stress-responsive pathways and increasing antioxidant systems.

2.5 Impact on Medicinal Plants

- *Improved growth and yield*: Nutrient availability and hormonal stimulation by rhizosphere bacteria increase biomass and production in medicinal plants such as *Withania somnifera* (ashwagandha) and *Ocimum sanctum* (holy basil).
- *Enhanced secondary metabolite production*: The metabolic interactions between roots and bacteria encourage the creation of bioactive chemicals, which improves the therapeutic value of plants like *Curcuma longa* (turmeric) and *Azadirachta indica* (neem).
- *Disease resistance*: Microbes induce systemic resistance and local resistance in medicinal plants, protecting them from soilborne and foliar diseases.
- *Sustainability in cultivation*: The use of rhizosphere bacteria in biofertilizers and biocontrol agents minimizes the need for chemical inputs, making medicinal plant agriculture more sustainable (Gupta *et al.*, 2015).

2.6 Conclusion

A diverse range of microorgansims such as bacteria, fungus, actinomycetes, protozoa, and archaea enhance the growth, health, and therapeutic potential of medicinal plants by upregulation of bioactive chemicals (alkaloids, flavonoids, and terpenoids). Therefore, the rhizosphere serves as the foundation for plant–microbe interactions and as a dynamic habitat in which plants interact with a varied range of microorganisms. Microbial activities in the rhizosphere guarantee a consistent supply of critical minerals and nutrients, which help the plants to combat biotic and abiotic stress such as salt, drought, and heavy metal toxicity, resulting in improved plant development and yields.

Sustainable agricultural practices and development of natural medicines will enhance the soil microbiome, resulting in better plant health and increased production. In addition, the involvement of local communities in sustainable agriculture practices can promote the use of microbial diversity, benefiting medicinal plant production. Therefore, comprehensive understanding is critical for both sustainable agriculture practices and the future of plant-based natural medicines.

References

Baker, B.J., Hyde, E. and Leão, P. (2024) Nature should be the model for microbial sciences. *Journal of Bacteriology* 206(9), e0022824. DOI: 10.1128/jb.00228-24.

Berendsen, R.L., Pieterse, C.M.J. and Bakker, P.A.H.M. (2012) The rhizosphere microbiome and plant health. *Trends in Plant Science* 17(8), 478–486. DOI: 10.1016/j.tplants.2012.04.001.

Berg, G. and Smalla, K. (2009) Plant species and soil type cooperatively shape the structure and function of microbial communities in the rhizosphere. *FEMS Microbiology Ecology* 68(1), 1–13. DOI: 10.1111/j.1574-6941.2009.00654.x.

Bhattacharyya, P.N. and Jha, D.K. (2012) Plant growth-promoting rhizobacteria (PGPR): Emergence in agriculture. *World Journal of Microbiology and Biotechnology* 28, 1327–1350.

Chaparro, J.M., Sheflin, A.M., Manter, D.K. and Vivanco, J.M. (2012) Manipulating the soil microbiome to increase soil health and plant fertility. *Biology and Fertility of Soils* 48(5), 489–499. DOI: 10.1007/s00374-012-0691-4.

Chaurasia, A., Meena, B.R., Tripathi, A.N., Pandey, K.K., Rai, A.B. *et al.* (2018) Actinomycetes: An unexplored microorganisms for plant growth promotion and biocontrol in vegetable crops. *World Journal of Microbiology and Biotechnology* 34(9), 132. DOI: 10.1007/s11274-018-2517-5.

Clarholm, M. (1985) Interactions of bacteria, protozoa and plants leading to mineralization of soil nitrogen. *Soil Biology and Biochemistry* 17(2), 181–187. DOI: 10.1016/0038-0717(85)90113-0.

El-Tarabily, K.A. and Sivasithamparam, K. (2006) Non-streptomycete actinomycetes as biocontrol agents of soil-borne fungal plant pathogens and as plant growth promoters. *Soil Biology and Biochemistry* 38(7), 1505–1520. DOI: 10.1016/j.soilbio.2005.12.017.

Foissner, W. (1999) Soil protozoa as bioindicators: Prospects and limitations. *Biology and Fertility of Soils* 29, 111–122. DOI: 10.1016/S0167-8809(99)00032-8.

Gupta, G., Parihar, S.S., Ahirwar, N.K., Snehi, S.K. and Singh, V. (2015) Plant growth promoting rhizobacteria (PGPR): Current and future prospects for development of sustainable agriculture. *Journal of Microbial and Biochemical Technology* 7(2), 096–102. DOI: 10.4172/1948-5948.1000188.

Khan, A.L. and Lee, I.J. (2013) Endophytic fungal associations of medicinal plants and their biotechnological applications. *Fungal Diversity* 54, 1–17.

Ling, N., Wang, T. and Kuzyakov, Y. (2022) Rhizosphere bacteriome structure and functions. *Nature Communications* 13(1), 836. DOI: 10.1038/s41467-022-28448-9.

Lugtenberg, B. and Kamilova, F. (2009) Plant-growth-promoting rhizobacteria. *Annual Review of Microbiology* 63, 541–556. DOI: 10.1146/annurev.micro.62.081307.162918.

Mendes, R., Garbeva, P. and Raaijmakers, J.M. (2013) The rhizosphere microbiome: Significance of plant beneficial, plant pathogenic, and human pathogenic microorganisms. *FEMS Microbiology Reviews* 37(5), 634–663. DOI: 10.1111/1574-6976.12028.

Nalini, M.S. and Prakash, H.S. (2017) Diversity and bioprospecting of actinomycete endophytes from the medicinal plants. *Letters in Applied Microbiology* 64(4), 261–270. DOI: 10.1111/lam.12718.

Oren, A. (2010) Industrial and environmental applications of halophilic microorganisms. *Environmental Technology* 31(8–9), 825–834. DOI: 10.1080/09593330903370026.

Oren, A. (2011) Thermodynamic limits to microbial life at high salt concentrations. *Environmental Microbiology* 13(8), 1908–1923. DOI: 10.1111/j.1462-2920.2010.02365.x.

Pathan, S.I., Ceccherini, M.T., Sunseri, F. and Lupini, A. (2020) Rhizosphere as hotspot for plant-soil-microbe interaction. *Carbon and Nitrogen Cycling in Soil* 17–43. DOI: 10.1007/978-981-13-7264-3_2.

Pester, M., Rattei, T., Flechl, S., Gröngröft, A., Richter, A. *et al.* (2012) AmoA-based consensus phylogeny of ammonia-oxidizing archaea and deep sequencing of AmoA genes from

soils of four different geographic regions. *Environmental Microbiology* 14(2), 525–539. DOI: 10.1111/j.1462-2920.2011.02666.x.
Prosser, J.I. and Nicol, G.W. (2012) Archaeal and bacterial ammonia-oxidisers in soil: The quest for niche specialisation and differentiation. *Trends in Microbiology* 20(11), 523–531. DOI: 10.1016/j.tim.2012.08.001.
Selim, M.S.M., Abdelhamid, S.A. and Mohamed, S.S. (2021) Secondary metabolites and biodiversity of actinomycetes. *Journal, Genetic Engineering and Biotechnology* 19(1), 72. DOI: 10.1186/s43141-021-00156-9.
Singh, A., Dubey, S. and Pandey, K. (2018) Actinomycetes as bio-inoculants for medicinal plants. *Journal of Applied Microbiology* 125(2), 276–292.
Smith, S.E. and Read, D.J. (2008) *Mycorrhizal Symbiosis*. Academic Press, New York, USA.
Sreevidya, M., Gopalakrishnan, S. and Kudapa, H. (2016) Role of actinomycetes in sustainable agricultural systems and biotechnological innovations. *Plant Soil and Environment* 62(4), 140–146.
Stetter, K.O., Huber, R., Blöchl, E., Kurr, M., Eden, R.D. *et al.* (1993) Hyperthermophilic archaea are thriving in deep North Sea and Alaskan oil reservoirs. *Nature* 365(6448), 743–745. DOI: 10.1038/365743a0.
Thauer, R.K., Kaster, A.-K., Seedorf, H., Buckel, W. and Hedderich, R. (2008) Methanogenic archaea: Ecologically relevant differences in energy conservation. *Nature Reviews Microbiology* 6(8), 579–591. DOI: 10.1038/nrmicro1931.
Thauer, R.K., Kaster, A.K., Goenrich, M., Schick, M., Hiromoto, T. *et al.* (2010) Hydrogenases from methanogenic archaea, nickel, a novel cofactor, and H2 storage. *Annual Review of Biochemistry* 79(1), 507–536. DOI: 10.1146/annurev.biochem.030508.152103.
Tshikhudo, P.P., Ntushelo, K. and Mudau, F.N. (2023) Sustainable applications of endophytic bacteria and their physiological/biochemical roles on medicinal and herbal plants: Review. *Microorganisms* 11(2), 453. DOI: 10.3390/microorganisms11020453.
Vessey, J.K. (2003) Plant growth promoting rhizobacteria as biofertilizers. *Plant and Soil* 255(2), 571–586. DOI: 10.1023/A:1026037216893.
Wink, M. (2010) Functions and biotechnology of plant secondary metabolites. In: *Annual Plant Reviews*, Vol. 39. Wiley Blackwell, Oxford, UK. DOI: 10.1002/9781444318876.
Wu, B., Liu, F., Fang, W., Yang, T., Chen, G.H. *et al.* (2021a) Microbial sulfur metabolism and environmental implications. *Science of the Total Environment* 778, 146085. DOI: 10.1016/j.scitotenv.2021.146085.
Wu, W., Chen, W., Liu, S., Wu, J., Zhu, Y. *et al.* (2021b) Beneficial relationships between endophytic bacteria and medicinal plants. *Frontiers in Plant Science* 12, 646146. DOI: 10.3389/fpls.2021.646146.
Zhang, Z., Fan, J., Wu, J., Zhang, L., Wang, J. *et al.* (2020) Alleviating effect of silicon on melon seed germination under autotoxicity stress. *Ecotoxicology and Environmental Safety* 188, 109901. DOI: 10.1016/j.ecoenv.2019.109901.

3 Exploration of Microbial Diversity in the Internal Tissues of Plants, and Their Interactions with the Plant

Shilpy Singh[1*], Varun Kumar Sharma[1], Pooja Singh[1], Shraddha Sahu[2], Renu Verma[2], Abhishek Awasthi[3], Ved Vrat Verma[4] and Dharmsheel Shrivastav[1*]

[1]*Department of Biotechnology and Microbiology, School of Sciences, Noida International University, Noida, India*
[2]*Department of Biotechnology, Faculty of Engineering and Technology, Rama University, Mandhana, India*
[3]*Department of Biotechnology, Shaheed Mahendra Karma Vishwavidyalaya, Bastar, India*
[4]*Department of Biotechnology, School of Engineering and Technology, Noida International University, Noida, India*

3.1 Introduction: Plant Endosphere and Microbiome

The plant endosphere represents one of the most selective, protected, and biologically active niches within the plant holobiont, the collective ecological unit comprising a plant and all its associated microorganisms (Compant *et al.*, 2021). This internal environment includes the tissues of roots, stems, leaves, seeds, and reproductive organs, each providing unique microhabitats that support diverse microbial consortia. These resident microorganisms, collectively known as endophytes, engage in close biological interactions with their host throughout its life cycle. Endophytic communities are dominated primarily by bacteria, fungi, and actinomycetes, though certain protists and viruses may also be present in specific contexts. Unlike pathogenic microbes, endophytes colonize plant tissues asymptomatically, meaning they do not cause visible disease symptoms or detrimental effects under normal conditions (Vimal *et al.*, 2024).

*Corresponding authors: shilpy.singh24@gmail.com and Dharmsheel.shrivastav1@gmail.com

DOI: 10.1079/9781836990475.0003

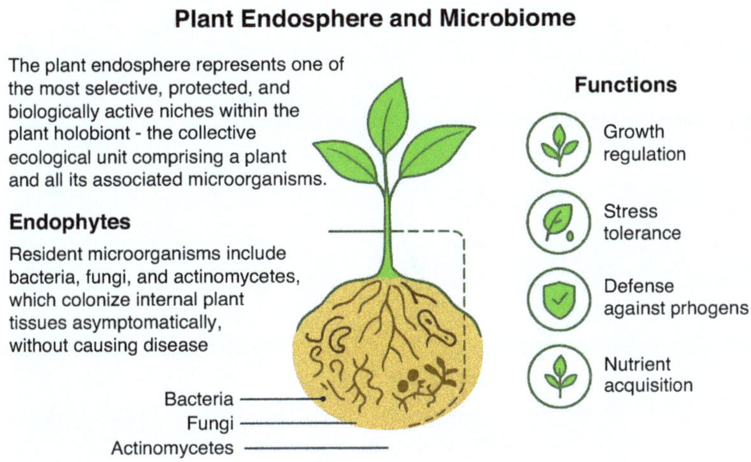

Fig. 3.1. Plant endosphere microbiome and its functions.

Endophytes infiltrate various internal plant compartments via natural openings, root hairs, stomata, wounds, or during seed development, eventually establishing themselves in intercellular (apoplastic) spaces and, in some cases, intracellular niches. Their ability to persist within these internal tissues differentiates them from epiphytes, which colonize plant surfaces, and from rhizospheric microbes, which inhabit the soil–root interface (Kandel *et al.*, 2017). The endosphere is therefore considered a highly selective environment, shaped by plant-derived chemical signals, immune responses, and nutrient availability. These factors, along with environmental and ecological pressures, determine which microbes successfully colonize and persist within plant tissues. As a result, the microbiome residing inside the plant is often less diverse but more functionally specialized compared to external microbial communities (Vandenkoornhuyse *et al.*, 2015).

Increasing evidence shows that plants and their endospheric microbiota form coevolved, mutualistic partnerships, suggesting long-term evolutionary adaptation. Endophytes significantly influence plant physiology by directly participating in processes such as nutrient acquisition, growth regulation, stress adaptation, and defense modulation (Fig. 3.1) (Adedayo and Olorunkosebi, 2025). Many bacterial endophytes produce phytohormones such as indole-3-acetic acid (IAA), cytokinins, and gibberellins, which regulate root architecture and overall plant development. Others enhance nutrient uptake by fixing atmospheric nitrogen, solubilizing phosphorus, or mobilizing essential micronutrients (Chaudhary *et al.*, 2022). Endophytic fungi and actinomycetes, meanwhile, are known for synthesizing a broad range of secondary metabolites, including antibiotics, antifungals, antioxidants, and plant-protective compounds that contribute to disease suppression and improved stress resilience (Gao *et al.*, 2025).

One of the most significant roles of endophytes is their ability to enhance plant tolerance to biotic and abiotic stressors. Through mechanisms such as induced systemic resistance (ISR), reactive oxygen species (ROS) regulation, and osmolyte production, endophytes equip plants with additional layers of protection against pathogens, drought, salinity, heavy metals, and temperature extremes. This makes endophytic communities crucial for plant survival in challenging or changing environments (Berg *et al.*, 2020). Moreover, the internal localization of these microbes gives them a functional advantage over rhizospheric or phyllospheric organisms, as they remain shielded from environmental fluctuations and can continuously influence plant metabolic and defense pathways.

Recent advances in multi-omics technologies, including genomics, metagenomics, transcriptomics, proteomics, and metabolomics, have revolutionized our understanding of endospheric microbiomes. These tools enable detailed characterization of microbial diversity, functional potential, metabolic interactions, and ecological dynamics within plant tissues. Multi-omics studies have uncovered complex networks of microbial cooperation, competition, and signaling within the endosphere, revealing that endophyte communities are not passive occupants but active contributors to plant health and ecosystem functioning (López-Mondéjar *et al.*, 2017). Through these approaches, researchers have been able to identify novel microbial taxa, previously unknown metabolic pathways, and regulatory mechanisms underlying endophyte–plant associations.

Our rapidly expanding knowledge of endosphere microbiomes has far-reaching implications for sustainable agriculture, crop improvement, ecosystem restoration, and biotechnology. Understanding how these microbial communities assemble, interact, and function opens possibilities for harnessing endophytes as biofertilizers, biocontrol agents, stress mitigation tools, phytoremediation enhancers, and sources of pharmaceutically valuable metabolites. Furthermore, insight into endophyte-mediated plant resilience offers promising avenues for developing climate-resilient crops and microbiome-informed breeding strategies aimed at enhancing plant performance under unpredictable environmental conditions (Liu *et al.*, 2020).

3.2 Microbial Diversity in the Internal Tissues of Plants

The internal tissues of plants harbour a highly specialized and diverse microbiome composed of bacteria, fungi, actinomycetes, archaea, and several lesser known microbial taxa. Unlike rhizospheric or phyllospheric microorganisms that reside on plant surfaces, these endophytic communities inhabit internal tissues, occupying intercellular apoplastic spaces and, in some cases, intracellular niches. Their capacity to persist asymptomatically reflects a finely tuned balance between microbial activity and host immune regulation, enabling stable associations without eliciting disease responses (Sivakumar *et al.*, 2020). Recent advances in cultivation techniques, together with high-resolution metagenomics, amplicon sequencing, and shotgun sequencing, have greatly expanded knowledge of endosphere diversity, revealing a microbial richness far exceeding what culture-dependent methods alone detect (Pérez-Cobas *et al.*, 2020).

Bacteria form a major component of this diversity, with Proteobacteria (*Pseudomonas*, *Burkholderia*, *Rhizobium*, and *Enterobacter*), Actinobacteria (*Streptomyces* and *Micromonospora*), and Firmicutes (*Bacillus* and *Bacteroidetes*) contributing to nutrient mobilization, stress tolerance, and biocontrol activities (Tshikhudo *et al.*, 2023) (Table 3.1). Fungal endophytes, primarily members of Ascomycota and Basidiomycota, also play essential roles by producing secondary metabolites such as alkaloids, terpenoids, polyketides, and phenolics that enhance plant defense and resilience. Other groups including actinomycetes, archaeal lineages, protists, endophytic viruses, and diverse uncultured taxa add further complexity to the endophytic landscape (Nisa *et al.*, 2015).

Moreover, endophyte diversity varies across plant organs, with roots typically supporting the highest microbial richness due to their role as the interface with soil microbiota, while stems, leaves, flowers, fruits, and seeds harbor distinct communities shaped by organ physiology, host genotype, developmental stage, and environmental conditions such as soil type, climate, and stress exposure (Delaux and Schornack, 2021). Some endophytes are vertically inherited through seeds, whereas others are acquired horizontally from the environment, contributing to dynamic and tissue-specific microbial colonization patterns. Functionally, endophytic microbes provide multiple benefits to the plant host, including nitrogen fixation, phosphate solubilization, micronutrient mobilization, siderophore production, phytohormone synthesis, and the generation of antimicrobial and antioxidant metabolites. They further enhance plant tolerance to drought, salinity, and temperature stress while also activating defense responses such as induced systemic resistance (ISR), collectively supporting plant growth, immunity, and resilience across diverse ecological settings (Delaux and Schornack, 2021).

3.3 Entry Routes and Colonization Processes

Endophytic colonization typically begins at the rhizoplane, a dynamic interface where plant roots interact directly with the surrounding soil environment. Plants continuously release a diverse and chemically rich mixture of root exudates, including amino acids, simple and complex sugars, organic acids, phenolics, flavonoids, and other secondary metabolites. These compounds serve not only as nutrient sources but also as chemosensory signals that attract beneficial soil microorganisms. Many rhizobacteria exhibit chemotaxis, enabling them to migrate along concentration gradients of root exudates. This targeted movement is the first step in the transition of microorganisms from free-living soil dwellers to potential endophytic symbionts (Chen and Liu, 2024).

Once microorganisms arrive at the root surface, they gain entry into internal plant tissues through natural openings and structural weak points.

Table 3.1. Examples of microbial diversity in plant internal tissues and their major ecological functions.

Host plant	Microbial group	Dominant taxa/genera	Tissue type	Functional traits	References
Arabidopsis thaliana	Bacteria	Proteobacteria (*Pseudomonas*, *Burkholderia*), Actinobacteria	Roots, leaves	Growth promotion, pathogen suppression	(Vimal *et al*., 2024)
Oryza sativa (rice)	Bacteria, fungi	Proteobacteria, Firmicutes; Ascomycota (*Fusarium*, *Trichoderma*)	Roots, stems	Phosphate solubilization, stress tolerance	(Fu *et al*., 2024)
Zea mays (maize)	Bacteria	Firmicutes (*Bacillus*), Actinobacteria (*Streptomyces*)	Roots, leaves	Nitrogen fixation, biocontrol agents	(Anand *et al*., 2023)
Solanum lycopersicum (tomato)	Bacteria, fungi	Proteobacteria, Actinobacteria, Ascomycota	Roots, stems, leaves	Phytohormone production, pathogen antagonism	(Vimal *et al*., 2024)
Medicago truncatula	Bacteria	*Rhizobium*, *Bacillus*, *Pseudomonas*	Roots, nodules	Nitrogen fixation, growth promotion	(Zancarini *et al*., 2025)
Triticum aestivum (wheat)	Bacteria, fungi	Ascomycota (*Fusarium*, *Alternaria*), Proteobacteria	Roots, leaves	Stress mitigation, nutrient cycling	(Kimotho and Maina, 2024)

In root systems, common portals of entry include root hairs, whose thin cell walls allow relatively easy penetration, as well as sites of lateral root emergence where temporary disruptions in epidermal integrity naturally occur (Mendes *et al.*, 2013). Microcracks formed during primary and secondary root growth also serve as important access points. In aerial plant organs, endophytes can enter through stomata, which regularly open for gas exchange, and through hydathodes, the specialized structures involved in guttation. Even minor wounds or microscopic injuries can provide additional opportunities for microbial infiltration. Together, these varied entry routes enable a wide range of microorganisms, including bacteria, fungi, actinomycetes, and even certain archaeal groups, to breach external plant barriers and initiate successful colonization (Kandel *et al.*, 2017).

Upon reaching an entry point, successful endophytes employ an arsenal of cell wall-modifying enzymes, such as cellulases, pectinases, xylanases, and hemicellulases. Rather than causing extensive tissue damage, these enzymes allow controlled and localized degradation of cell wall components, enabling microbes to navigate through the apoplast (intercellular spaces). This delicate enzymatic activity distinguishes endophytes from pathogens, which typically provoke aggressive host responses through extensive cell wall disruption (Dogan and Taskin, 2021). Once inside, microbes expand their niche primarily within intercellular spaces, where nutrient availability, moisture conditions, and reduced immune surveillance facilitate establishment. The apoplast acts as a favorable habitat that supports microbial growth while providing access to plant-derived compounds (Fig. 3.2). Some specialized endophytes go a step further and adopt an intracellular lifestyle, entering living plant cells without triggering hypersensitive responses. These intracellular colonizers must carefully modulate host cellular processes to maintain compatibility (Roussin-Léveillée *et al.*, 2024).

To ensure long-term survival without triggering strong host defense responses, endophytes employ a variety of strategies that help maintain a balanced, nonpathogenic relationship with their plant hosts. One key mechanism involves the production of antioxidants that neutralize ROS generated during early immune signaling, thereby preventing oxidative damage and limiting defense escalation (Praveen *et al.*, 2023). Endophytes also secrete phytohormone mimics or modulators, such as auxins, cytokinins, and gibberellins which help adjust plant growth–defense trade-offs and promote favorable physiological conditions for colonization. In addition, many endophytes release effector molecules capable of transiently suppressing, bypassing, or fine-tuning innate immune pathways, allowing them to persist within tissues without eliciting strong defense reactions. Modifications of surface structures, including alterations to lipopolysaccharides, further enable these microbes to evade recognition by host pattern recognition receptors. Together, these finely coordinated strategies illustrate the high degree of coevolution and interdependence between plants and their associated endophytes (Hardoim *et al.*, 2008; Pandey *et al.*, 2023).

Ultimately, colonization success relies on the microbe's capacity to establish a stable niche, proliferate at controlled densities, and remain

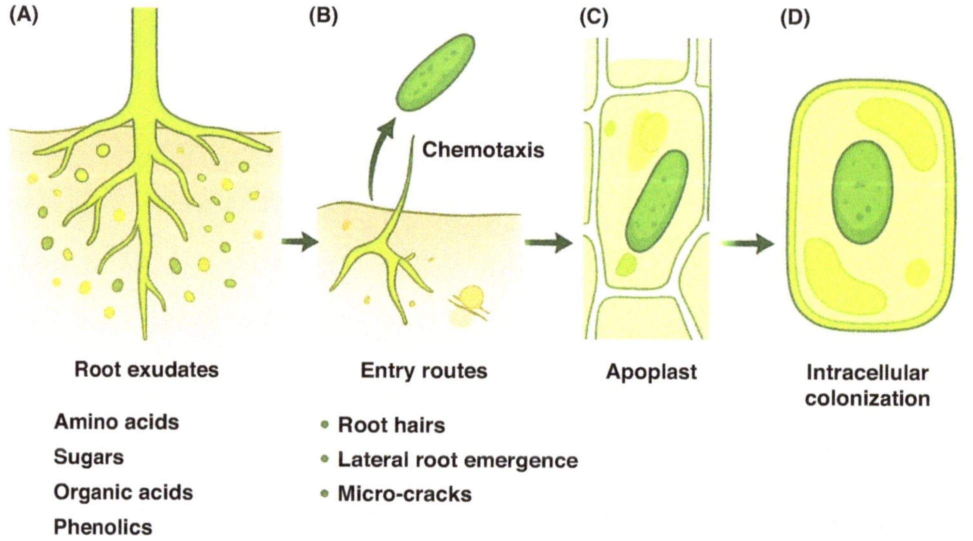

Fig. 3.2. Entry routes and colonization processes of microbes in plant roots. (A) Root exudates: Plant roots release compounds such as amino acids, sugars, organic acids, and phenolics that attract microbes. (B) Entry routes: Microbes enter roots through root hairs, sites of lateral root emergence, or micro-cracks in the root surface, guided by chemotaxis. (C) Apoplast colonization: Microbes move and multiply in the apoplastic space (the cell wall and intercellular spaces) of root tissues. (D) Intracellular colonization: Microbes penetrate plant cells, establishing intracellular colonization.

asymptomatic. This equilibrium shapes the assembly of functional endosphere communities, contributing to plant health, stress tolerance, and environmental adaptability. The subtlety and complexity of these processes underscore the intricately evolved nature of plant endophyte symbioses.

3.4 Assembly and Determinants of Endosphere Communities

The formation of endosphere microbiomes is a multistage, highly selective ecological process shaped by interactions among environmental reservoirs, plant physiological processes, and microbe–microbe dynamics. Unlike the relatively open and diverse microbial habitats surrounding plants, the endosphere represents a highly constrained internal niche where only a fraction of environmental microbes can successfully colonize, survive, and function. This hierarchical filtering ensures that internal plant tissues host microbial communities that are not only taxonomically distinct but functionally adapted to the plant environment (Vimal *et al.*, 2024).

3.4.1 Environmental reservoirs and initial recruitment

The establishment of endosphere microbiomes begins with the recruitment of microorganisms from diverse environmental reservoirs that act as primary microbial pools for potential colonizers. Bulk soil, one of Earth's most diverse microbial habitats, provides a vast array of bacteria, fungi, archaea, and protists, although only a small fraction possess the traits required for internal colonization. As roots grow, they modify this soil environment to create the rhizosphere, a nutrient-rich zone shaped by root exudates such as amino acids, sugars, phenolics, fatty acids, and signaling compounds (Xia *et al.*, 2021). These exudates attract motile and chemotactic microbes, select metabolically compatible taxa, and function as a critical ecological filter that narrows the diverse soil community into organisms capable of adhering to and interacting with the root surface.

Beyond soil, the phyllosphere, including leaves, stems, and aerial plant tissues, receives constant microbial inputs from airborne particulates, dust, water droplets, and atmospheric aerosols. Despite fluctuating UV, humidity, and temperature conditions, these aerial communities serve as important sources of stress-tolerant microbes that can enter plant tissues through stomata, hydathodes, microwounds, and trichome bases (Wu *et al.*, 2024) (Fig. 3.3). Rainwater and irrigation systems further contribute microbial influxes, serving as conduits for both local and long-distance taxa introduced during rain events or transported through agricultural water infrastructure; these waterborne microbes may infiltrate plants via hydathodes, moisture-facilitated stomatal openings, and water-induced microcracks (Bell-Dereske *et al.*, 2017).

Seeds provide another foundational reservoir, enabling vertical transmission of microbes associated with the embryo, endosperm, and seed coat. These seedborne endophytes often influence germination, shape early root and shoot microbiomes, prime immune responses, and establish priority effects that guide microbial succession throughout plant development. Additionally, insects and arthropods act as biological vectors that transfer microbes among plants through feeding wounds, deposition of frass or saliva, and passive transport on their exoskeletons, with some endophytes exhibiting coevolved tripartite relationships involving both plants and insect hosts (Shahzad *et al.*, 2018).

Ultimately, while each environmental reservoir contributes unique microbial lineages, only a subset possesses the necessary abilities such as immune evasion, adhesion, stress tolerance, and metabolic adaptability to move from external environments into internal tissues. Seedborne microbes typically dominate early developmental stages, whereas soil-derived taxa become more prominent during root expansion and vegetative growth, creating a dynamic integration of vertical inheritance and horizontal environmental acquisition that drives endosphere community assembly.

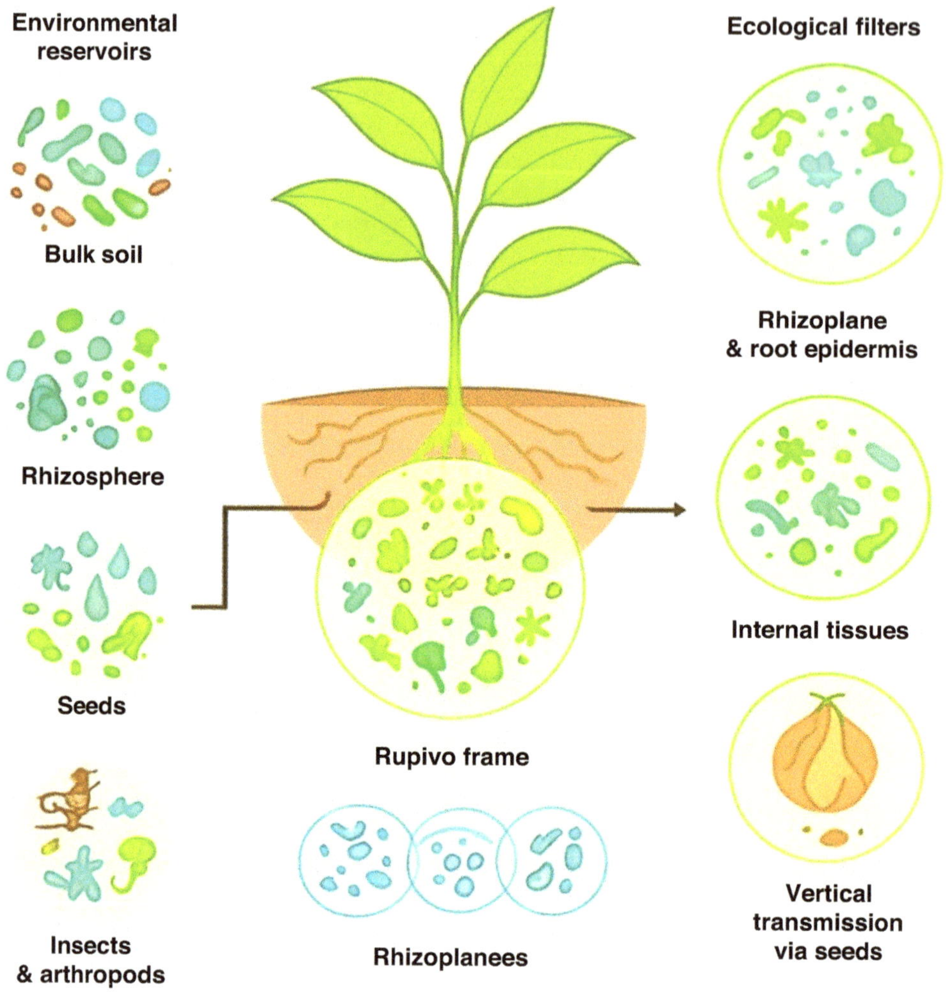

Fig. 3.3. Assembly and determinants of plant endosphere microbiomes. Plant microbiomes are shaped by microbial inputs from environmental reservoirs, including bulk soil, rhizosphere, seeds, and insects/arthropods. Microbes encounter ecological filters at the rhizoplane and root epidermis, internal tissues, and through vertical transmission via seeds. The 'Rupivo Frame' illustrates the microbial communities within the plant, while Rhizoplanees represent microbial populations specifically associated with root surfaces. Microbiome composition is influenced by both abiotic factors (pH, temperature, moisture, salinity) and biotic interactions (competition, cooperation, antagonism, viral effects).

3.4.2 Rhizosphere: The first level of ecological filtering

The rhizosphere serves as the first and most critical ecological filter that shapes the transition of environmental microbes toward potential endophytic lifestyles, functioning as a highly dynamic interface where plant–soil–microbe interactions exert strong selective pressures. This biologically active zone is enriched with root exudates including sugars, amino acids, organic and fatty acids, phenolics, flavonoids, alkaloids, vitamins, and diverse signaling molecules which collectively represent 20–40% of photosynthetically fixed carbon and create nutrient-rich chemical gradients that attract and structure microbial populations (Pantigoso *et al.*, 2022).

Microbial recruitment in the rhizosphere is strongly influenced by metabolic compatibility, with only those microbes capable of efficiently utilizing exudate components such as glucose, sucrose, malate, citrate, amino acids, flavonoids, and coumarins becoming enriched near root surfaces. Chemotactic responsiveness further refines this selection, as exudates act as chemoattractants that enable motile taxa, including *Pseudomonas*, *Burkholderia*, *Bacillus*, and *Azospirillum*, to migrate toward the root through flagellar movement and specialized chemotaxis gene pathways. At the same time, the rhizosphere presents fluctuating conditions characterized by oxygen gradients, variable moisture, pH shifts, and intense microbial competition, requiring incoming taxa to possess stress resilience traits such as oxidative stress tolerance, biofilm formation capacity, and efficient nutrient acquisition systems.

These combined pressures transform the rhizosphere into a structured and selectively enriched microbiome distinct from bulk soil, dominated by microbes adapted to root-associated lifestyles, adhesion to root surfaces, antibiotic and siderophore production, and potential for endophytic entry (Yang *et al.*, 2024). Ultimately, this refined rhizosphere community determines which microbes progress to the rhizoplane, root epidermis, and natural entry sites, meaning that only those microorganisms capable of meeting metabolic, chemotactic, and stress-related criteria successfully pass through this ecological sieve to become endophytes. Thus, the rhizosphere is not simply a microbial recruitment zone but a fundamental determinant of endosphere assembly, shaping the diversity, composition, and functional capacities of microorganisms capable of colonizing internal plant tissues.

3.4.3 Rhizoplane and root epidermis: Selective entry barriers

As microbes move from the rhizosphere toward the plant interior, they encounter the rhizoplane, the root surface which functions as a critical selective barrier regulating microbial access to internal tissues. This interface forms a highly competitive microenvironment where structural defenses, biochemical secretions, and molecular recognition systems intensify microbial filtering, allowing only a small subset of organisms to advance to the root epidermis and cortex (Bhadrecha *et al.*, 2023a; Grzyb and Szulc, 2024).

Successful endophytes often possess finely regulated enzymatic capabilities, including cellulases, pectinases, polygalacturonases, xylanases, and other hemicellulases, which facilitate localized softening or microchannel formation in cell walls without causing the extensive tissue maceration characteristic of pathogens. At the same time, they must navigate plant immune responses initiated by pattern recognition receptors (PRRs) that detect microbial features such as flagellin, chitin, or lipopolysaccharides; thus, endophytes employ immune-modulating strategies such as effector secretion to suppress pattern-triggered immunity, structural modification of microbe-associated molecular patterns (MAMPs) to avoid recognition, antioxidant production to neutralize reactive oxygen species, and modulation of phytohormone pathways, for example lowering ethylene levels via ACC deaminase. Compatibility with plant PRRs further influences colonization success, as microorganisms with reduced surface immunogenicity or those capable of masking recognizable molecules through extracellular polysaccharides progress more readily toward endophytic lifestyles (García-Latorre *et al.*, 2023).

The rhizoplane also serves as a biofilm-enriched niche where adhesion and persistence are essential, enabling endophytes to utilize extracellular polymeric substances for stable attachment, protection from oxidative or antimicrobial compounds, creation of buffered microenvironments, and coordinated colonization through quorum sensing (Riseh *et al.*, 2025). Once at the epidermis, only microbes that avoid strong immune activation, move through tissues without causing damage, tolerate apoplastic nutrient limitations, and appropriately modulate or respond to plant signaling pathways penetrate deeper into the cortex and eventually the vascular system (Zhang *et al.*, 2025). Collectively, the rhizoplane and root epidermis act as finely tuned selective gateways that integrate structural barriers, biochemical defenses, molecular recognition events, and microbial adaptability to ensure that only organisms capable of controlled entry, immune evasion, and mutualistic or neutral behavior successfully establish within the plant endosphere.

3.4.4 Internal colonization and niche establishment

Once endophytes successfully traverse the rhizoplane and root epidermis, they enter the internal plant tissues, encountering a highly selective, resource-limited environment where stable niche establishment is critical for functional endosphere communities. Within intercellular spaces and the apoplast, endophytes must compete with resident microbiota for limited nutrients such as sugars, amino acids, and iron, employing strategies including siderophore-mediated scavenging, antimicrobial compound production, and spatial partitioning across roots, stems, and leaves (Kandel *et al.*, 2017). Concurrently, they modulate host immune responses to avoid triggering detrimental reactions by suppressing reactive oxygen species bursts, secreting effectors that transiently attenuate pattern-triggered immunity, producing antioxidants and detoxifying enzymes, and fine-tuning phytohormone signaling pathways, including ethylene, jasmonic acid, and salicylic acid. Endophytes must also adapt to heterogeneous microenvironments characterized by oxygen and nutrient gradients, localized pH variations, and the presence of antimicrobial secondary metabolites, which necessitate tolerance or detoxification mechanisms. By integrating into host hormonal networks, including auxins, cytokinins, and gibberellins, endophytes influence colonization success, microbial community stability, and plant development, including root architecture, shoot growth, and reproductive tissue formation (Wu *et al.*, 2021; Bhadrecha *et al.*, 2023b).

Microbes that successfully navigate these selective pressures establish structured, tissue-specific communities within intercellular or, in some cases, intracellular niches, forming stable consortia that contribute to plant growth promotion, stress tolerance, and pathogen suppression. Overall, internal colonization represents a highly selective, dynamic process shaped by microbial competition, immune evasion, environmental adaptation, and hormonal interactions, ensuring that only compatible and functionally beneficial microbes become persistent members of the plant endosphere (Fadiji *et al.*, 2025).

3.4.5 Core microbiomes and host-specific assembly

Through successive ecological filtering and selective recruitment, plants establish tissue-specific and developmentally regulated endophytic communities that collectively contribute to growth, immunity, and environmental adaptability. These selective processes give rise to core microbiomes, defined as microbial taxa consistently present across individuals of the same species or genotype and maintained throughout particular developmental stages. Endophyte composition varies among organs, reflecting distinct microenvironments and functional requirements: roots harbour the most diverse communities enriched in nitrogen-fixing bacteria (e.g. *Azospirillum*, *Herbaspirillum*), phosphate solubilizers, and stress-tolerant taxa; stems and leaves support microbes that produce antimicrobial compounds, volatile organic compounds (VOCs), and secondary metabolites, while participating in systemic signaling that primes distal tissues; seeds carry vertically transmitted endophytes that establish heritable lineages influencing germination, early seedling vigor, and subsequent root and shoot microbiomes (Dadi *et al.*, 2025). The core microbiome is characterized by consistency across host individuals, persistence through developmental stages, and functional redundancy, ensuring nutrient acquisition, stress tolerance, and immune modulation even under fluctuating environmental conditions.

Plant ontogeny further shapes microbial assembly; early growth stages are dominated by seedborne taxa that establish priority effects and modulate immune development, whereas vegetative and reproductive stages incorporate additional microbes from soil, rhizosphere, and aerial sources, enhancing diversity and functional capacity. Stress conditions selectively enrich endophytes that confer tolerance or protection, illustrating the adaptive flexibility of core communities. Collectively, these tissue- and host-specific microbiomes perform essential roles in nitrogen fixation, phosphate solubilization, micronutrient mobilization, phytohormone production (IAA, gibberellins, cytokinins), stress resilience, and intergenerational inheritance via seeds, forming stable, functionally indispensable consortia that underpin plant health, productivity, and coevolution with their microbial partners (Zhou *et al.*, 2024).

3.4.6 Influence of abiotic and biotic factors

The assembly, composition, and stability of endophytic communities are shaped by a complex interplay of abiotic factors and biotic interactions, which collectively determine microbial recruitment, colonization efficiency, and long-term persistence within plant tissues.

Abiotic determinants, such as soil texture, pH, organic matter content, temperature, moisture regimes, salinity, heavy metals, and agricultural practices, influence both the availability of potential endophytes and the physiological state of the host plant, thereby modulating selective filtering at the rhizosphere, rhizoplane, and internal tissues (Vimal *et al.*, 2024). Soil structure affects water retention and microbial mobility, while pH and nutrient availability select for specific microbial taxa. Environmental stresses, including drought, temperature extremes, salinity, or heavy metal contamination, favor microbes with adaptive traits such as osmoprotectant synthesis, metal-chelating siderophores, or stress-tolerant enzymes, which in turn enhance host resilience. Agricultural interventions, including fertilizer application, pesticide use, crop rotation, and intercropping, further shape microbial composition by altering resource availability, competitive dynamics, and environmental reservoirs.

Biotic interactions within the endosphere refine these assemblages by mediating competition, cooperation, and antagonism among cocolonizing microbes (Abdul Rahman *et al.*, 2021; Bhawana *et al.*, 2024). Resource competition for limited carbon, nitrogen, phosphorus, and micronutrients drives niche partitioning, while syntrophic exchanges and metabolic crossfeeding stabilize communities under nutrient-limited conditions. Quorum sensing coordinates biofilm formation, enzyme secretion, and secondary metabolite production, facilitating cooperative behavior and defense against competitors. Additionally, antagonistic interactions through antibiotics, bacteriocins, and lytic enzymes suppress rival taxa and maintain microbial homeostasis, while viral infections (mycoviruses and bacteriophages) can alter population dynamics, horizontal gene transfer, and competitive outcomes.

Together, abiotic and biotic factors act synergistically to structure the endosphere microbiome, determining which taxa are recruited, persist, and contribute functional capabilities, ultimately shaping plant health, resilience, and productivity (Wang *et al.*, 2025).

3.5 Spectrum of Plant–Microbe Interaction Outcomes

Plant–microbe interactions span a dynamic continuum ranging from mutualistic and commensal relationships to latent pathogenicity and outright parasitism. This continuum, often referred to as "behavioral dualism," highlights that the outcome of a plant–endophyte association is context dependent, influenced by host genotype, microbial genotype, environmental conditions, and microbial population density (Schulz and Boyle, 2005; Anand *et al.*, 2023). Understanding this spectrum is essential for elucidating the ecological roles of endophytes and their potential applications in agriculture and biotechnology.

3.5.1 Mutualistic and beneficial associations

Mutualistic endophytes form beneficial associations with their host plants, enhancing growth, development, and resilience to environmental stresses. These microbes contribute to nutrient acquisition by fixing atmospheric nitrogen through diazotrophic bacteria such as *Azospirillum* and *Herbaspirillum*, solubilizing phosphate and mobilizing micronutrients including iron and zinc. Siderophore production further facilitates nutrient availability (Chaudhary *et al.*, 2022). Endophytes also synthesize phytohormones: IAA promotes root proliferation, gibberellins stimulate stem elongation and seed germination, and cytokinins regulate cell division and organ differentiation, thereby directly influencing plant development. In addition, they enhance tolerance to abiotic stresses: osmoprotectants like proline and trehalose mitigate drought and salinity effects, heat-shock or antifreeze proteins confer temperature resilience, and metal-chelating compounds aid in heavy metal detoxification (Table 3.2, Fig. 3.4).

Biotic stress protection is provided through induction of plant defense pathways such as induced systemic resistance (ISR) and systemic acquired resistance (SAR), as well as by producing antimicrobial secondary metabolites and VOCs that suppress pathogens. These mutualistic interactions are often accompanied by immune modulation, allowing endophytes to fine-tune host defenses and establish stable, asymptomatic colonization (Hacquard *et al.*, 2017).

3.5.2 Commensal and neutral interactions

Commensal endophytes inhabit plant tissues with minimal or often undetectable effects on host physiology, maintaining a largely neutral relationship. These microbes passively reside within intercellular spaces without significantly altering plant growth, development, or immune responses. Despite their apparent neutrality, commensal endophytes play important indirect ecological roles: they compete with potential pathogens for nutrients and space, modulate microbial community dynamics through signaling and quorum sensing, and contribute to functional redundancy within the endosphere. By maintaining community stability and buffering against perturbations, these organisms indirectly support plant health, resilience, and productivity (Hardoim *et al.*, 2008; Compant *et al.*, 2021; Xia *et al.*, 2022).

Table 3.2. Spectrum of plant–microbe interaction outcomes and their key characteristics.

Interaction type	Description	Mechanisms	Examples	Reference
Mutualistic/ beneficial	Endophytes enhance plant growth, development, and stress resilience	Nutrient acquisition: N_2 fixation, phosphate solubilization, micronutrient mobilization (Fe, Zn) Phytohormone production: IAA, gibberellins, cytokinins Abiotic stress tolerance: drought, salinity, temperature, heavy metal detoxification Biotic defense: ISR, SAR, antimicrobial compounds, VOCs Immune modulation for asymptomatic colonization	*Azospirillum*, *Herbaspirillum*	(Tariq *et al*., 2025)
Commensal/ neutral	Endophytes inhabit plant tissues with minimal or no effect on host physiology	Passive residence in intercellular spaces Neutral impact on growth and immune response Competes with pathogens for space and nutrients Contributes to community stability and functional redundancy	*Bacillus subtilis* and *Streptomyces* species	(Xia *et al*., 2022)
Latent pathogenic/ opportunistic	Endophytes remain asymptomatic under normal conditions but can become pathogenic under stress	Pathogenic traits expressed in response to environmental or host stress (nutrient imbalance, oxidative stress, impaired immunity, high microbial density) Context-dependent pathogenicity	*Fusarium* spp., *Colletotrichum* spp.	(Chaudhary *et al*., 2022)
Parasitic/ pathogenic	Endophytes cause disease and damage to the host	Obligate or facultative pathogens Direct harm to plant tissues Often symptomatic infection	Classic plant pathogens (e.g. certain strains of *Fusarium*, *Phytophthora*)	(de Lamo and Takken, 2020)

IAA, indole-3-acetic acid; ISR, induced systemic resistance; SAR, systemic acquired resistance; VOC, volatile organic compound.

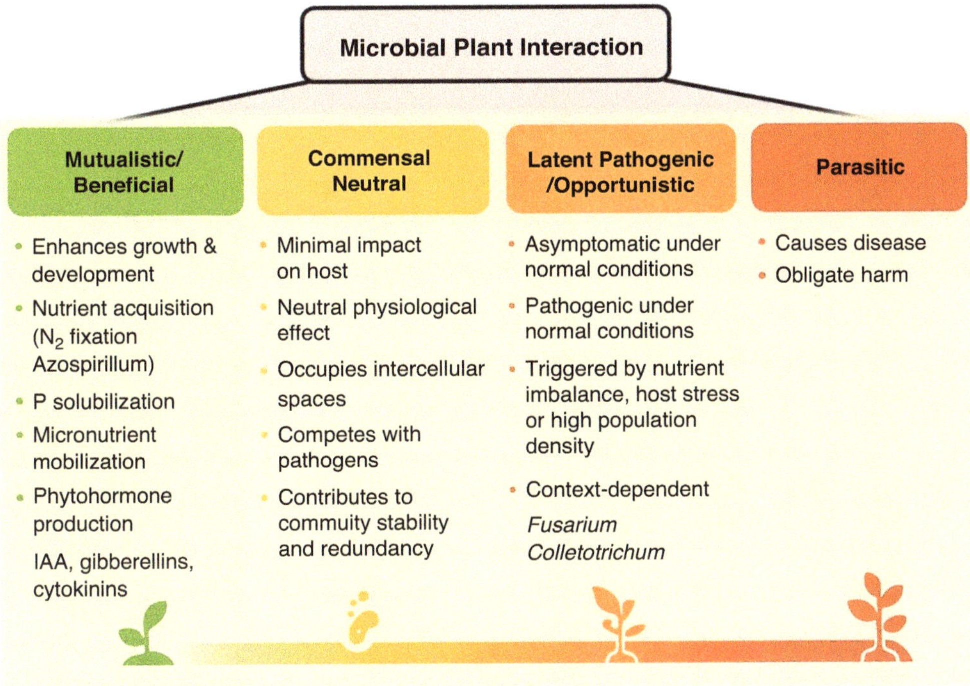

Fig. 3.4. Spectrum of plant–microbe interaction outcomes.

3.5.3 Latent pathogenicity and opportunism

Some endophytes exhibit latent pathogenicity or opportunistic behavior, where they remain asymptomatic under normal conditions but can express pathogenic traits in response to specific environmental or host-related triggers. Factors that can induce this pathogenic shift include nutrient imbalances, carbon limitation, oxidative stress, impaired host immunity, or high microbial population density and competitive pressure. Under such conditions, otherwise benign endophytes may cause disease, as observed with certain *Fusarium* or *Colletotrichum* species that behave as latent pathogens in some host plants. This context-dependent pathogenicity highlights the dynamic and flexible nature of plant–microbe interactions, demonstrating that microbial behavior is not fixed but emerges from the interplay between microbial traits, host physiology, and environmental conditions (Hardoim *et al.*, 2008; Compant *et al.*, 2021; Chaudhary *et al.*, 2022).

3.6 Molecular Dialogue and Plant Immunity

The interaction between plants and endophytic microorganisms is orchestrated through an intricate molecular dialogue that determines whether colonization results in mutualism, commensalism, or disease. Plants constantly monitor microbial presence via PRRs located on the plasma membrane, which detect conserved MAMPs such as flagellin, chitin, peptidoglycan, and lipopolysaccharides. Recognition of these molecular signatures causes MAMP-triggered immunity (MTI), initiating a rapid cascade of defense responses, including ROS bursts, cell wall reinforcement through callose and lignin deposition, calcium influx, and the induction of defense-related genes (Berg *et al.*, 2020; Liu *et al.*, 2020). MTI represents the first immune barrier preventing excessive microbial proliferation within plant tissues.

Successful endophytes must therefore overcome or modulate host immunity to achieve stable colonization while maintaining host health. Beneficial endophytes secrete a repertoire of cell wall-modifying enzymes such as pectinases, cellulases, and xylanases facilitating controlled tissue penetration without triggering hypersensitive reactions. They additionally release effectors, antioxidants, and phytohormone modulators, including IAA and ACC deaminase, which fine-tune defense pathways and suppress excessive MTI activation (Liu *et al.*, 2020). This immunomodulatory behavior enables coexistence and distinguishes beneficial or neutral endophytes from pathogens, which typically induce strong immune activation and tissue damage.

Beyond localized immune interactions, many beneficial endophytes enhance whole-plant immunity by inducing SAR and ISR, which are mediated primarily through salicylic acid (SA) and jasmonic acid/ethylene (JA/ET) signaling pathways, respectively. These systemic responses prime the plant immune system, enabling faster and stronger activation upon subsequent pathogen attack while avoiding the energetic cost of continuous defense activation. Species of *Bacillus*, *Pseudomonas*, and *Trichoderma* are widely reported to enhance systemic immunity, stimulating defense enzymes such as peroxidases, phenylalanine ammonia-lyase, chitinases, and β-1,3-glucanases, contributing to broad-spectrum disease suppression (Khavkin, 2021; Srivastava *et al.*, 2025).

3.7 Functional Roles in Plant Growth and Nutrition

The endophytic microbiome plays a crucial role in modulating plant growth and nutrition, acting as an integral component of the holobiont that influences host development and productivity. Vandenkoornhuyse *et al.* (2015) emphasize that microbes inhabiting internal plant tissues contribute significantly to nutrient cycling, hormone regulation, and stress mitigation, which collectively boost plant fitness and adaptation. Endophytes facilitate nutrient acquisition through diverse mechanisms such as nitrogen fixation, phosphate solubilization, potassium mobilization, and siderophore production. These processes increase the bioavailability of essential macro- and micronutrients, often poorly accessible to plants under natural soil conditions. For example, nitrogen-fixing bacteria within the endosphere supplement host nitrogen requirements without the need for costly external fertilizers, helping sustain growth in nutrient-limited environments (Ramírez-López *et al.*, 2025). Phytohormone production is another salient feature of beneficial endophytes. Many produce auxins (IAA), cytokinins, gibberellins, and enzymes like ACC deaminase, which modulate ethylene stress hormone levels, leading to improved root architecture, enhanced cell division, and better adaptation under abiotic stresses such as drought and salinity. This hormone modulation results in increased root surface area, which enhances nutrient and water uptake efficiency (Vandana *et al.*, 2021).

Furthermore, endophytes contribute to maintaining nutrient homeostasis by decomposing complex organic matter within plant tissues and recycling nutrients critical for growth. Microbial metabolites can also elicit expression of host genes involved in nutrient transport and metabolism, indicating a sophisticated crosstalk that fine-tunes plant physiological processes. The functional benefits extend beyond single strains; synergistic interactions within microbial consortia amplify positive effects on the host. Such microbial networks modulate metabolic pathways, enhance stress resilience, and stabilize the plant microbiome, underscoring the need to view plant nutrition as a product of integrated host–microbe systems rather than isolated components.

Understanding and harnessing these microbial functions holds promise for developing sustainable agricultural practices that reduce dependence on chemical fertilizers and pesticides, while improving crop productivity and resistance to environmental stresses (Vandenkoornhuyse *et al.*, 2015; Vandana *et al.*, 2021; Jadhav *et al.*, 2025).

3.7.1 Roles in resistance to biotic and abiotic stress

Endophytic microbes are key players in enhancing plant resistance to both biotic stresses (caused by pathogens, insects, and other organisms) and abiotic stresses (environmental pressures such as drought, salinity, heavy metals, and extreme temperatures). This dual protective role positions endophytes as critical components of integrated plant defense strategies for sustainable agriculture (Ramatsitsi and Manyevere, 2025).

3.7.1.1 Resistance to biotic stress

Endophytes combat biotic stress through multiple direct and indirect mechanisms. Directly, endophytic bacteria and fungi produce diverse antimicrobial compounds, including antibiotics, phenolic acids, lytic enzymes (chitinases, glucanases, proteases), and VOCs that actively inhibit or eliminate plant pathogens (Fig. 3.5). For instance, secondary metabolites such as polyketides and nonribosomal peptides, synthesized via polyketide synthase (PKS) and nonribosomal peptide synthetase (NRPS) pathways, demonstrate potent antibacterial and antifungal activities against multidrug-resistant pathogens. Additionally, endophytes compete for ecological niches and nutrients within plant tissues, creating a hostile microenvironment that deters pathogenic colonization and proliferation (Nazir *et al.*, 2024).

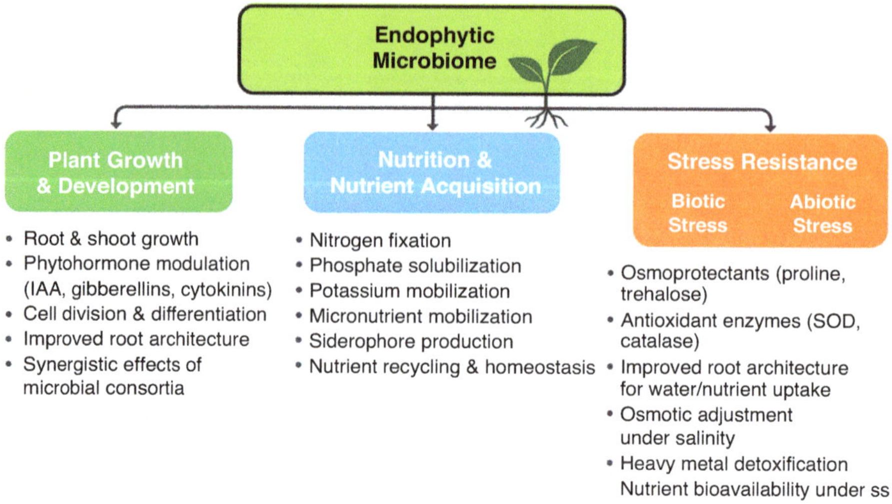

Fig. 3.5. Functional roles of the endophytic microbiome in plant growth, nutrition, and stress resistance.

Indirectly, endophytes fortify plant immunity by inducing SAR and ISR, priming plants for enhanced and faster defense responses upon pathogen encounter. These priming mechanisms involve modulation of salicylic acid and jasmonic acid signaling pathways, leading to upregulation of PR genes without incurring significant fitness costs associated with constitutive defense activation. Furthermore, endophytes produce siderophores that sequester iron, making it unavailable to iron-scavenging pathogens and thereby inhibiting their growth. Many endophytes also modulate host hormone levels by producing auxins, ACC deaminase, and other phytohormones, reducing plant stress and strengthening the capacity to mount robust immune responses (Chaudhary *et al.*, 2022). Research has demonstrated that endophytic inoculation significantly reduces disease incidence and severity caused by bacterial, fungal, and viral pathogens in numerous crop species. Endophytes are therefore recognized as promising eco-friendly biocontrol agents, reducing dependence on synthetic pesticides and contributing to sustainable crop management.

3.7.1.2 Resistance to abiotic stress

Endophytes enhance plant tolerance to abiotic stresses including drought, salinity, heavy metals, temperature extremes, and nutrient deficiency through both chemical and physiological mechanisms. Under drought stress, endophytes modulate plant osmoprotectant accumulation (e.g. proline, trehalose, glycine betaine) and enhance antioxidant enzyme production (superoxide dismutase, catalase, peroxidase), mitigating oxidative damage and maintaining cellular water status. Additionally, endophytes promote beneficial root architecture alterations that improve water and nutrient acquisition efficiency in water-limited environments (Kamran *et al.*, 2022).

For salinity stress, endophytic fungi such as *Trichoderma* species improve plant tolerance by increasing plant biomass, enhancing nutrient acquisition rates, promoting osmotic adjustments, and alleviating oxidative stress in host cells. These fungi alter root architecture to maximize resource uptake and produce stress-related enzymes and phytohormones that stabilize plant growth under high salt concentrations. A metaanalysis demonstrated that endophytic bacterial inoculation significantly improves plant biomass and stress resilience under heavy metal exposure (cadmium, copper, chromium, nickel, arsenic, aluminum), reducing bioaccumulation and toxicity (Verma *et al.*, 2022).

Endophytes also produce antioxidant metabolites and enzymes that neutralize ROS generated during abiotic stress, preventing oxidative damage to cellular lipids, proteins, and nucleic acids. Phytohormone production by endophytes, particularly ACC deaminase activity that reduces ethylene levels, directly attenuates stress responses and maintains plant development even under adverse conditions. Furthermore, endophytes enhance nutrient bioavailability (nitrogen fixation, phosphate solubilization, micronutrient mobilization) under stress, enabling sustained plant growth when external nutrient availability is compromised (Sena *et al.*, 2024).

3.7.1.3 Integrated stress resilience

Recent evidence indicates that endophytes provide superior protection when plants face combined biotic and abiotic stresses, a scenario increasingly common under climate change. The symbiotic relationship between plant and endophytic microbiome enables dynamic adjustment of defense strategies, resource allocation, and stress response pathways tailored to specific environmental challenges. This integrated approach to stress management represents a paradigm shift toward microbiome-informed sustainable agriculture that reduces chemical inputs while enhancing crop productivity and climate resilience (Verma *et al.*, 2022; Sena *et al.*, 2024).

3.8 Multi-omics Approaches to Endosphere Exploration

Understanding the complexity of plant endosphere microbiomes requires approaches that extend beyond traditional culture-dependent microbiology. Multi-omics technologies including genomics, transcriptomics, proteomics, metabolomics, metagenomics, metatranscriptomics, metaproteomics, and integrated systems biology have emerged as transformative platforms for deciphering the structure, function, and ecological roles of internal plant microbiomes (López-Mondéjar *et al.*, 2017). These multilayered analytical strategies provide a holistic understanding of gene content, gene expression dynamics, protein functional profiles, and metabolite signaling networks, thereby uncovering the molecular mechanisms that underlie plant–endophyte interactions (Younas *et al.*, 2025).

Advances in next-generation sequencing (NGS) technologies have revolutionized plant microbiome research by enabling high-resolution profiling of microbial communities and their functions (Table 3.3). Metagenomics,

Table 3.3. Multi-omics approaches for exploring plant endosphere microbiomes: platforms, insights, and applications.

Omics platform	Primary output/data type	Key analytical insights	Applications in endosphere microbiome research	References
Genomics/whole genome sequencing (WGS)	Complete genome sequences, gene annotations	Identifies functional genes, biosynthetic clusters, virulence and symbiosis factors	Discovery of plant growth-promoting genes, mapping biosynthetic pathways for secondary metabolites, taxonomic characterization	(Battu *et al*., 2017)
Metagenomics (shotgun/ amplicon 16S, ITS)	Microbial community structure and functional gene potential	Reveals diversity of culturable and unculturable microbes, predicts metabolic functions	Profiling community composition across plant tissues, identifying core microbiomes and beneficial strains	(Fadiji *et al*., 2021; Mehra *et al*., 2022)
Transcriptomics (RNA-Seq)	Differential gene expression profiles	Host and microbial transcriptional responses during colonization	Identifying regulatory genes, signaling pathways, stress response networks	(Chen *et al*., 2022)
Metatranscriptomics	Active microbiome expression datasets	Identifies actively expressed microbial genes in plants	Understanding metabolic activity and colonization competitiveness	(Vannier *et al*., 2023)
Proteomics/ metaproteomics	Protein abundance and functional protein profiles	Identifies enzymes, transporters, stress-responsive proteins	Understanding mechanisms of plant immunity modulation and nutrient transport	(Starr *et al*., 2018)
Metabolomics (LC-MS, GC-MS, NMR)	Small-molecule metabolites, biochemical fingerprints	Signaling molecules, phytohormones, defense compounds, antimicrobial metabolites	Mapping chemical dialogue between plant and microbes, discovering novel bioactive compounds	(Farag *et al*., 2012)

Epigenomics	DNA methylation, histone modifications	Plant–microbe interaction-induced epigenetic remodelling	Understanding heritable resistance traits and microbiome-induced gene regulation	(Wang and Chang, 2018)
Phenomics/imaging-based omics	Quantitative phenotypic trait data	Phenotypic expression of genotypes and stress responses	Linking microbiome functions to growth performance and stress tolerance	(Hériché *et al*., 2019)
Systems biology/integrated omics	Multilayer correlation networks and computational models	Holistic understanding of regulatory interactions and metabolic flux	Microbiome engineering, predictive modelling, designing synthetic consortia for agriculture	(Fukushima *et al*., 2009)

transcriptomics, proteomics, and metabolomics now link microbial taxonomy with metabolic pathways, regulatory signaling networks, and bioactive compound production (Liu *et al.*, 2020). These approaches help elucidate how endophytes contribute to nutrient mobilization, biosynthesis of antimicrobial metabolites, and enhancement of plant tolerance under abiotic and biotic stress conditions.

Emerging developments such as single-cell omics and spatial omics, combined with high-resolution microscopy, allow researchers to visualize colonization patterns and microbe–host interactions at tissue and even subcellular scales (Compant *et al.*, 2021). These tools offer unprecedented insights into spatial organization, niche specialization, and microbial behavior inside plant tissues, facilitating precise mapping of symbiotic zones and cellular entry routes. Such innovations underpin future efforts to rationally manage and engineer plant microbiomes, accelerating the development of targeted microbial inoculants and synthetic communities for improved crop health, productivity, and biotechnology applications (López-Mondéjar *et al.*, 2017; Jadhav *et al.*, 2025; Afoma *et al.*, 2026).

Integrative multi-omics frameworks, supported by systems biology and computational modelling, are now crucial for constructing functional interaction networks that describe how microbial consortia and plant hosts coordinate metabolism and signaling (Kumar *et al.*, 2023; Verma *et al.*, 2025). By correlating molecular data layers, these platforms identify regulatory modules, metabolic dependencies, and response pathways that sustain stable endosphere communities. Ultimately, multi-omics pipelines enable the transition from descriptive diversity analysis to predictive, mechanistic understanding, advancing microbiome-engineering strategies for sustainable agriculture (Younas *et al.*, 2025).

3.9 Future Directions and Applications

The expanding knowledge of plant endophytic microbiomes presents significant opportunities for sustainable agriculture, biotechnology, and environmental management. Future research and applications will focus on rational microbiome engineering, crop improvement, phytoremediation, and climate resilience. Development of engineered synthetic microbial consortia (SynComs) designed with complementary metabolic functions will enable more stable and effective inoculants compared to single-strain approaches. These consortia leverage functional redundancy and syntrophic interactions to ensure robust colonization and sustained benefits within the host (Afridi *et al.*, 2022; Malik *et al.*, 2025; Afoma *et al.*, 2026).

Endophytes will serve as living biofertilizers and biocontrol agents, reducing chemical input dependence while maintaining crop productivity. Integration of endophyte inoculants into microbiome-assisted breeding programs will select plant genotypes that preferentially recruit beneficial microbes, enhancing inherent resilience and stress tolerance. Endophytes capable of degrading pollutants and chelating heavy metals offer sustainable alternatives for soil and water remediation, particularly when combined with hyperaccumulator plants. Multi-omics profiling of plant–endophyte interactions will accelerate discovery of novel antimicrobial and therapeutic compounds from endophytic biosynthetic pathways, addressing the growing challenge of antimicrobial resistance (Najjar, 2025).

Endophyte-enhanced crops exhibiting superior drought, salinity, and temperature tolerance will support agricultural sustainability under climate change scenarios. However, regulatory frameworks for microbial inoculants, standardized quality control protocols, cost-effective formulation methods, and comprehensive field-scale validation studies remain critical priorities for translating endophyte research into widespread agricultural adoption.

3.10 Conclusion

The plant endosphere is a highly selective, dynamic, and biologically active niche where intricate interactions between host plants and endophytic microorganisms shape growth, health, and ecological adaptability. Endophytes, including bacteria, fungi, actinomycetes, archaea, protists, and viruses, colonize internal tissues asymptomatically, occupying intercellular and, in some cases, intracellular spaces. Their establishment is governed by hierarchical ecological filters, from recruitment in environmental reservoirs through selective passage via the rhizosphere, rhizoplane, and root epidermis to stable niche formation within plant tissues. Each stage imposes metabolic, structural, chemotactic, and immune-mediated constraints, ensuring that only functionally compatible microbes persist.

Endophytic communities are taxonomically and functionally diverse, with roots typically harbouring the richest assemblages and aerial tissues, reproductive organs, and seeds hosting specialized, sometimes heritable microbes. Core microbiomes—tissue-specific, host-adapted groups consistently maintained across developmental stages—perform critical roles in nutrient acquisition, phytohormone production, stress tolerance, and defense, establishing mutualistic equilibria that support plant fitness while limiting opportunistic pathogenicity.

Plant–microbe interactions span a continuum from mutualism and commensalism to latent pathogenicity, with outcomes influenced by host genotype, microbial traits, environmental factors, and population density. Successful endophytes modulate plant immunity, prime systemic defenses, and enhance resilience to biotic and abiotic stresses, including drought, salinity, heavy metals, and temperature extremes, providing integrated protection under challenging conditions.

Advances in multi-omics technologies, including metagenomics, transcriptomics, proteomics, metabolomics, and spatial single-cell analyses, have elucidated the structural complexity, functional potential, and ecological dynamics of endophytic communities. These insights reveal active microbial contributions to nutrient cycling, metabolic crossfeeding, stress adaptation, and plant defense, enabling rational microbiome engineering for sustainable agriculture, climate-resilient crops, phytoremediation, and biotechnological applications.

In summary, the plant endosphere is a coevolved, structured ecosystem in which microbial communities play essential roles in host fitness, stress resilience, and ecological adaptability. Leveraging these endophytic microbiomes offers transformative opportunities for sustainable food production, environmental management, and biotechnological innovation, addressing global challenges in food security, climate resilience, and ecosystem sustainability.

References

Abdul Rahman, N.S.N., Abdul Hamid, N.W. and Nadarajah, K. (2021) Effects of abiotic stress on soil microbiome. *International Journal of Molecular Sciences* 22(16), 9036. DOI: 10.3390/ijms22169036.

Adedayo, A.A. and Olorunkosebi, M.T. (2025) Coevolution dynamics of beneficial and pathogenic microbes in plant-microbe interactions. *Biology* 14(11), 1505. DOI: 10.3390/biology14111505.

Afoma, U.M., Singh, S., Mishra, A.K., Sharma, C.K., Gupta, K. *et al.* (2026) Integrating artificial intelligence in environmental monitoring: A paradigm shift in data-driven sustainability. *EcoHealth* 23, 38–57. DOI: 10.1007/s10393-025-01752-8.

Afridi, M.S., Javed, M.A., Ali, S., De Medeiros, F.H.V., Ali, B. *et al.* (2022) New opportunities in plant microbiome engineering for increasing agricultural sustainability under stressful conditions. *Frontiers in Plant Science* 13, 899464. DOI: 10.3389/fpls.2022.899464.

Anand, U., Pal, T., Yadav, N., Singh, V.K., Tripathi, V. *et al.* (2023) Current scenario and future prospects of endophytic microbes: Promising candidates for abiotic and biotic stress management for agricultural and environmental sustainability. *Microbial Ecology* 86(3), 1455–1486. DOI: 10.1007/s00248-023-02190-1.

Battu, L., Reddy, M.M., Goud, B.S., Ulaganathan, K. and Kandasamy, U. (2017) Genome inside genome: NGS based identification and assembly of endophytic *Sphingopyxis granuli* and *Pseudomonas aeruginosa* genomes from rice genomic reads. *Genomics* 109(3–4), 141–146. DOI: 10.1016/j.ygeno.2017.02.002.

Bell-Dereske, L., Takacs-Vesbach, C., Kivlin, S.N., Emery, S.M. and Rudgers, J.A. (2017) Leaf endophytic fungus interacts with precipitation to alter belowground microbial communities in primary successional dunes. *FEMS Microbiology Ecology* 93(6), fix036. DOI: 10.1093/femsec/fix036.

Berg, G., Rybakova, D., Fischer, D., Cernava, T., Vergès, M.-C.C. *et al.* (2020) Microbiome definition re-visited: Old concepts and new challenges. *Microbiome* 8(1), 103. DOI: 10.1186/s40168-020-00875-0.

Bhadrecha, P., Singh, S. and Assumi, R. (2023a) Investigation of plant–microbe interactions using CRISPR/Cas for sustainable agriculture. In: Singh, S. (ed.) *Plant Microbiome Engineering*. CRC Press, Boca Raton, USA, pp. 1–20. DOI: 10.1201/9781003331759-16.

Bhadrecha, P., Singh, S. and Dwibedi, V. (2023b) 'A plant's major strength in rhizosphere': The plant growth promoting rhizobacteria. *Archives of Microbiology* 205(5), 165. DOI: 10.1007/s00203-023-03502-2.

Bhawana, S.S., Singh, S. and Bhadrecha, P. (2024) Endophytic microorganisms as a source of bioactive compounds. In: Singh, S. (ed.) *Plant Microbiome Engineering*. Elsevier, pp. 247–274. DOI: 10.1016/B978-0-443-15199-6.00006-3.

Chaudhary, P., Agri, U., Chaudhary, A., Kumar, A. and Kumar, G. (2022) Endophytes and their potential in biotic stress management and crop production. *Frontiers in Microbiology* 13, 933017. DOI: 10.3389/fmicb.2022.933017.

Chen, L. and Liu, Y. (2024) The function of root exudates in the root colonization by beneficial soil rhizobacteria. *Biology* 13(2), 95. DOI: 10.3390/biology13020095.

Chen, X., Sun, M., Chong, S., Si, J. and Wu, L. (2022) Transcriptomic and metabolomic approaches deepen our knowledge of plant–endophyte interactions. *Frontiers in Plant Science* 12, 700200. DOI: 10.3389/fpls.2021.700200.

Compant, S., Cambon, M.C., Vacher, C., Mitter, B., Samad, A. *et al.* (2021) The plant endosphere world – bacterial life within plants. *Environmental Microbiology* 23(4), 1812–1829. DOI: 10.1111/1462-2920.15240.

Dadi, F.A., Muthusamy, S., Ghosh, S., Muleta, D., Tesfaye, K. *et al.* (2025) Plant development influences dynamic shifts in the root compartment microbiomes of wild and domesticated finger millet cultivars. *BMC Microbiology* 25(1), 259. DOI: 10.1186/s12866-025-03976-8.

de Lamo, F.J. and Takken, F.L.W. (2020) Biocontrol by *Fusarium oxysporum* using endophyte-mediated resistance. *Frontiers in Plant Science* 11, 37. DOI: 10.3389/fpls.2020.00037.

Delaux, P.M. and Schornack, S. (2021) Plant evolution driven by interactions with symbiotic and pathogenic microbes. *Science* 371(6531), eaba6605. DOI: 10.1126/science.aba6605.

Dogan, G. and Taskin, B. (2021) Hydrolytic enzymes producing bacterial endophytes of some poaceae plants. *Polish Journal of Microbiology* 70(3), 297–304. DOI: 10.33073/pjm-2021-026.

Fadiji, A.E., Ayangbenro, A.S. and Babalola, O.O. (2021) Shotgun metagenomics reveals the functional diversity of root-associated endophytic microbiomes in maize plant. *Current Plant Biology* 25, 100195. DOI: 10.1016/j.cpb.2021.100195.

Fadiji, A.E., Adeniji, A., Lanrewaju, A.A., Adedayo, A.A., Chukwuneme, C.F. *et al.* (2025) Key challenges in plant microbiome research in the next decade. *Microorganisms* 13(11), 2546. DOI: 10.3390/microorganisms13112546.

Farag, M.A., Porzel, A. and Wessjohann, L.A. (2012) Comparative metabolite profiling and fingerprinting of medicinal licorice roots using a multiplex approach of GC-MS, LC-MS and 1D NMR techniques. *Phytochemistry* 76, 60–72. DOI: 10.1016/j.phytochem.2011.12.010.

Fukushima, A., Kusano, M., Redestig, H., Arita, M. and Saito, K. (2009) Integrated omics approaches in plant systems biology. *Current Opinion in Chemical Biology* 13(5–6), 532–538. DOI: 10.1016/j.cbpa.2009.09.022.

Fu, Z., Xiao, R., Hu, Z., Zhang, M., Wu, S. *et al.* (2024) Characterization of endophytic streptomyces rhizosphaericola Ahn75 and its potential for biocontrol against rice blast. *Journal of Microbiology and Biotechnology* 34(12), 2516–2526. DOI: 10.4014/jmb.2407.07018.

Gao, Y., Xu, Y., Dong, Z., Guo, Y., Luo, J. *et al.* (2025) Endophytic fungal diversity and its interaction mechanism with medicinal plants. *Molecules* 30(5), 1028. DOI: 10.3390/molecules30051028.

García-Latorre, C., Rodrigo, S. and Santamaría, O. (2023) Potential of fungal endophytes isolated from pasture species in Spanish dehesas to produce enzymes under salt conditions. *Microorganisms* 11(4), 908. DOI: 10.3390/microorganisms11040908.

Grzyb, T. and Szulc, J. (2024) Deciphering molecular mechanisms and diversity of plant holobiont bacteria microhabitats, community ecology, and nutrient acquisition. *International Journal of Molecular Sciences* 25(24), 13601. DOI: 10.3390/ijms252413601.

Hacquard, S., Spaepen, S., Garrido-Oter, R. and Schulze-Lefert, P. (2017) Interplay between innate immunity and the plant microbiota. *Annual Review of Phytopathology* 55(1), 565–589. DOI: 10.1146/annurev-phyto-080516-035623.

Hardoim, P.R., van Overbeek, L.S. and van Elsas, J.D. (2008) Properties of bacterial endophytes and their proposed role in plant growth. *Trends in Microbiology* 16(10), 463–471. DOI: 10.1016/j.tim.2008.07.008.

Hériché, J.K., Alexander, S. and Ellenberg, J. (2019) Integrating imaging and omics: Computational methods and challenges. *Annual Review of Biomedical Data Science* 2(1), 175–197. DOI: 10.1146/annurev-biodatasci-080917-013328.

Jadhav, A.V., Pati, P., Patel, K.G., Sharma, V.K., Shewale, M. *et al.* (2025) Biotechnological approaches for enhancing biological control agents against crop pests. *Journal of Entomological Research* 49(3), 788–795. DOI: 10.5958/0974-4576.2025.00139.7.

Kamran, M., Imran, Q.M., Ahmed, M.B., Falak, N., Khatoon, A. *et al.* (2022) Endophyte-mediated stress tolerance in plants: A sustainable strategy to enhance resilience and assist crop improvement. *Cells* 11(20), 3292. DOI: 10.3390/cells11203292.

Kandel, S.L., Joubert, P.M. and Doty, S.L. (2017) Bacterial endophyte colonization and distribution within plants. *Microorganisms* 5(4), 77. DOI: 10.3390/microorganisms5040077.

Khavkin, E.E. (2021) Plant–pathogen molecular dialogue: Evolution, mechanisms and agricultural implementation. *Russian Journal of Plant Physiology* 68(2), 197–211. DOI: 10.1134/S1021443721020072.

Kimotho, R.N. and Maina, S. (2024) Unraveling plant–microbe interactions: Can integrated omics approaches offer concrete answers. *Journal of Experimental Botany* 75(5), 1289–1313. DOI: 10.1093/jxb/erad448.

Kumar, R., Yadav, G., Kuddus, M., Ashraf, G. and Singh, R. (2023) Unlocking the microbial studies through computational approaches: How far have we reached? *Environmental Science and Pollution Research* 30, 48929–48947. DOI: 10.1007/s11356-023-26220-0.

Liu, H., Brettell, L.E. and Singh, B. (2020) Linking the phyllosphere microbiome to plant health. *Trends in Plant Science* 25(9), 841–844. DOI: 10.1016/j.tplants.2020.06.003.

López-Mondéjar, R., Kostovčík, M., Lladó, S., Carro, L. and García-Fraile, P. (2017) Exploring the plant microbiome through multi-omics approaches. In: Kumar, V., Kumar, M., Sharma, S. and Prasad, R. (eds) *Probiotics in Agroecosystems*. Springer, Singapore, pp. 233–268.

Malik, J., Naushad, Z., Singh, S., Shrivastav, D., Verma, V.V. *et al.* (2025) Emerging contaminants in healthcare effluents: Antibiotic resistance and virulence factor dissemination. *Folia Microbiologica* 70, 1201–1216. DOI: 10.1007/s12223-025-01356-3.

Mehra, S., Tyagi, C., Dudha, N., Reddy, Y.P., Manjhi, J. *et al.* (2022) Molecular characterization and phylogenetic analysis of *Pseudomonas aeruginosa* obtained from wound infection. *Arab Gulf Journal of Scientific Research* 39(3), 221–235. DOI: 10.51758/AGJSR-03-2021-0027.

Mendes, R., Garbeva, P. and Raaijmakers, J.M. (2013) The rhizosphere microbiome: Significance of plant beneficial, plant pathogenic, and human pathogenic microorganisms. *FEMS Microbiology Reviews* 37(5), 634–663. DOI: 10.1111/1574-6976.12028.

Najjar, A.A. (2025) Therapeutic potential of endophytic microbes: Emphasizing both fungal and bacterial endophytes. *Applied Microbiology* 5(1), 5. DOI: 10.3390/applmicrobiol5010005.

Nazir, A., Puthuveettil, A.R., Hussain, F.H.N., Hamed, K.E. and Munawar, N. (2024) Endophytic fungi: Nature's solution for antimicrobial resistance and sustainable agriculture. *Frontiers in Microbiology* 15, 1461504. DOI: 10.3389/fmicb.2024.1461504.

Nisa, H., Kamili, A.N., Nawchoo, I.A., Shafi, S., Shameem, N. *et al.* (2015) Fungal endophytes as prolific source of phytochemicals and other bioactive natural products: A review. *Microbial Pathogenesis* 82, 50–59. DOI: 10.1016/j.micpath.2015.04.001.

Pandey, P., Tripathi, A., Dwivedi, S., Lal, K. and Jhang, T. (2023) Deciphering the mechanisms, hormonal signaling, and potential applications of endophytic microbes to mediate stress tolerance in medicinal plants. *Frontiers in Plant Science* 14, 1250020. DOI: 10.3389/fpls.2023.1250020.

Pantigoso, H.A., Newberger, D. and Vivanco, J.M. (2022) The rhizosphere microbiome: Plant-microbial interactions for resource acquisition. *Journal of Applied Microbiology* 133(5), 2864–2876. DOI: 10.1111/jam.15686.

Pérez-Cobas, A.E., Gomez-Valero, L. and Buchrieser, C. (2020) Metagenomic approaches in microbial ecology: An update on whole-genome and marker gene sequencing analyses. *Microbial Genomics* 6(8), mgen000409. DOI: 10.1099/mgen.0.000409.

Praveen, A., Dubey, S., Singh, S. and Sharma, V.K. (2023) Abiotic stress tolerance in plants: A fascinating action of defense mechanisms. *3 Biotech* 13(3), 102. DOI: 10.1007/s13205-023-03519-w.

Ramatsitsi, N. and Manyevere, A. (2025) Roles of endophytic fungi in plant resilience under abiotic stress: A mechanistic review with implications for climate-smart agriculture. *Plant Signaling and Behavior* 20(1), 2578712. DOI: 10.1080/15592324.2025.2578712.

Ramírez-López, M., Bautista-Cruz, A., Toledo-López, A. and Aquino-Bolaños, T. (2025) Rhizospheric and endophytic plant growth-promoting bacteria associated with *Coffea arabica* L. and *Coffea canephora* Pierre ex Froehner: A review of their agronomic potential. *Microorganisms* 13(11), 2567. DOI: 10.3390/microorganisms13112567.

Riseh, R.S., Fathi, F., Vazvani, M.G. and Tarkka, M.T. (2025) Plant colonization by biocontrol bacteria and improved plant health: A review. *Frontiers in Bioscience* 30(1), p23223. DOI: 10.31083/FBL23223.

Roussin-Léveillée, C., Mackey, D., Ekanayake, G., Gohmann, R. and Moffett, P. (2024) Extracellular niche establishment by plant pathogens. *Nature Reviews Microbiology* 22(6), 360–372. DOI: 10.1038/s41579-023-00999-8.

Schulz, B. and Boyle, C. (2005) The endophytic continuum. *Mycological Research* 109(6), 661–686.

Sena, L., Mica, E., Valè, G., Vaccino, P. and Pecchioni, N. (2024) Exploring the potential of endophyte-plant interactions for improving crop sustainable yields in a changing climate. *Frontiers in Plant Science* 15, 1349401.

Shahzad, R., Khan, A.L., Bilal, S., Asaf, S. and Lee, I.J. (2018) What is there in seeds? Vertically transmitted endophytic resources for sustainable improvement in plant growth. *Frontiers in Plant Science* 9, 24. DOI: 10.3389/fpls.2018.00024.

Sivakumar, N., Sathishkumar, R., Selvakumar, G., Shyamkumar, R. and Arjunekumar, K. (2020) Phyllospheric microbiomes: Diversity, ecological significance, and biotechnological applications. *Plant Microbiomes for Sustainable Agriculture* 25, 113–172. DOI: 10.1007/978-3-030-38453-1_5.

Srivastava, A.K., Singh, R.D., Pandey, G.K., Mukherjee, P.K. and Foyer, C.H. (2025) Unravelling the molecular dialogue of beneficial microbe–plant interactions. *Plant, Cell and Environment* 48(4), 2534–2548. DOI: 10.1111/pce.15245.

Starr, A.E., Deeke, S.A., Li, L., Zhang, X., Daoud, R. *et al.* (2018) Proteomic and metaproteomic approaches to understand host–microbe interactions. *Analytical Chemistry* 90(1), 86–109.

Tariq, A., Tanvir, A., Barasarathi, J. and Nazir, A. (2025) Endophytes: Key role players for sustainable agriculture. *Plant Growth Regulation* 105, 1969–1990. DOI: 10.1007/s10725-025-01370-y.

Tshikhudo, P.P., Ntushelo, K. and Mudau, F.N. (2023) Sustainable applications of endophytic bacteria and their physiological/biochemical roles on medicinal and herbal plants: Review. *Microorganisms* 11(2), 453. DOI: 10.3390/microorganisms11020453.

Vandana, U.K., Rajkumari, J., Singha, L.P., Satish, L., Alavilli, H. *et al.* (2021) The endophytic microbiome as a hotspot of synergistic interactions, with prospects of plant growth promotion. *Biology* 10(2), 101. DOI: 10.3390/biology10020101.

Vandenkoornhuyse, P., Quaiser, A., Duhamel, M., Le Van, A. and Dufresne, A. (2015) The importance of the microbiome of the plant holobiont. *New Phytologist* 206(4), 1196–1206. DOI: 10.1111/nph.13312.

Vannier, N., Mesny, F., Getzke, F., Chesneau, G., Dethier, L. *et al.* (2023) Genome-resolved metatranscriptomics reveals conserved root colonization determinants in a synthetic microbiota. *Nature Communications* 14(1), 8274. DOI: 10.1038/s41467-023-43688-z.

Verma, A., Shameem, N., Jatav, H.S., Sathyanarayana, E., Parray, J.A. *et al.* (2022) Fungal endophytes to combat biotic and abiotic stresses for climate-smart and sustainable agriculture. *Frontiers in Plant Science* 13, 953836. DOI: 10.3389/fpls.2022.953836.

Verma, V.V., Vimal, S., Mishra, M.K. and Sharma, V.K. (2025) A comprehensive review on structural insights through molecular visualization: Tools, applications, and limitations. *Journal of Molecular Modeling* 31(6), 173. DOI: 10.1007/s00894-025-06402-y.

Vimal, S.R., Singh, J.S., Kumar, A. and Prasad, S.M. (2024) The plant endomicrobiome: Structure and strategies to produce stress resilient future crop. *Current Research in Microbial Sciences* 6, 100236. DOI: 10.1016/j.crmicr.2024.100236.

Wang, K.C. and Chang, H.Y. (2018) Epigenomics: Technologies and applications. *Circulation Research* 122(9), 1191–1199. DOI: 10.1161/CIRCRESAHA.118.310998.

Wang, Z., Wang, S., He, Q., Yang, X., Zhao, B. *et al.* (2025) Ecological design of high-performance synthetic microbial communities: From theoretical foundations to functional optimization. *ISME Communications* 5(1), ycaf133. DOI: 10.1093/ismeco/ycaf133.

Wu, W., Chen, W., Liu, S., Wu, J., Zhu, Y. *et al.* (2021) Beneficial relationships between endophytic bacteria and medicinal plants. *Frontiers in Plant Science* 12, 646146. DOI: 10.3389/fpls.2021.646146.

Wu, D., He, X., Jiang, L., Li, W., Wang, H. *et al.* (2024) Root exudates facilitate the regulation of soil microbial community function in the genus *Haloxylon*. *Frontiers in Plant Science* 15, 1461893. DOI: 10.3389/fpls.2024.1461893.

Xia, Q., Rufty, T. and Shi, W. (2021) Predominant microbial colonizers in the root endosphere and rhizosphere of turfgrass systems: *Pseudomonas veronii*, *Janthinobacterium lividum*, and *Pseudogymnoascus* spp. *Frontiers in Microbiology* 12, 643904. DOI: 10.3389/fmicb.2021.643904.

Xia, Y., Liu, J., Chen, C., Mo, X., Tan, Q. *et al.* (2022) The multifunctions and future prospects of endophytes and their metabolites in plant disease management. *Microorganisms* 10(5), 1072. DOI: 10.3390/microorganisms10051072.

Yang, L., Qian, X., Zhao, Z., Wang, Y., Ding, G. *et al.* (2024) Mechanisms of rhizosphere plant-microbe interactions: Molecular insights into microbial colonization. *Frontiers in Plant Science* 15, 1491495. DOI: 10.3389/fpls.2024.1491495.

Younas, M.U., Zuo, S., Qasim, M., Ahmad, I., Feng, Z. *et al.* (2025) Multi-omics approaches in plant biology: Decoding agronomic traits for sustainable agriculture. *Plant Stress* 18, 101118. DOI: 10.1016/j.stress.2025.101118.

Zancarini, A., Le Signor, C., Terrat, S., Aubert, J., Salon, C. *et al.* (2025) Medicago truncatula genotype drives the plant nutritional strategy and its associated rhizosphere bacterial communities. *New Phytologist* 245(2), 767–784. DOI: 10.1111/nph.20272.

Zhang, H., Li, Y., Wang, J., Chen, X. and Liu, Q. (2025) Plant–rhizosphere microbe interactions and their roles in nitrogen cycles under periodic flooding: From cooperation mechanisms to regulation pathways. *Critical Reviews in Environmental Science and Technology* 55, 1–25.

Zhou, Y., Liu, D., Li, F., Dong, Y., Jin, Z. *et al.* (2024) Superiority of native soil core microbiomes in supporting plant growth. *Nature Communications* 15(1), 6599. DOI: 10.1038/s41467-024-50685-3.

4 Microbial Communities on the Leaves and Stems of Plants: Their Ecological Roles and Significance

Ahmad Mahmood[1,2*], Waleed Asghar[2,3], Md. Tareq Bin Salam[2,4], Syeda Irsa Mazhar Kazmi[1], Muhammad Arshad[5], Ryota Kataoka[2] and Zhongli Chen[6]

[1]School of Interdisciplinary Engineering and Sciences (SINES), National University of Sciences and Technology (NUST), Islamabad, Pakistan
[2]Department of Environmental Sciences, University of Yamanashi, Yamanashi, Japan
[3]Department of Biochemistry and Molecular Biology, Oklahoma State University, Stillwater, USA
[4]Soil, Water and Environment Discipline, Khulna University, Khulna, Bangladesh
[5]Institute of Environmental Science and Engineering (IESE), National University of Sciences and Technology (NUST), Islamabad, Pakistan
[6]Key Laboratory of the Three Gorges Reservoir Region's Eco-Environment, Chongqing University, Chongqing, People's Republic of China

4.1 Introduction

Plants exist as holobionts, closely intertwined with diverse microbial communities that colonize every part of the plant, from the roots to the aerial tissues (Dastogeer *et al.*, 2020; Lüttge, 2023). While the rhizosphere has long received attention for its impacts on plant nutrition and health, the above-ground microbiomes of plants are equally vital yet historically understudied (Pantigoso *et al.*, 2022; Han *et al.*, 2025). The phyllosphere, encompassing all aerial plant surfaces including leaves and stems, represents one of the largest microbial habitats on Earth (Sarver *et al.*, 2025). It is estimated that on a global scale, leaf surfaces may harbour in the order of 10^{26} bacterial cells, rivaling

*Corresponding author: ahmad.mahmood@sines.nust.edu.pk

 Medicinal Plants and the Rhizosphere (eds M.K. Mishra and S. Tiwari)
DOI: 10.1079/9781836990475.0004

or exceeding the microbial abundance in soils (Remus-Emsermann and Vorhold, 2014). These microbes include bacteria, the most numerous colonizers, fungi (yeasts and filamentous fungi), archaea, algae, and even viruses living on or inside the plant's aerial tissues (Bashir *et al.*, 2022). Collectively, the communities inhabiting leaf surfaces—epiphytes—and the interior of leaves and stems—endophytes—form a critical component of the plant's microbiome (Zeng *et al.*, 2025).

Microbes in the phyllosphere face a harsh and dynamic environment. Leaves and stems are exposed to fluctuating temperature and humidity, intense UV radiation, and limited nutrients, mostly dilute leaf exudates and atmospheric deposits (Rangel and Leveau, 2024). Despite these challenges, a diverse microbiota thrives on healthy plant foliage. Proteobacteria, especially Alpha- and Gammaproteobacteria, often dominate leaf surface communities, followed by members of Firmicutes, Actinobacteria, and Bacteroidota phyla. Fungal residents include yeasts, e.g. *Cryptococcus*, *Symmetrospora*, and filamentous fungi, e.g. *Cladosporium*, that can live as commensals (Perazzolli *et al.*, 2014). Endophytic microorganisms reside within stem and leaf tissues without causing disease, engaging in largely symbiotic or neutral relationships (Mahmood and Kataoka, 2020).

The exact composition of these communities is influenced by plant species, genotype, leaf age, and environmental conditions. For example, closely related plant species tend to host more similar microbiomes than distantly related species, indicating an evolutionary component in microbiome assembly. Within a single plant, the communities also differ by organ, as roots typically harbor the highest microbial diversity, and leaves often the lowest, with stem communities intermediate (Wang *et al.*, 2023). This reflects the gradient in resource availability and exposure: nutrient-rich roots support more copious growth, whereas nutrient-poor, exposed leaves select for a more specialized microbiota.

Crucially, these above-ground microbial communities are not passive occupants; they play integral ecological and functional roles for the plant host and ecosystem. Phyllosphere microbes participate in global biogeochemical cycles by metabolizing carbon, nitrogen, and other elements (Sivakumar *et al.*, 2020; Bashir *et al.*, 2022). For instance, leaf-colonizing bacteria, such as methylotrophs in genera like *Methylobacterium*, consume plant-emitted methanol, a common volatile by-product of plant cell wall metabolism, thereby moderating plant methanol emissions to the atmosphere (Sy *et al.*, 2005; Tani *et al.*, 2023). Some phyllospheric bacteria can fix atmospheric nitrogen or solubilize mineral nutrients, effectively supplying the plant with essential nutrients outside of the root–soil pathway (Fürnkranz *et al.*, 2008; Yang *et al.*, 2025). Phyllosphere microbes also contribute to degradation of environmental pollutants that settle on leaves, such as degrading aromatic hydrocarbons or pesticide residues, effectively performing a "phylloremediation" service (Wei *et al.*, 2017).

On the plant itself, leaf and stem microbes often act as a frontline defense against pathogens and stress. They compete with or antagonize foliar pathogens, produce antifungal and antibacterial compounds, induce the plant's own immune responses, and can even form protective biofilms on surfaces. Endophytic microbes inside stems and leaves can similarly bolster plant defenses by producing defensive metabolites and priming host immunity. In addition, some endophytes are known to produce the same secondary metabolites as their medicinal plant hosts, a fascinating convergence that has implications for natural product sourcing. For example, numerous endophytic fungi isolated from *Taxus* (yew) trees can biosynthesize paclitaxel, Taxol®, the renowned anticancer compound originally obtained from yew bark (Gärditz and Czesnick, 2024). More broadly, endophytes in many medicinal plants have been shown to exhibit pharmacological activities similar to their hosts', producing or modulating bioactive compounds. This symbiotic production may enhance the plant's chemical defenses and offers potential biotechnological avenues for drug production.

Given these critical functions, the ecological significance of phyllosphere and endosphere communities cannot be overstated. They contribute not only to the health and resilience of individual plants, including medicinal species of high therapeutic value, but also to broader ecosystem processes like nutrient cycling, air quality regulation, and climate interactions, through trace gas and aerosol dynamics. Yet, as will be discussed, modern agricultural practices have often ignored these above-ground microbiomes, sometimes inadvertently eroding their diversity and functions. In this chapter, we provide an extensive analysis of microbial communities on plant leaves and stems, covering their composition, ecological roles, and significance, with a special emphasis on medicinal plants.

4.2 Microbial Communities of Stems and Their Roles

Inside plant stems resides a hidden world of microbes. The stem endosphere, internal tissues of stems, including vascular tissue, pith, cortex, etc., harbors diverse endophytic bacteria and fungi that live within the plant's architecture. Stems, being the conduits between roots and leaves, form a highway that many microorganisms can travel, some entering from the soil via roots and moving upward, others entering through aerial parts, e.g. via stomata or wounds, and establishing systemic colonization.

Culturable surveys and DNA sequencing have revealed that stem endophyte communities are often a subset of the soil and root microbiome, but selectively filtered by the plant's defenses and the unique stem environment (Shelake *et al.*, 2019). Typically, stem endophytes include taxa in Proteobacteria, e.g. *Pseudomonas*, *Enterobacter*, *Sphingomonas*, Actinobacteria, e.g. *Streptomyces*, *Microbacterium*, Firmicutes, notably *Bacillus*, and various fungi, e.g. *Fusarium* and *Colletotrichum* species as endophytes, and many yeast-like fungi (Yadav and Meena, 2021). The diversity of endophytic microbes tends to be lower in stems than in roots, but higher than in leaves. For example, a recent high-throughput study on sweet potato found that bacterial community diversity follows a Root > Stem > Leaf pattern under healthy conditions (Gao *et al.*, 2021). This gradient likely reflects the diminishing nutrient availability and increasing plant immune scrutiny from root to stem to leaf. Interestingly, when plants are diseased or stressed, these patterns can shift; in the sweet potato study, fungal diversity was highest in roots in healthy plants, but dropped in infected plants such that leaves/stems of diseased plants showed higher fungal diversity, due to proliferation of the pathogen and associated microbes. Thus, stem microbiomes are dynamic and responsive to plant health status.

From an ecological perspective, stem microbial communities contribute to plant fitness in several ways. A central role is in disease protection. The interior of stems, especially the xylem vessels, can serve as avenues for pathogenic invasion, as seen in wilts and systemic infections, but endophytes act as resident antagonists and competitors in this space. Many studies document endophytic bacteria and fungi that inhibit pathogens and suppress vascular diseases from within (Eljounaidi *et al.*, 2016; Muthu Narayanan *et al.*, 2022; Sallam *et al.*, 2023; Tao *et al.*, 2024; Das *et al.*, 2025; Najjar, 2025). For instance, in apple and potato, certain endophytic *Bacillus* strains colonizing stems significantly reduced incidence of scab diseases caused by pathogenic fungi (Ebrahimi *et al.*, 2022). In one trial, inoculation of potato tubers with an endophytic *Bacillus* sp. resulted in a 70.4% inhibition of potato scab lesions on progeny tubers, along with a 12.4% increase in yield, compared to controls (Shuang *et al.*, 2022). These endophytes likely produce antifungal compounds and induce local resistance, thereby protecting the stem and subterranean tissues. Likewise, endophytic fungi isolated from tree stems have shown activity against wood decay pathogens. Endophytes can also physically occupy niches, such as xylem vessels or cortical intercellular spaces, forming a barrier to pathogen movement—"colonization resistance." In woody medicinal shrubs, endophytic microbes have been observed to form biofilm-like networks inside stems, which may impede the spread of pathogenic fungi and also help the host by retaining moisture and nutrients (Hong *et al.*, 2024).

Beyond disease resistance, growth promotion and stress tolerance are key services provided by stem endophytes (Table 4.1). Many bacterial endophytes in stems are known to produce phytohormones or modulate plant hormone levels, for example by producing indole-3-acetic acid (IAA), an auxin, or gibberellins, or by lowering ethylene levels via ACC deaminase activity. In sweet potato, an endophytic *Bacillus amyloliquefaciens*, strain YTB1407, isolated from stems was found to secrete IAA, and its inoculation led to enhanced plant growth and vigor. The same strain also induced the plant's salicylic acid (SA)-dependent defense pathway, effectively priming systemic resistance against pathogens (Wang *et al.*, 2020).

Some stem endophytes improve the host's nutrient status by fixing atmospheric nitrogen internally or by solubilizing nutrients *in situ*. Sugarcane is a notable example, as it harbors nitrogen-fixing endophytic bacteria, e.g. *Gluconacetobacter diazotrophicus*, in stems and vascular tissues, which can contribute significantly to the plant's nitrogen budget, explaining how sugarcane can thrive in low-fertility soils with reduced fertilizer input (Carvalho *et al.*, 2014). Similarly, *Herbaspirillum* and *Azospirillum* species have been identified

Table 4.1. Examples of microbial communities in plant stems, stem endosphere, and their documented roles and benefits.

Plant/crop	Microbial taxa	Association type	Functional/ ecological role	Observed effect/ benefit	Reference
Taxus (yew)	*Fusarium*, *Penicillium* and other fungal endophytes	Endophyte	Secondary metabolite production	Some endophytic fungi produce Taxol (paclitaxel) in culture, suggesting potential contribution to host metabolite pool	(Strobel *et al.*, 1992; El-Sayed *et al.*, 2020)
Sweet potato (*Ipomoea batatas*)	*Bacillus amyloliquefaciens*	Endophyte	Hormone production (IAA), disease resistance induction	Inoculation led to enhanced growth and primed SA-dependent systemic resistance	(Wang *et al.*, 2025)
Potato (*Solanum tuberosum*)	*Bacillus* sp.	Endophyte	Antagonism to scab pathogen	Up to ~70% reduction in scab lesions, yield increase ~12%	(Shuang *et al.*, 2022)

IAA, indole-3-acetic acid; SA, salicylic acid.

in the stems of cereals and grasses, where they likely deliver fixed nitrogen or stimulate root growth via hormones (Cortés-Patiño *et al.*, 2021).

In the context of medicinal plants, an intriguing role of endophytes is their influence on secondary metabolites. Endophytic fungi and bacteria can activate or enhance the biosynthetic pathways for the plant's own chemical defenses, often the compounds of medicinal value. For example, endophytes in *Catharanthus roseus*, Madagascar periwinkle, have been reported to increase alkaloid production, like vincristine, by the plant (Pandey *et al.*, 2016; Khan *et al.*, 2024). They may do so by inducing mild stress or by providing precursor molecules and signals that upregulate the plant's metabolite pathways. In some cases, the endophytes themselves produce the compounds directly. A famous case is the production of the anticancer drug Taxol (paclitaxel) by endophytic fungi in yew trees; since the initial discovery in the 1990s, numerous endophytic fungal isolates from *Taxus* species and even unrelated plants have been shown to synthesize Taxol in culture. Although yields are often low, this phenomenon suggests a symbiotic collaboration in nature; the endophyte may help the host by contributing Taxol, a deterrent to herbivores and pathogens, while benefiting from residing in a host that is less browsed by herbivores. Endophytes also produce a variety of other bioactive metabolites, e.g. antimicrobial peptides, antioxidants, immunomodulatory compounds, that can contribute to the medicinal properties attributed to the plant. Indeed, it is now thought that the overall therapeutic efficacy of some herbal medicines may result from the combined metabolites of the plant and its endophytes, rather than the plant alone.

Another ecological role of stem microbiomes is in abiotic stress tolerance. Endophytes can enhance drought and heat tolerance by multiple mechanisms: producing ACC deaminase, to lower ethylene and prevent premature senescence under stress, improving water retention, some produce extracellular polysaccharides, and adjusting the plant's osmolytes. Certain *Serendipita*, formerly *Piriformospora*, fungal endophytes in stems/roots are known to increase host drought tolerance by triggering stress response genes (Jangir *et al.*, 2021). Under cold stress, some endophytes produce antifreeze proteins or promote accumulation of cryoprotectants in the plant (Satyakam *et al.*, 2022; Zhang *et al.*, 2024). In saline environments, endophytes have been shown to help sequester or exclude sodium in plant tissues (Guo *et al.*, 2023). While these roles have been more documented for root endophytes, stem and leaf endophytes likely contribute similarly for above-ground stress responses, ensuring the delicate vascular system of the plant continues to function under harsh conditions.

Finally, stem microbial communities play a part in the larger ecosystem by influencing litter decomposition and nutrient cycling after plant tissues senesce. Many foliar and cortical endophytes switch to saprophytic modes when the stem or leaf dies, hastening the decay of organic matter (Wolfe and Ballhorn, 2020). This releases nutrients back to the soil and can influence the soil microbiome. Endophytes thus form a continuum from living plant to decomposing litter, integrating the plant's life cycle with nutrient cycles in the environment.

4.3 Microbial Communities of Leaves and Their Ecological Roles

Leaves constitute the largest component of the phyllosphere and present a unique habitat often described as a "microbial jungle at the plant–air interface." The leaf surface, also termed the phylloplane, is analogous to an expansive landscape for microbes, replete with microhabitats: ridges of veins, depressions around stomata, and vast flat plains of epidermis, all with variable moisture and nutrient conditions. On a healthy leaf, microbial populations typically stabilize at around 10^6–10^7 cells per square centimeter, forming communities that can be highly specific to the plant species and even genotype.

The composition of leaf microbial communities is characteristically dominated by bacteria. Studies using both culture and sequencing consistently find that Proteobacteria, especially Alpha-proteobacterial genera like *Sphingomonas*, *Methylobacterium*, and *Pseudomonas*, are abundant on most plant leaves (Yashiro *et al.*, 2011; Ding and Melcher, 2016; Alibrandi *et al.*, 2020; Glick and Gamalero, 2021). These bacteria are well adapted to oligotrophic, UV-exposed conditions; for instance, many *Sphingomonas* have pigments that protect against UV and can utilize leaf hydrocarbons, and *Methylobacterium* famously consumes methanol emitted from plant cells. Other common phyla include Firmicutes, e.g. *Bacillus* species forming hardy spores to survive desiccation, Actinobacteria, e.g. *Micrococcus*, *Curtobacterium*, and Bacteroidota, formerly Bacteroidetes, e.g. *Chryseobacterium*. The relative proportions of these groups can vary with leaf age and conditions; for example, Actinobacteria often increase on older or senescing leaves, possibly due to better tolerance of low nutrients or higher pH from leaching.

Fungi on leaf surfaces include yeasts, such as *Cryptococcus* and *Sporobolomyces*, that can utilize simple sugars and tolerate sunlight, and filamentous fungi often present as epiphyllous molds or latent pathogens, e.g. *Cladosporium*, *Alternaria* are common phyllosphere fungi. Unlike roots, where arbuscular mycorrhizal fungi dominate, leaf fungal communities are more limited and often pigmented, to withstand UV, or melanized. Some filamentous fungi exist as endophytes within leaves, only visibly sporulating after leaf senescence or under stress. These endophytic fungi, e.g. species of *Colletotrichum*, *Penicillium*, and *Xylaria*, quietly inhabit the leaf interior and can have mutualistic roles like deterring herbivores by producing antifeedant compounds.

An interesting aspect of leaf microbiomes is the spatial heterogeneity on the leaf itself. The upper and lower leaf surfaces can host different communities. The adaxial, upper, surface typically receives more UV and is drier due to direct sun, so microbes there often aggregate in protected microsites, along veins or in microscopic depressions. The abaxial, lower, surface, being more shaded and often richer in stomata and hairs, can support larger populations and more water films, from transpiration droplets. Studies have observed that certain bacteria preferentially colonize stomatal cavities or the bases of trichomes on the abaxial side, where humidity and nutrient exudates are higher. For example, rods of *Pseudomonas* syringae, a common epiphyte and opportunistic pathogen, often lodge around stomata on the lower side, awaiting an opportunity to enter, whereas *Sphingomonas* cells might spread more thinly over the upper cuticle (Arnold and Preston, 2019). Leaf topology influences colonization; microbes tend to cluster along leaf veins, perhaps attracted by nutrients leaking from veins or the slight topographical ridge that shelters them. On smooth areas between veins, bacteria may form small microcolonies encased in self-produced polysaccharides, biofilms, to resist desiccation. Imaging studies, Leaf-FISH and electron microscopy, have even shown bacteria colonizing fungal hyphae on leaves, essentially using fungal networks as "highways" or nutrient sources (Grobelski *et al.*, 2025). Thus, the phyllosphere is a rich tapestry of microbe–microbe and microbe–host interactions at microscopic scales.

Ecologically, phyllosphere microbes fill many functional niches that benefit the host plant and the environment. One well-recognized role is in plant protection. Nonpathogenic epiphytes act as a barrier against pathogens simply by occupying sites on the leaf, space preemption, and consuming available nutrients, resource competition, thereby limiting pathogen establishment. This concept of competitive exclusion is supported by experiments; when researchers inoculated leaves with beneficial commensal strains, they observed that pathogens were less able to colonize because the resident microbes outcompeted them for sugars and amino acids on the leaf surface (Lee *et al.*, 2022; Vuolo *et al.*, 2022; Etesami, 2025; Sarver *et al.*, 2025). Some epiphytes produce specific antibiotics or antifungal compounds that directly inhibit pathogens. For example, certain *Pseudomonas fluorescens* and *Pantoea agglomerans* strains on apple and pear blossoms produce antibiotics that suppress the fire blight pathogen *Erwinia amylovora* (Anderson *et al.*, 2004). In other cases, *Bacillus* species on leaves secrete lipopeptides, surfactin, iturin, etc., that can kill fungal spores of pathogens like *Botrytis cinerea* (Cawoy *et al.*, 2015). Indeed, commercial biocontrol products, e.g. those based on *Bacillus subtilis*, are sprayed on leaves to protect crops; these work by a combination of antibiosis and niche exclusion. Phyllosphere microbes can also form biofilms that coat the leaf, which not only physically blocks pathogen attachment but also retains moisture for the microbiota and plant. Biofilm-forming strains help maintain a stable presence on the leaf, ensuring longer-term protection as the biofilm matrix can resist wash-off by rain.

Perhaps one of the most fascinating services of phyllosphere communities is the ability to induce systemic resistance in the plant. Just as root-associated beneficial microbes, PGPR, can trigger immune responses, foliar microbes can elicit plant defense pathways. Epiphytic bacteria, or even their cell surface molecules, can be perceived by the leaf's immune system, leading to priming of defenses not only locally but throughout the plant, a phenomenon known as induced systemic resistance (ISR). For instance, a consortium of phyllosphere bacteria applied to tomato leaves was shown to activate the plant's salicylic acid and jasmonic acid defense genes, bolstering resistance against multiple foliar pathogens (Gowtham *et al.*, 2024). Notably, this priming often occurs without causing the direct defense, e.g. no hypersensitive response or significant stress on the plant; the plant is simply on "high alert," able to react faster when a real pathogen attack comes. This is a form of vaccination by the microbiome. Some studies have demonstrated that when the native leaf microbiome is disrupted, for example by antibiotics or gnotobiotic growth, plants become more susceptible to leaf diseases, suggesting that a well-assembled leaf microbiome actively suppresses pathogen colonization and possibly secretes signals that keep the plant's immunity primed.

Beyond plant health, phyllosphere microbes have important environmental functions. They are key players in the biogeochemical cycling of elements at the interface of biosphere and atmosphere (Table 4.2). One well-documented function is the metabolism of volatile organic compounds (VOCs). Leaves emit various VOCs like methanol, especially during pectin degradation in cell wall growth, and terpenes (Voyard *et al.*, 2024). Phyllosphere bacteria, particularly methylotrophs, consume methanol as a carbon and energy source, thereby reducing its emission to the atmosphere (Yurimoto *et al.*, 2021). This is globally significant because methanol is a precursor to formaldehyde and ozone in the atmosphere; by consuming methanol, phyllosphere microbes may slightly mitigate ozone formation and the atmospheric lifetime of methane. Phyllosphere microbes also contribute to nitrogen cycling; some bacteria, e.g. certain *Klebsiella*, *Enterobacter*, or cyanobacteria on leaves, can fix N_2 gas into ammonia, part of which could be available to the plant. Although leaf nitrogen fixation is relatively minor compared to that in root nodules, it can be significant in ecosystems where leaves intercept wet/dry nitrogen deposition (Fürnkranz *et al.*, 2008).

Additionally, as mentioned, leaf bacteria can solubilize mineral nutrients. Researchers have isolated phyllosphere strains capable of phosphate solubilization, secreting organic acids that mobilize inorganic phosphate from dust or from leaf surface deposits (Abadi *et al.*, 2020). In tropical forests, epiphytic microbes on canopy leaves might provide nutrients by capturing aerosols or mineral particles and solubilizing them for leaf uptake, essentially acting as an aerial nutrient input system. Some leaf microbes can even oxidize carbon monoxide or other pollutants, contributing to atmospheric cleaning (Wei *et al.*, 2017). Another intriguing role is in the formation of rain and snow; certain bacteria like *Pseudomonas syringae* have ice-nucleating proteins so when they become airborne from leaf surfaces, they can catalyze ice crystal formation in clouds at higher temperatures than usual (Huang *et al.*, 2021), possibly influencing precipitation; this is an area of active research. Thus, the collective microbial life on leaves forms a thin but impactful "biofilm" on the planet's vegetation, mediating exchanges between plants and the atmosphere.

Given their many beneficial roles, it is concerning that these communities can be disrupted by human activities. Heavy use of agrochemicals, air pollution, and other anthropogenic factors can shift phyllosphere community composition. For instance, spraying copper-based fungicides on leaves may kill pathogens but also drastically reduces beneficial phyllosphere

Table 4.2. Ecological functions of leaf, phyllosphere, microbial communities and supporting examples from the literature.

Plant/crop	Microbial taxa	Association type	Functional/ecological role	Observed effect/benefit	Reference
Tomato (*Solanum lycopersicum*)	*Bacillus amyloliquefaciens*, *Pseudomonas chlororaphis*, *P. azotoformans*, *Trichoderma harzianum*	SynCom (synthetic community) on leaf surface	Induced systemic resistance (ISR), biofilm formation, antimicrobial compound production	50–70% reduction in lesion areas from foliar pathogens when SynCom sprayed	(Sarver *et al*., 2025)
Tobacco (*Nicotiana tabacum*)	*Sphingomonas*, *Methylobacterium*	Natural phyllosphere / after biocontrol treatment	Modulation of microbial network, activation of plant defense (SA, JA, ET, ABA)	Biocontrol agents promoted these taxa, which then helped inhibit *Pseudomonas syringae* and enhanced defense gene expression	(Yang *et al*., 2024)
Pseudostellaria heterophylla (a medicinal plant)	*Pseudomonas palleroniana* (strain B-BH16-1) plus other keystone genera including *Chryseobacterium*, *Flavobacterium*	Pathogen-induced phyllosphere recruitment	Antagonism (via VOCs), restructuring community to suppress foliar disease	VOCs emitted by B-BH16-1 inhibited several pathogens; induced beneficial community assembly and improved disease resistance in monocropped plants	(Yuan *et al*., 2024)
Maize (*Zea mays*)	*Bacillus*, *Pseudomonas*, *Tumebacillus*	Naturally resident phyllosphere bacteria	Biocontrol, shaping community via plant genotype	Higher bacterial diversity in resistant maize hybrids; biomarker species isolated with known biocontrol potential against *Setosphaeria turcica*	(Xueliang *et al*., 2020)

ABA, abscisic acid; ET, ethylene; JA, jasmonic acid; SA, salicylic acid; VOC, volatile organic compound.

microbes, sometimes allowing secondary pathogens or opportunists to flourish. Similarly, elevated atmospheric pollutants, like particulate matter, can alter leaf surface pH and UV exposure, affecting microbial colonists. However, there is growing interest in manipulating phyllosphere communities positively, for example by designing synthetic microbial communities (SynComs) to inoculate onto leaves and enhance plant traits. Recent trials have successfully used tailor-made SynCom sprays to protect crops (mixtures of beneficial strains have reduced foliar disease severity in tomato and other plants by over 50%) and even to promote yields, as one study showed that a phyllosphere SynCom increased tomato fruit production under greenhouse conditions (Mehlferber *et al.*, 2023). These promising results emphasize that understanding and managing the phyllosphere microbiome could become a key aspect of future sustainable agriculture.

4.4 Neglect of Phyllosphere and Stem Microbiomes in Modern Agriculture

Despite mounting evidence of the crucial roles played by leaf and stem microbiota, conventional industrial agriculture has largely neglected these communities in both practice and research until recently. This neglect has often been inadvertent; the primary focus in crop management for decades was on maximizing yield through inputs, fertilizers, irrigation, and controlling pests with chemicals, under the paradigm that plants are essentially independent entities. The result has been agricultural practices that frequently disrupt or diminish the natural microbial assemblages on plants.

4.4.1 Intensive pesticide and fungicide use

The regular application of broad-spectrum chemicals to foliage is a cornerstone of high-input farming for pest and disease control. However, these chemicals do not distinguish between pathogenic targets and beneficial or benign residents. A clear example is the use of copper-based fungicides and bactericides on fruit and vegetable crops. Studies on grapevine and tobacco have shown that after spraying such chemicals, there is a sharp decline in the overall phyllosphere microbial populations, especially the beneficial bacteria (Perazzolli *et al.*, 2014; Chen *et al.*, 2021; Feng *et al.*, 2023). In a 2023 study on tobacco, copper hydroxide application significantly reduced the pathogenic fungi *Alternaria* and *Cladosporium*, which is good, but also decreased beneficial bacteria like actinomycetes and *Pantoea*, while allowing some opportunistic fungi, e.g. *Boeremia*, *Stagonosporopsis*, to increase (Feng *et al.*, 2023). This indicates a microbial imbalance, dysbiosis, where, after chemical treatment, the checks and balances provided by the native microbiota are disturbed.

Repeated pesticide use can lead to chronically simplified foliar microbiomes that lack the richness and functional guilds of an untouched state. For instance, insecticides that leave residues on leaves may also inhibit certain bacteria or fungi. There is emerging evidence that fungicide residues can alter the leaf microbiome enough to affect processes like nitrogen fixation or VOC degradation. In short, the "scorched earth" approach to killing pests often collaterally damages the helpful phyllosphere tenants. Over time, this can make crops even more chemically dependent, because the natural suppression of pathogens by epiphytes is lost, potentially leading to a vicious cycle of more disease outbreaks and more chemical use.

4.4.2 Monoculture and low plant diversity

Modern agriculture often involves large monocultures of genetically uniform crops. While efficient for mechanization, this uniformity can reduce the diversity of microbial inocula that colonize the plants. In natural settings or polycultures, plants exchange microbes via air, insects, and rain splash; a diversity of neighboring plant species can "seed" each other's phyllospheres with a broader range of microbes. Monocultures, however, create a more homogeneous microbial pool. Moreover, studies comparing wild vs domesticated plant microbiomes suggest that domestication and breeding have altered the way plants recruit and maintain microbiota (Li *et al.*, 2024; Soldan *et al.*, 2024). For example, a study comparing wild and cultivated rice found that the wild rice phyllosphere had greater community-level niche breadth and more deterministic assembly processes, whereas the cultivated rice had a more constrained set of microbes, likely due to breeding and uniform management (Yin *et al.*, 2023). This implies that as humans selected for certain traits such as yield, aesthetics, and disease resistance genes, we may have unintentionally also selected for certain microbiome profiles or lost associations that wild plants had.

Domesticated plants might have less variable surface chemistry or immune responses that inadvertently exclude some of the microbes their wild ancestors hosted. Additionally, monocultures amplify the spread of specialized pathogens, since a pathogen finds the same susceptible host everywhere, often leading farmers to intensify chemical controls, further harming nontarget microbes. In contrast, plant diversity is known to enhance microbial diversity, one study showing that in mixed-species crop stands, phyllosphere microbial richness was higher and no single pathogen dominated, as opposed to monocultures. Yet the prevailing agricultural model often does not leverage this, resulting in an "impoverished" phyllosphere microbiota on crops compared to plants in natural ecosystems.

4.4.3 Propagation and planting practices

The way we propagate crops can also lead to loss of microbiota. Many crops are grown from sterilized seeds or cuttings. For instance, vegetable seeds are often treated with bleach or fungicides to ensure no pathogens are carried, but this also kills beneficial endophytes that would naturally reside in seeds, and vertical transmission of microbes from parent to offspring is an important mechanism for maintaining symbioses. Research on seed microbiomes indicates that domesticated cereal seeds have lower endophytic diversity than wild relatives, possibly due to cleaning and breeding practices (Abdullaeva *et al.*, 2021). In horticulture, cuttings of ornamentals or medicinal plants are typically dipped in fungicide and grown in sterile media, which can result in essentially microbiologically naïve shoots that then lack a rich phyllosphere until random environmental microbes colonize or disease strikes due to lack of competition (Gikas *et al.*, 2022).

The postharvest handling of produce also reflects the disregard for native microbiota. Fruits and leaves are often washed with sanitizers, which is necessary for food safety but eliminates microbes that might otherwise have been studied for their roles or retained for probiotic benefits. There is some evidence that produce from organic systems with intact epiphytes may have a longer shelf-life due to microbial antagonism of rot organisms.

4.4.4 Focus on pathogens in research and extension

Historically, the phyllosphere was studied mostly in the context of disease, identifying the bad actors, pathogens like mildews, blights, etc., and finding ways to eliminate them. The concept of a beneficial phyllosphere microbiome is relatively new to mainstream agronomy. Consequently, crop breeding programs did not include microbiome traits as selection criteria. If a wild variety harbored lots of endophytes but also had lower yield, breeders might discard it without realizing that the endophytes could be gold. Likewise, disease-resistant cultivars were selected for genetic resistance, not for the ability to recruit protective microbes. This narrow focus has likely led to cultivars that *only* rely on their genetics and chemical protection, lacking the multilayered defense that wild plants employ—genes + symbionts.

4.4.5 Environmental pollution and climate factors in intensive systems

High-input agriculture can create microclimatic and chemical conditions unfavorable for a healthy microbiome. For example, ammonia from heavy fertilization can deposit on leaves and select for ammonia-tolerant but not necessarily beneficial microbes, possibly skewing community composition. Greenhouse cultivation, while controlled, often involves UV-filtering films that alter the radiation regime, which might let more pathogen-susceptible microbes proliferate; in nature, UV keeps some populations in check. Also, climate change intertwined with agriculture, such as extreme heat waves in monocultures, can cause phyllosphere dysbiosis, e.g. beneficial populations crash and latent opportunists take over under heat stress. But in a diverse wild ecosystem, resilient microbiomes might buffer those extremes.

The net effect of these factors is that in many modern agricultural contexts, the natural phyllosphere and endosphere communities are simplified, less diverse, and functionally diminished compared to their potential. Studies comparing organic with conventional farming systems illustrate this well; organically managed crop leaves often harbor significantly higher microbial richness and a more balanced community, whereas conventionally managed leaves may show dominance of a few tolerant species, or sometimes a high load of one opportunistic pathogen kept in check only by chemicals. For instance, a study on apple orchards found that organically grown apples had a richer and more connected epiphytic bacterial community on leaves/fruits, including many beneficial taxa, while conventional orchards, treated with fungicide sprays, had lower diversity and networks (Matthews *et al.*, 2025). Another study on wheat reported that organic farming increased the richness of phyllosphere fungal operational taxonomic units (OTUs) and altered community composition significantly, likely because no fungicides were used to suppress the natural foliar fungi (Karlsson *et al.*, 2017).

The loss of these communities matters because it can lead to increased vulnerability. Without a robust microbiome, plants may experience more severe disease outbreaks, pathogens face no competition, and there is possibly reduced stress tolerance and even subtle deficiencies like poorer nutrient uptake or slower decomposition of leaf litter. There is also an unrealized cost in terms of lost opportunities, as many endophytes or epiphytes with plant growth-promoting or medicinal compound-producing abilities might be undiscovered or underutilized due to their depletion in the very systems we depend on for food and medicine.

In summary, industrial and commercialized agriculture has often treated the phyllosphere and endosphere as blanks to be controlled or sanitized, rather than as integral parts of the crop holobiont to be nurtured (Table 4.3). However, a shift in perspective is under way. Recognizing what has been lost is the first step; researchers are now cataloguing how farming practices affect the plant microbiome, and finding that practices which preserve biodiversity, like reduced chemicals, crop rotation, polycultures, and organic inputs, correlate with richer above-ground microbiomes. This sets the stage for reintegrating these microbial allies into crop production, which we will explore in the next section, focusing on how we can conserve, restore, and even enhance the leaf and stem microbiomes for the benefit of plant and environmental health.

4.5 Conserving and Enriching Leaf and Stem Microbial Communities for Plant and Environmental Health

In light of the clear benefits conferred by phyllosphere and endosphere microbiota, and the losses incurred when they are diminished, there is a growing movement toward conservation and intentional management of these microbial communities in agriculture, horticulture, and ecosystem

Table 4.3. Examples of how industrial agricultural practices impact leaf and stem microbial communities.

Plant/crop	Agricultural practice	Microbial taxa affected	Association type	Functional role affected	Observed impact	Reference
Tobacco (*Nicotiana tabacum*)	Foliar application of copper hydroxide fungicide	*Actinomycetes*, *Pantoea*, *Pseudomonas*, *Sphingomonas*, *Ralstonia*	Epiphyte/ phyllosphere	Beneficial bacteria, antagonists	After copper treatment: decrease in beneficial bacteria (e.g. *Pantoea*), increase of potential pathogens (*Boeremia*, *Stagonosporopsis*)	(Feng *et al.*, 2023)
Pepper (*Capsicum* species)	Repeated applications of fungicide iprodione (foliar)	*Paenarthrobacter* strains, among others	Epiphyte/ phyllosphere	Biodegradation, community structure alteration	Accelerated fungicide degradation on leaves after repeated application; significant shifts in epiphytic bacterial and fungal communities	(Katsoula *et al.*, 2020)
Cucumber (*Cucumis sativus*)	Application of synthetic fungicides (e.g. tebuconazole) vs environmentally friendly fungicides	Diverse phyllosphere bacteria and fungi	Phyllosphere	Diversity, resilience, functional guilds	Synthetic fungicide tended to reduce fungal richness; eco-friendly alternatives (sulfur, NPA, biopesticides) preserved or had minimal negative impacts on microbial diversity	(Wu *et al.*, 2023)

restoration. This shift aligns with the broader concept of viewing plants not just as individual organisms but as metaorganisms, holobionts, composed of the plant and its symbiotic microbes. Here, we discuss strategies to conserve the native diversity of leaf and stem microbiomes, as well as active interventions to enrich and restore these communities, ultimately aiming to support both plant health and environmental sustainability.

4.5.1 Adoption of microbiome-friendly agricultural practices

One of the simplest ways to conserve above-ground microbiomes is to reduce or modify practices that harm them. This means judicious use of pesticides—for example switching to targeted biopesticides or using threshold-based spraying rather than routine calendar sprays, to minimize collateral damage to nontarget microbes. Many farmers are turning to integrated pest management (IPM) which combines biological controls and cultural practices, thereby reducing reliance on harsh chemicals. By cutting down broad-spectrum fungicides, the beneficial phyllosphere fungi and bacteria are allowed to persist. Moreover, practices such as crop rotation, polyculture, intercropping, and maintaining refugial plants around fields can increase microbial inoculum diversity. For instance, intercropping a medicinal plant with a companion crop known to harbor rich phyllosphere communities might "spill over" beneficial microbes to the medicinal plant's leaves, much like mixed forests share microbiota via air currents. Planting hedgerows or strips of wildflowers around crop fields similarly can provide reservoirs of beneficial foliar microbes that colonize crops. Another practice is reduced tillage in perennial systems, which indirectly helps phyllosphere microbiomes by promoting soil health and a steady influx of soil-derived beneficials onto the plant via aerosols or insect vectors.

Crucially, organic farming methods, avoiding synthetic chemicals and using organic inputs, have been shown to maintain higher microbial diversity on plants. In an organic citrus orchard study, the leaf bacterial community was not only more diverse but also formed a more complex interaction network than in a conventional orchard. Such complex networks are thought to be more resilient and better at suppressing diseases. While organic farming is not universally feasible, incorporating some of its principles, like using compost teas, which often contain beneficial epiphytes, as foliar sprays, can enrich the phyllosphere even in conventional systems.

4.5.2 Utilizing microbial inoculants and synthetic communities

A proactive approach to enriching phyllosphere/stem microbiomes is to introduce beneficial microorganisms to the plant. This concept, often likened to a "probiotic for plants," has gained traction with the development of foliar bio-inoculants. These can be single strains or, increasingly, SynComs, carefully selected multispecies cocktails. Researchers design SynComs to include microbes with complementary functions; for example, a mix might contain one strain that strongly inhibits fungal pathogens, another that fixes nitrogen or solubilizes phosphorus, another that induces plant systemic resistance, and perhaps a yeast that occupies the same niche a pathogen would. By combining these, the SynCom can provide broad benefits. An illustrative case is a recent tomato trial where a SynCom of six strains, including *Bacillus amyloliquefaciens*, *Pseudomonas chlororaphis*, *P. azotoformans*, and even a beneficial *Trichoderma* fungus, was sprayed on leaves; this consortium reduced gray mold disease, caused by *Botrytis cinerea*, by 50–70%, outperforming most single-strain treatments (Minchev *et al.*, 2021). The SynCom strains produced multiple antibiotics and formed biofilms, creating a multilayered defense on the leaf. Similarly, a study on *Arabidopsis* identified about 8% of tested leaf isolates that could singly confer full protection against a bacterial pathogen, and mixing these effective strains led to even better outcomes (Vogel *et al.*, 2021).

These examples indicate that we can engineer microbiomes for specific goals. There are now commercial products emerging, often labeled as "biologicals," that are meant to be foliar applied, acting essentially as seedings of beneficial epiphytes. For medicinal plants, one could imagine inoculating seedlings with a SynCom of endophytes known to enhance the production of desired phytochemicals. For instance, endophytic *Piriformospora* or *Bacillus* strains that induce higher alkaloid levels could be applied to young plants destined for medicinal harvest (Watts *et al.*, 2023). Importantly, successful inoculation requires the introduced microbes to establish on/in the plant (Mahmood and Kataoka, 2018). Techniques like formulating microbes in protective carriers, oils or polymers, have improved their survival on leaf surfaces (Ali *et al.*, 2023). Biofilm-forming strains are particularly useful since they stick to and proliferate on leaves, as noted with certain epiphytic *Pseudomonas* which achieve high populations and effectively exclude pathogens. In essence, we are learning to restore lost microbial functions by adding the right players back to the phyllosphere.

A striking demonstration of microbiome restoration comes from microbiota transplantation experiments. In a 2023 study, scientists took the phyllosphere microbiome from wild coffee species, which are resistant to coffee leaf rust disease, and transplanted it onto susceptible arabica coffee plants (de Sousa *et al.*, 2021). The treated arabica plants showed increased resistance to coffee leaf rust, presumably because the transplanted microbes either directly antagonized the rust fungus or induced the plant's defenses. This highlights that even when a plant lacks genetic resistance, its associated microbiota can fill that gap if properly managed. Similarly, researchers have undertaken "cotransplantation" of phyllosphere and rhizosphere communities from healthy to diseased plants, observing enhanced recovery and pathogen suppression in the recipients (Khoiri *et al.*, 2024). While still experimental, these approaches signal a future where farmers and horticulturalists might inoculate entire plant microbiomes, like a smear from a healthy plant's leaf microbiota, to protect or invigorate their crops.

4.5.3 Breeding and biotechnology for microbiome synergy

Another strategy is to incorporate microbiome considerations into plant breeding programs. This could involve selecting plant varieties not just for yield or disease resistance genes, but for their ability to recruit and sustain beneficial microbiomes. For example, if one genotype consistently harbors a rich phyllosphere community and has fewer diseases, that trait could be selected. There is evidence that plants have genetic determinants that influence which microbes colonize them, through leaf exudate composition, surface structures, immune tolerance, etc. By identifying those traits, breeders could develop crop lines that naturally attract "probiotic" microbes or at least do not impede them. The concept of the plant holobiont suggests that future high-performing cultivars may be ones that come paired with specific microbes, e.g. a tomato variety sold with its recommended microbial partners. In fact, seed companies are already coating seeds with beneficial endophytes, mainly for roots, but it could extend to foliar microbes that later migrate up.

Biotechnological interventions could also help; for instance, if a crop lacks a certain microbial symbiont, one might engineer the microbe to be more easily taken up by the plant or engineer the plant to produce compounds that favor the microbe. CRISPR and synthetic biology tools are being explored to create microbes that are "evolution proof" and tailored for plant protection, e.g. by inserting circuits that trigger antimicrobial production when a pathogen is sensed. One could imagine a leaf bacterium engineered to produce a burst of fungicide only when a pathogenic fungus attempts infection, thereby minimizing impact on other microbiota. While such futuristic solutions need careful risk assessment, they highlight the possibilities for enhancing microbiome functions beyond what evolution has provided.

4.5.4 Conservation and biobanking of microbial diversity

On the conservation side, it is crucial to preserve the rich diversity of plant-associated microbes found in wild ecosystems and traditional agroecosystems. This involves efforts such as creating microbial culture collections, biobanks, of endophytes and epiphytes from diverse plant species, especially medicinal plants which might host unique taxa. Open-access biobanks and databases are being called for, to allow researchers and growers to tap into a library of beneficial strains. Such resources would facilitate finding the right microbe for a given plant and environment. For example, a biobank could hold dozens of foliar *Bacillus* strains, each isolated from different plant species, with metadata on what conditions they thrive in and what pathogens they inhibit. A farmer could consult this and choose a strain to spray before a known disease season. Likewise, storing the microbiome of wild relatives of crops ensures that if domestication or breeding lost some microbes, they are not gone forever. Some scientists have proposed microbiome reserves akin to seed banks, preserving not just the plant genetic material but also the living microbial community, via frozen stocks, lyophilized cultures, or DNA/RNA sequence data to revive communities synthetically. Table 4.4 presents the different strategies used in various crops to conserve and enhance leaf and stem microbiomes.

Field-level conservation can also be pursued; maintaining strips of native vegetation or medicinal plant refuges in farms can conserve the ambient pool of microbes. For medicinal plant cultivation, using fewer intensive practices can help retain the natural endophytes that often contribute to the plants' medicinal properties. For instance, wild-harvested medicinal herbs often have a rich endophytic community that may be tied to their efficacy; when those herbs are cultivated intensively with chemicals, they might lose some of those endophytes. Recognizing this, some growers now use mild stress or organic conditions in cultivation to encourage secondary metabolite production, possibly via endophyte activity.

4.5.5 Monitoring and guided intervention

In the future, we will likely see tools for monitoring phyllosphere health, perhaps quick DNA-based sensors that tell a farmer if the leaf microbiome is out of balance, for example low diversity or an overload of one species. This could trigger an intervention, like applying a microbial cocktail or changing a practice, such as reducing irrigation at night to avoid washing off beneficials. There is also potential to use prebiotics, substances that specifically feed beneficial foliar microbes. For instance, a dilute spray of certain sugars or amino acids might selectively boost the growth of commensal bacteria over pathogens, which might need different nutrients. Early studies show that adding small amounts of trehalose on leaves can promote *Pantoea* and *Sphingomonas* growth, which in turn suppress pathogens (Yang *et al.*, 2025). Another angle is phyllosphere habitat engineering, for example developing crop varieties with more leaf surface microsites, like grooved wax structures, to harbor more microbes, or antitranspirant sprays that maintain a slight moisture film to support microbial life during dry periods.

In restoring lost communities, one must also consider context and safety. Introducing microbes into new regions or plant hosts requires ensuring they will not become invasive or pathogenic. Generally, using microbes isolated from the target plant species, or close relatives, is best, as they coevolved and are unlikely to become pests. Regulatory frameworks for microbial inoculants are evolving, hopefully to accommodate the complexity of multistrain products.

4.6 Conclusion

Formerly overlooked, the microbial communities of leaves and stems are now coming into focus as indispensable partners of plants. For medicinal plants, prized for their biochemical riches, these microbes can be coproducers

Table 4.4. Strategies to conserve and enhance leaf/stem microbiomes, with examples of their implementation and effects.

Plant/crop	Strategy	Microbial taxa targeted	Association type	Functional role	Outcome/benefit	Reference
Tomato (*Solanum lycopersicum*)	Foliar inoculation with a synthetic microbial community (SynCom)	*Bacillus megaterium*, *B. subtilis*, *Pseudomonas aeruginosa*, *Ralstonia pickettii*	Epiphyte/phyllosphere	Induce ISR, compete with pathogens, form biofilm	Significant reduction in foliar pathogen damage, improved plant health; demonstrated long-term colonization	(Sarver *et al*., 2025)
Pseudostellaria heterophylla	Pathogen-driven recruitment via probiotics	*Pseudomonas palleroniana* and other beneficial genera (e.g. *Flavobacterium*)	Epiphyte/phyllosphere	Antagonism (via VOCs), community rearrangement	Enhanced disease resistance under continuous monocropping by suppressing pathogens and reshaping phyllosphere community	(Yuan *et al*., 2024)
Willow (*Salix* spp.)	Natural enrichment via herbivory	*Pseudomonas* spp. (entomopathogenic)	Epiphyte/phyllosphere	Insecticidal activity	After beetle damage, enrichment of insecticidal *Pseudomonas* in the phyllosphere increased resistance to further herbivory	(Wang *et al*., 2024)

ISR, induced systemic resistance; VOC, volatile organic compound.

of therapeutic compounds and guardians of plant health. For crops, they are unpaid sentinels and aides, carrying out services from pathogen suppression to nutrient provisioning. The phyllosphere and stem endosphere, once regarded as relatively barren or irrelevant compared to soil, are in fact teeming with life that holds significance from the molecular scale, e.g. influencing plant gene expression, to the global scale, affecting atmospheric chemistry and climate interactions.

As discussed, microbial communities on leaves and stems play diverse ecological roles: they protect plants from disease through competitive and antagonistic interactions; they enhance plant growth and resilience by fixing nitrogen, solubilizing nutrients, producing growth hormones, and inducing systemic resistance; they contribute to nutrient cycling and pollutant degradation in the ecosystem; and they can even augment the medicinal properties of plants by boosting the production of bioactive metabolites. However, the status quo in industrial agriculture has not valued these roles, often eroding the very communities that perform them. The good news is that a paradigm shift is under way. There is a conscious move toward practices and interventions that conserve and enrich phyllosphere and endosphere microbiomes. These include gentle farming methods that work with natural microbial processes, and cutting-edge biotechnological solutions that reintroduce or even redesign microbiomes for optimal plant performance. In essence, we are coming to manage not just crops but crop hologenomes, the plant plus its suite of microbes.

Looking ahead, a holistic approach to plant cultivation will likely become the norm. In the context of medicinal plants and the rhizosphere, the broader theme of this volume, it will not be enough to optimize soil conditions and root symbionts; we must also tend to the above-ground microbiota that influences the plant's phenotype and phytochemistry. This could mean inoculating tissue-cultured medicinal plantlets with endophytes before field planting, or formulating foliar feeds that include both nutrients and microbial amendments. It could also mean protecting wild medicinal plant populations not only by preventing overharvest but by preserving their habitat quality so their microbiomes remain intact, since a medicinal plant grown in its native environment with its native microbiome might have different, potentially superior, phytochemical profiles than one grown in sterile greenhouse conditions.

In summary, the leaves and stems of plants host complex microbial ecosystems integral to plant ecology. By exhaustively surveying these communities, as we have, we gain insight into an entire frontier of plant science that merges microbiology, ecology, and agronomy. The evidence is clear that embracing this knowledge can lead to more sustainable and robust agricultural systems and aid in the conservation of medicinal plants and their unique attributes. The challenge now is to apply this understanding in practice: to adapt agricultural protocols, train farmers and horticulturalists in microbiome-aware techniques, and continue exploring the rich vein of symbiotic biodiversity that lies upon every leaf and within every stem.

By fostering and restoring the phyllosphere and endosphere microbiomes, we not only support healthier plants, with reduced chemical inputs and improved yields/quality, but also promote environmental health, reducing chemical run-off, enhancing biodiversity, and potentially mitigating some climate-related effects through natural microbial activity. The plant and its microbes are a package deal forged by eons of evolution; our agriculture and plant usage stand to benefit immensely by honoring and harnessing that partnership.

References

Abadi, V.A.J.M., Sepehri, M., Rahmani, H.A., Zarei, M., Ronaghi, A. *et al.* (2020) Role of dominant phyllosphere bacteria with plant growth-promoting characteristics on growth and nutrition of maize (*Zea mays* L.). *Journal of Soil Science and Plant Nutrition* 20(4), 2348–2363. DOI: 10.1007/s42729-020-00302-1.

Abdullaeva, Y., Ambika Manirajan, B., Honermeier, B., Schnell, S. and Cardinale, M. (2021) Domestication affects the composition, diversity, and co-occurrence of the cereal seed microbiota. *Journal of Advanced Research* 31, 75–86. DOI: 10.1016/j.jare.2020.12.008.

Alibrandi, P., Schnell, S., Perotto, S. and Cardinale, M. (2020) Diversity and structure of the endophytic bacterial communities associated with three terrestrial orchid species as revealed by 16S rRNA gene metabarcoding. *Frontiers in Microbiology* 11, 604964. DOI: 10.3389/fmicb.2020.604964.

Ali, M., Cybulska, J., Frąc, M. and Zdunek, A. (2023) Application of polysaccharides for the encapsulation of beneficial microorganisms for agricultural purposes: A review. *International Journal of Biological Macromolecules* 244, 125366. DOI: 10.1016/j.ijbiomac.2023.125366.

Anderson, L.M., Stockwell, V.O. and Loper, J. (2004) An extracellular protease of *Pseudomonas fluorescens* inactivates antibiotics of *Pantoea agglomerans*. *Phytopathology* 94, 1228–1234. DOI: 10.1094/PHYTO.2004.94.11.1228.

Arnold, D.L. and Preston, G.M. (2019) *Pseudomonas syringae*: Enterprising epiphyte and stealthy parasite. *Microbiology* 165(3), 251–253. DOI: 10.1099/mic.0.000715.

Bashir, I., War, A.F., Rafiq, I., Reshi, Z.A., Rashid, I. *et al.* (2022) Phyllosphere microbiome: Diversity and functions. *Microbiological Research* 254, 126888. DOI: 10.1016/j.micres.2021.126888.

Carvalho, T.L., Pires, E., Saraiva, R., Vargas, L., Bomfim, A.C.J. *et al.* (2014) Nitrogen fixation in grasses – gluconacetobacter activates genes in sugarcane. *BMC Proceedings* 8(S4), 20. DOI: 10.1186/1753-6561-8-S4-O20.

Cawoy, H., Debois, D., Franzil, L., De Pauw, E. and Thonart, P. (2015) Lipopeptides as main ingredients for inhibition of fungal phytopathogens by *Bacillus subtilis/amyloliquefaciens*. *Microbial Biotechnology* 8, 281–295. DOI: 10.1111/1751-7915.12238.

Chen, X., Wicaksono, W.A., Berg, G. and Cernava, T. (2021) Bacterial communities in the plant phyllosphere harbour distinct responders to a broad-spectrum pesticide. *Science of the Total Environment* 751, 141799. DOI: 10.1016/j.scitotenv.2020.141799.

Cortés-Patiño, S., Vargas, C., Álvarez-Flórez, F., Bonilla, R. and Estrada-Bonilla, G. (2021) Potential of *Herbaspirillum* and *Azospirillum* consortium to promote growth of perennial ryegrass under water deficit. *Microorganisms* 9(1), 91. DOI: 10.3390/microorganisms9010091.

Das, D., Sharma, P.L., Paul, P., Baruah, N.R., Choudhury, J. *et al.* (2025) Harnessing endophytes: Innovative strategies for sustainable agricultural practices. *Discover Bacteria* 2(1), 1. DOI: 10.1007/s44351-025-00011-z.

Dastogeer, K.M.G., Tumpa, F.H., Sultana, A., Akter, M.A. and Chakraborty, A. (2020) Plant microbiome – an account of the factors that shape community composition and diversity. *Current Plant Biology* 23, 100161. DOI: 10.1016/j.cpb.2020.100161.

de Sousa, L.P., Guerreiro Filho, O. and Costa Mondego, J.M. (2021) Differences between the leaf mycobiome of *Coffea arabica* and wild *Coffee* species and their modulation by caffeine/chlorogenic acid content. *Microorganisms* 9(11), 2296. DOI: 10.3390/microorganisms9112296.

Ding, T. and Melcher, U. (2016) Influences of plant species, season and location on leaf endophytic bacterial communities of non-cultivated plants. *PLoS ONE* 11(3), e0150895. DOI: 10.1371/journal.pone.0150895.

Ebrahimi, L., Hatami Rad, S. and Etebarian, H.R. (2022) Apple endophytic fungi and their antagonism against apple scab disease. *Frontiers in Microbiology* 13, 1024001. DOI: 10.3389/fmicb.2022.1024001.

Eljounaidi, K., Lee, S.K. and Bae, H. (2016) Bacterial endophytes as potential biocontrol agents of vascular wilt diseases – review and future prospects. *Biological Control* 103, 62–68. DOI: 10.1016/j.biocontrol.2016.07.013.

El-Sayed, A.S.A., El-Sayed, M.T., Rady, A., Zein, N., Enan, G. *et al.* (2020) Exploiting the biosynthetic potency of taxol from fungal endophytes of conifers plants. *Genome Mining and Metabolic Manipulation Molecules* 25, 3000. DOI: 10.3390/molecules25133000.

Etesami, H. (2025) The dual nature of plant growth-promoting bacteria: Benefits, risks, and pathways to sustainable deployment. *Current Research in Microbial Sciences* 9, 100421. DOI: 10.1016/j.crmicr.2025.100421.

Feng, R., Wang, H., Liu, T., Wang, F., Cai, L. *et al.* (2023) Response of microbial communities in the phyllosphere ecosystem of tobacco exposed to the broad-spectrum copper hydroxide. *Frontiers in Microbiology* 14, 1229294. DOI: 10.3389/fmicb.2023.1229294.

Fürnkranz, M., Wanek, W., Richter, A., Abell, G., Rasche, F. *et al.* (2008) Nitrogen fixation by phyllosphere bacteria associated with higher plants and their colonizing epiphytes of a tropical lowland rainforest of Costa Rica. *ISME Journal* 2(5), 561–570. DOI: 10.1038/ismej.2008.14.

Gao, Z., Hu, Y., Han, M., Xu, J., Wang, X. *et al.* (2021) Effects of continuous cropping of sweet potatoes on the bacterial community structure in rhizospheric soil. *BMC Microbiology* 21(1), 102. DOI: 10.1186/s12866-021-02120-6.

Gärditz, K.F. and Czesnick, H. (2024) Paclitaxel – a product of fungal secondary metabolism or an artefact? *Planta Medica* 90(9), 726–735. DOI: 10.1055/a-2309-6298.

Gikas, G.D., Parlakidis, P., Mavropoulos, T. and Vryzas, Z. (2022) Particularities of fungicides and factors affecting their fate and removal efficacy: A review. *Sustainability* 14(7), 4056. DOI: 10.3390/su14074056.

Glick, B.R. and Gamalero, E. (2021) Recent developments in the study of plant microbiomes. *Microorganisms* 9(7), 1533. DOI: 10.3390/microorganisms9071533.

Gowtham, H.G., Murali, M., Shilpa, N., Amruthesh, K.N., Gafur, A. *et al.* (2024) Harnessing abiotic elicitors to bolster plant's resistance against bacterial pathogens. *Plant Stress* 11, 100371. DOI: 10.1016/j.stress.2024.100371.

Grobelski, G., Kułakowski, M. and Górzyńska, K. (2025) Hidden highways: Fungi associated with glossiphoniid leeches. *International Journal for Parasitology: Parasites and Wildlife* 28, 101158.

Guo, X., Peng, W., Xu, X., Xie, K. and Yang, X. (2023) The potential of endophytes in improving salt–alkali tolerance and salinity resistance in plants. *International Journal of Molecular Sciences* 24(23), 16917. DOI: 10.3390/ijms242316917.

Han, M., Han, Y., Liu, X., Li, G. and Li, P. (2025) Rhizosphere microbiomes altered by environmental stresses and agronomic practices: Implications for plant adaptation and soil biogeochemical processes. *Plant Stress* 18, 101062. DOI: 10.1016/j.stress.2025.101062.

Hong, L., Wang, Q., Zhang, J., Chen, X., Liu, Y. *et al.* (2024) Advances in the beneficial endophytic fungi for the growth and health of woody plants. *Forestry Research* 4, e028. DOI: 10.48130/forres-0024-0025.

Huang, S., Hu, W., Chen, J., Wu, Z., Zhang, D. *et al.* (2021) Overview of biological ice nucleating particles in the atmosphere. *Environment International* 146, 106197. DOI: 10.1016/j.envint.2020.106197.

Jangir, P., Shekhawat, P.K., Bishnoi, A., Ram, H. and Soni, P. (2021) Role of *Serendipita indica* in enhancing drought tolerance in crops. *Physiological and Molecular Plant Pathology* 116, 101691. DOI: 10.1016/j.pmpp.2021.101691.

Karlsson, I., Friberg, H., Kolseth, A., Steinberg, C. and Persson, P. (2017) Organic farming increases richness of fungal taxa in the wheat phyllosphere. *Molecular Ecology* 26(13), 3424–3436. DOI: 10.1111/mec.14132.

Katsoula, A., Vasileiadis, S., Sapountzi, M. and Karpouzas, D.G. (2020) The response of soil and phyllosphere microbial communities to repeated application of the fungicide iprodione: Accelerated biodegradation or toxicity? *FEMS Microbiology Ecology* 96(6), fiaa056. DOI: 10.1093/femsec/fiaa056.

Khan, U., Lata, R., Kumar, H. and Gond, S.K. (2024) Exploration of oncolytic drugs from endophytic fungi of *Catharanthus roseus*. *South African Journal of Botany* 173, 330–337. DOI: 10.1016/j.sajb.2024.08.014.

Khoiri, A.N., Wattanachaisaereekul, S., Jirakkakul, J., Sutheeworapong, S., Kusonmano, K. *et al.* (2024) Co-transplantation of phyllosphere and rhizosphere microbes promotes microbial colonization and enhances sugarcane growth. *Applied Soil Ecology* 201, 105469. DOI: 10.1016/j.apsoil.2024.105469.

Lee, M.D., Pedroso, A.A. and Maurer, J.J. (2022) Bacterial composition of a competitive exclusion product and its correlation with product efficacy at reducing *Salmonella* in poultry. *Frontiers in Physiology* 13, 1043383. DOI: 10.3389/fphys.2022.1043383.

Li, F., Lyu, H., Li, H., Xi, K., Yi, Y. *et al.* (2024) Domestication and genetic improvement alter the symbiotic microbiome structure and function of tomato leaf and fruit pericarp. *Microorganisms* 12(7), 1351. DOI: 10.3390/microorganisms12071351.

Lüttge, U. (2023) Holobionts in the plant kingdom. In: Lüttge, U., Cánovas, F.M., Risueño, M.-C., Leuschner, C. and Pretzsch, H. (eds) *Progress in Botany*, Vol. 83. Springer International Publishing, Cham, Switzerland.

Mahmood, A. and Kataoka, R. (2018) Potential of biopriming in enhancing crop productivity and stress tolerance. In: Rakshit, A. and Singh, H.B. (eds) *Advances in Seed Priming*. Springer, Singapore, pp. 127–145. DOI: 10.1007/978-981-13-0032-5_9.

Mahmood, A. and Kataoka, R. (2020) Metabolite profiling reveals a complex response of plants to application of plant growth-promoting endophytic bacteria. *Microbiological Research* 234, 126421. DOI: 10.1016/j.micres.2020.126421.

Matthews, K.E., Breed, M.F., Stirling, E., Macdonald, L.M. and Cavagnaro, T.R. (2025) Comparing apples and apples; evaluating the impact of conventional and organic management on the soil microbial communities of apple orchards. *Applied Soil Ecology* 215, 106470. DOI: 10.1016/j.apsoil.2025.106470.

Mehlferber, E.C., Debray, R., Conover, A.E., Sherman, J.K., Kaulbach, G. *et al.* (2023) Phyllosphere microbial associations improve plant reproductive success. *Frontiers in Plant Science* 14, 1273330. DOI: 10.3389/fpls.2023.1273330.

Minchev, Z., Kostenko, O., Soler, R. and Pozo, M.J. (2021) Microbial consortia for effective biocontrol of root and foliar diseases in tomato. *Frontiers in Plant Science* 12, 756368. DOI: 10.3389/fpls.2021.756368.

Muthu Narayanan, M., Ahmad, N., Shivanand, P. and Metali, F. (2022) The role of endophytes in combating fungal- and bacterial-induced stress in plants. *Molecules* 27(19), 27. DOI: 10.3390/molecules27196549.

Najjar, A.A. (2025) Therapeutic potential of endophytic microbes: Emphasizing both fungal and bacterial endophytes. *Applied Microbiology* 5(1), 5. DOI: 10.3390/applmicrobiol5010005.

Pandey, S.S., Singh, S., Babu, C.S.V., Shanker, K., Srivastava, N.K. *et al.* (2016) Fungal endophytes of *Catharanthus roseus* enhance vindoline content by modulating structural and regulatory genes related to terpenoid indole alkaloid biosynthesis. *Scientific Reports* 6(1), 26583. DOI: 10.1038/srep26583.

Pantigoso, H.A., Newberger, D. and Vivanco, J.M. (2022) The rhizosphere microbiome: Plant–microbial interactions for resource acquisition. *Journal of Applied Microbiology* 133(5), 2864–2876. DOI: 10.1111/jam.15686.

Perazzolli, M., Antonielli, L., Storari, M., Puopolo, G., Pancher, M. *et al.* (2014) Resilience of the natural phyllosphere microbiota of the grapevine to chemical and biological pesticides. *Applied and Environmental Microbiology* 80(12), 3585–3596. DOI: 10.1128/AEM.00415-14.

Rangel, L.I. and Leveau, J.H.J. (2024) Applied microbiology of the phyllosphere. *Applied Microbiology and Biotechnology* 108(1), 211. DOI: 10.1007/s00253-024-13042-4.

Remus-Emsermann, M.N.P. and Vorhold, J. (2014) Complexities of microbial life on leaf surfaces. *Microbe Magazine* 9(11), 448–452. DOI: 10.1128/microbe.9.448.1.

Sallam, N.M.A., AbdElfatah, H.-A.S., Khalil Bagy, H.M.M., Elfarash, A., Abo-Elyousr, K.A.M. *et al.* (2023) Exploring the mechanisms of endophytic bacteria for suppressing early blight disease in tomato (*Solanum lycopersicum* L.). *Frontiers in Microbiology* 14, 1184343. DOI: 10.3389/fmicb.2023.1184343.

Sarver, E., González-Morelo, K.J., Christensen, K.G., Lefevers, H.M. and Corbin, K.R. (2025) Phyllosphere synthetic microbial communities: A new frontier in plant protection. *BMC Plant Biology* 25(1), 949. DOI: 10.1186/s12870-025-06935-7.

Satyakam, Zinta, G., Singh, R.K. and Kumar, R. (2022) Cold adaptation strategies in plants – An emerging role of epigenetics and antifreeze proteins to engineer cold resilient plants. *Frontiers in Genetics* 13, 909007. DOI: 10.3389/fgene.2022.909007.

Shelake, R.M., Pramanik, D. and Kim, J.-Y. (2019) Exploration of plant-microbe interactions for sustainable agriculture in CRISPR era. *Microorganisms* 7(8), 269. DOI: 10.3390/microorganisms7080269.

Shuang, M., Wang, Y., Teng, W. and Jin, G. (2022) Isolation and identification of an endophytic bacteria *Bacillus* sp. K-9 exhibiting biocontrol activity against potato common scab. *Archives of Microbiology* 204(8), 483. DOI: 10.1007/s00203-022-02989-5.

Sivakumar, N., Sathishkumar, R., Selvakumar, G., Shyamkumar, R. and Arjunekumar, K. (2020) Phyllospheric microbiomes: Diversity, ecological significance, and biotechnological applications. In: Yadav, A., Singh, J., Rastegari, A. and Yadav, N. (eds) *Plant Microbiomes for Sustainable Agriculture*. Spring, Cham, Switzerland, pp. 113–172. DOI: 10.1007/978-3-030-38453-1_5.

Soldan, R., Fusi, M., Cardinale, M., Homma, F., Santos, L.G. *et al.* (2024) Consistent effects of independent domestication events on the plant microbiota. *Current Biology* 34(3), 557–567. DOI: 10.1016/j.cub.2023.12.056.

Strobel, G.A., Stierle, A. and van Kuijk, F.J.G.M. (1992) Factors influencing the *in vitro* production of radiolabeled taxol by pacific yew, *Taxus brevifolia*. *Plant Science* 84(1), 65–74. DOI: 10.1016/0168-9452(92)90209-5.

Sy, A., Timmers, A.C.J., Knief, C. and Vorholt, J.A. (2005) Methylotrophic metabolism is advantageous for *Methylobacterium extorquens* during colonization of *Medicago truncatula* under competitive conditions. *Applied and Environmental Microbiology* 71(11), 7245–7252. DOI: 10.1128/AEM.71.11.7245-7252.2005.

Tani, A., Masuda, S., Fujitani, Y., Iga, T., Haruna, Y. *et al.* (2023) Metabolism-linked methylotaxis sensors responsible for plant colonization in *Methylobacterium aquaticum* strain 22A. *Frontiers in Microbiology* 14, 1258452. DOI: 10.3389/fmicb.2023.1258452.

Tao, J., Gu, M., Yu, S., Shi, J., Cheng, L. *et al.* (2024) The beneficial endophytic microbes enhanced tobacco defense system to resist bacterial wilt disease. *Chemical and Biological Technologies in Agriculture* 11(1), 21. DOI: 10.1186/s40538-024-00542-8.

Vogel, C.M., Potthoff, D.B., Schäfer, M., Barandun, N. and Vorholt, J.A. (2021) Protective role of the *Arabidopsis* leaf microbiota against a bacterial pathogen. *Nature Microbiology* 6(12), 1537–1548. DOI: 10.1038/s41564-021-00997-7.

Voyard, A., Ciuraru, R., Lafouge, F., Decuq, C., Fortineau, A. *et al.* (2024) Emissions of volatile organic compounds from aboveground and belowground parts of rapeseed (*Brassica napus* L.) and tomato (*Solanum lycopersicum* L.). *Science of the Total Environment* 955, 177081. DOI: 10.1016/j.scitotenv.2024.177081.

Vuolo, F., Novello, G., Bona, E., Gorrasi, S. and Gamalero, E. (2022) Impact of plant-beneficial bacterial inocula on the resident bacteriome: Current knowledge and future perspectives. *Microorganisms* 10(12), 2462. DOI: 10.3390/microorganisms10122462.

Wang, C.-J., Wang, Y.-Z., Chu, Z.-H., Wang, P.-S., Liu, B.-Y. *et al.* (2020) Endophytic *Bacillus amyloliquefaciens* YTB1407 elicits resistance against two fungal pathogens in sweet potato (*Ipomoea batatas* (L.) Lam.). *Journal of Plant Physiology* 253, 153260. DOI: 10.1016/j.jplph.2020.153260.

Wang, H., Zhang, F., Zhang, Y., Wang, M., Zhang, Y. *et al.* (2024) Enrichment of novel entomopathogenic *Pseudomonas* species enhances willow resistance to leaf beetles. *Microbiome* 12(1). DOI: 10.1186/s40168-024-01884-z.

Wang, S., Ma, T., Yao, X., Yao, Z., Wang, Z. *et al.* (2023) Structure of endophytes in the root, stem and leaf tissues of sweetpotato and their response to sweetpotato scab disease caused by *Elsinoë batatas*. *Agronomy* 13(12), 2965. DOI: 10.3390/agronomy13122965.

Wang, Y., Hao, J., Gu, J., Wu, J., Zhang, Y. *et al.* (2025) Genome-driven functional validation of *Bacillus amyloliquefaciens* strain MEPW12: A multifunctional endophyte for sustainable sweet potato cultivation. *Microorganisms* 13(6), 1322. DOI: 10.3390/microorganisms13061322.

Watts, D., Palombo, E.A., Jaimes Castillo, A. and Zaferanloo, B. (2023) Endophytes in agriculture: Potential to improve yields and tolerances of agricultural crops. *Microorganisms* 11(5), 1276. DOI: 10.3390/microorganisms11051276.

Wei, X., Lyu, S., Yu, Y., Wang, Z., Liu, H. *et al.* (2017) Phylloremediation of air pollutants: Exploiting the potential of plant leaves and leaf-associated microbes. *Frontiers in Plant Science* 8. DOI: 10.3389/fpls.2017.01318.

Wolfe, E.R. and Ballhorn, D.J. (2020) Do foliar endophytes matter in litter decomposition? *Microorganisms* 8(3), 446. DOI: 10.3390/microorganisms8030446.

Wu, P.H., Chang, H.X. and Shen, Y.M. (2023) Effects of synthetic and environmentally friendly fungicides on powdery mildew management and the phyllosphere microbiome of cucumber. *PLoS ONE* 18(3), e0282809. DOI: 10.1371/journal.pone.0282809.

Xueliang, T., Dan, X., Tingting, S., Songyu, Z., Ying, L. *et al.* (2020) Plant resistance and leaf chemical characteristic jointly shape phyllosphere bacterial community. *World Journal of Microbiology and Biotechnology* 36(9), 139. DOI: 10.1007/s11274-020-02908-0.

Yadav, G. and Meena, M. (2021) Bioprospecting of endophytes in medicinal plants of Thar desert: An attractive resource for biopharmaceuticals. *Biotechnology Reports* 30, e00629. DOI: 10.1016/j.btre.2021.e00629.

Yang, Z., Liu, T., Fan, J., Chen, Y., Wu, S. *et al.* (2024) Biocontrol agents modulate phyllosphere microbiota interactions against pathogen *Pseudomonas syringae*. *Environmental Science and Ecotechnology* 21, 100431. DOI: 10.1016/j.ese.2024.100431.

Yang, H., Liu, J., Ma, M., Tan, Z., Zhang, K. *et al.* (2025) Leaf development and its interaction with phyllospheric microorganisms: Impacts on plant stress responses. *Plant Stress* 16, 100843. DOI: 10.1016/j.stress.2025.100843.

Yashiro, E., Spear, R.N. and McManus, P.S. (2011) Culture-dependent and culture-independent assessment of bacteria in the apple phyllosphere. *Journal of Applied Microbiology* 110(5), 1284–1296. DOI: 10.1111/j.1365-2672.2011.04975.x.

Yin, Y., Wang, Y.-F., Cui, H.-L., Zhou, R., Li, L. *et al.* (2023) Distinctive structure and assembly of phyllosphere microbial communities between wild and cultivated rice. *Microbiology Spectrum* 11(1), e0437122. DOI: 10.1128/spectrum.04371-22.

Yuan, Q.-S., Gao, Y., Wang, L., Wang, X., Wang, L. *et al.* (2024) Pseudomonas reshaped the phyllosphere microbiome in combination with *Pseudostellaria heterophylla* foliar disease resistance via the release of volatile organic compounds. *Environmental Microbiome* 19(1), 61. DOI: 10.1186/s40793-024-00603-3.

Yurimoto, H., Shiraishi, K. and Sakai, Y. (2021) Physiology of methylotrophs living in the phyllosphere. *Microorganisms* 9(4), 809. DOI: 10.3390/microorganisms9040809.

Zeng, Q., Hu, H., Ge, A., Xiong, C., Zhai, C. *et al.* (2025) Plant–microbiome interactions and their impacts on plant adaptation to climate change. *Journal of Integrative Plant Biology* 67(3), 826–844. DOI: 10.1111/jipb.13863.

Zhang, Z., Liu, W., Huang, Y. and Li, P. (2024) Research progress on plant anti-freeze proteins. *Phyton* 93(6), 1263–1274. DOI: 10.32604/phyton.2024.050755.

5 Deciphering the Rhizosphere Microbiome: Integrating High-throughput Sequencing and Multi-omics Approaches

Rajat Awasthi[1], Roshini Singh[2], Sanjay Mishra[2], Pankaj Gupta[2] and Manoj Kumar Mishra[2]*

[1]*Department of Biotechnology, Institute of Engineering and Technology, Lucknow, India*
[2]*Department of Biotechnology, S.R. Institute of Management and Technology, Lucknow, India*

5.1 Introduction

The rhizosphere represents a dynamic interface where plant roots actively modify the surrounding soil environment through exudation, nutrient uptake, and physical interactions. These processes influence biogeochemical cycling, soil quality, and plant stress tolerance. Recent advances in high-throughput sequencing (HTS) and integrated multi-omics approaches have enabled detailed investigation of plant–microbe interactions within this zone, providing new insights into microbial functions that support plant productivity and ecosystem stability.

5.1.1 Importance of the rhizosphere microbiome

The rhizosphere is the narrow region of soil directly influenced by root secretions and associated microbial communities. It forms a highly interactive environment in which complex relationships occur among plant roots, soil components, and diverse microorganisms (Mendes *et al.*, 2013). This ecological niche regulates numerous biological and biochemical processes essential for plant growth and development.

*Corresponding author: manoj30biotech@gmail.com

DOI: 10.1079/9781836990475.0005

A defining characteristic of the rhizosphere is its high microbial diversity, including bacteria, fungi, archaea, and other microorganisms. These communities are not only abundant but also functionally specialized, contributing significantly to plant health and yield. They participate in mutualistic and commensal interactions such as nitrogen transformation, phosphate solubilization, pathogen suppression, and phytohormone production (Mendes *et al.*, 2013).

Through rhizodeposition, plants stimulate microbial diversity that supports nutrient cycling, soil resilience, and stress tolerance. Modern HTS and multi-omics approaches have revealed the central role of rhizosphere microbes in nutrient turnover and agroecosystem functioning. Microbial processes regulate the transformation and mobilization of key nutrients such as nitrogen, phosphorus, and iron, thereby improving their bioavailability to plants (Fig. 5.1) (Li *et al.*, 2021). In addition, rhizosphere microbes enhance plant stress tolerance by promoting systemic resistance, improving water uptake and facilitating nutrient acquisition under adverse environmental conditions (Ling *et al.*, 2022).

Microbial metabolic activities also contribute to soil health by decomposing organic residues, detoxifying harmful compounds and improving soil structure, thus maintaining ecological stability (Mendes *et al.*, 2013). Owing to its biological richness and functional complexity, the rhizosphere is considered a hotspot of plant–microbe interactions. It operates as a coordinated signaling network in which chemical communication between roots and microorganisms shapes microbial community structure, plant fitness, and ecosystem performance. Overall, the rhizosphere microbiome plays a fundamental role in regulating nutrient cycles, enhancing resistance to biotic and abiotic stresses, and sustaining agricultural productivity.

5.1.2 Role of high-throughput sequencing and multi-omics in microbiome studies

The advent of HTS and integrated multi-omics approaches has substantially advanced microbiome research by enabling deeper insights into microbial diversity, functionality, and ecological interactions. Traditional methods such as culturing and microscopy provide limited resolution and fail to capture the complexity and functional dynamics of microbial communities. In contrast, modern approaches including metagenomics, metatranscriptomics, metaproteomics, and metabolomics allow culture-independent assessment of microbial composition and activity across diverse environments.

High-throughput sequencing technologies have become essential tools for analyzing plant-associated microbiomes, as they enable precise identification of microbial taxa, detection of uncultured or rare organisms, and evaluation of community responses to environmental changes (Kimotho and Maina, 2024). Such information is crucial for understanding plant–microbe mutualism and its implications for sustainable agriculture and crop improvement.

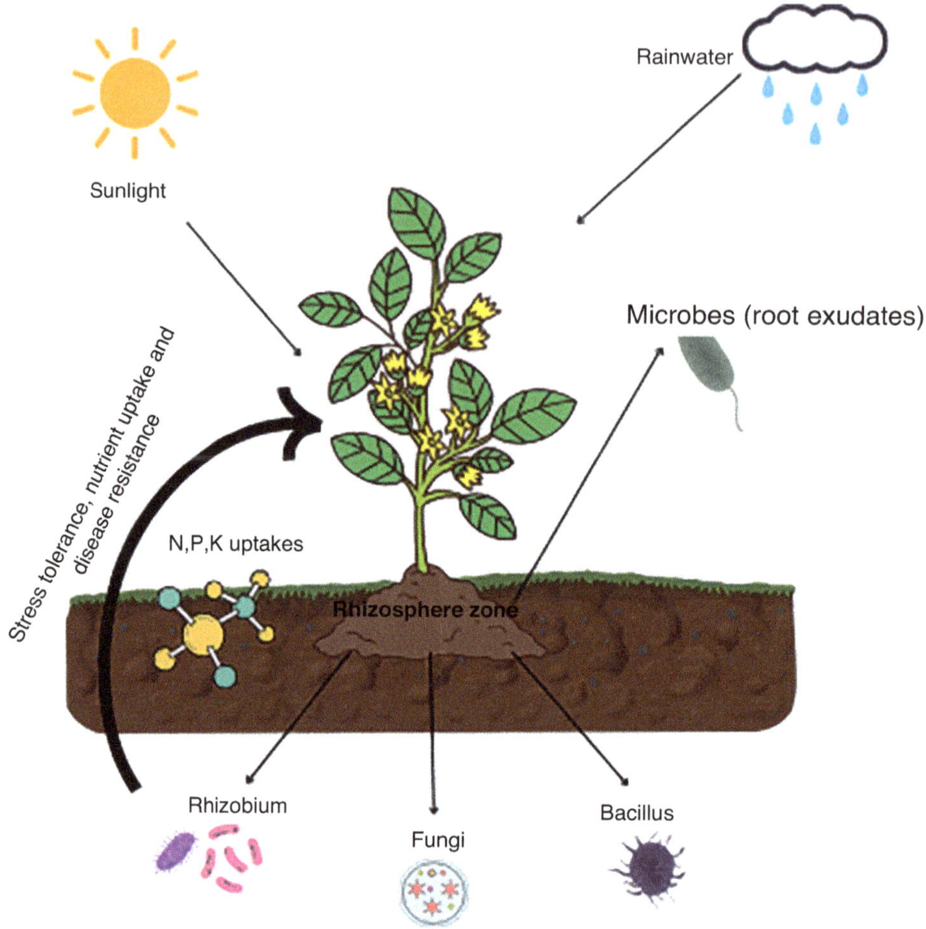

Fig. 5.1. Functional role of the rhizosphere microbiome. Schematic representation of interactions within the rhizosphere illustrating how plant roots, soil microorganisms, and environmental inputs collectively influence plant growth and nutrient dynamics. Root exudates recruit diverse microbial communities, including rhizobia, fungi, and beneficial bacteria such as *Bacillus*, which contribute to nutrient mobilization, biological interactions, and plant health. Microbial processes enhance the uptake of essential macronutrients (N, P, K) from soil, while external factors such as sunlight and rainwater regulate plant metabolism and soil moisture, thereby shaping rhizosphere activity and ecosystem functioning.

Multi-omics integration further strengthens microbiome analysis by linking genetic potential with functional expression. The combined use of genomics, transcriptomics, and proteomics facilitates reconstruction of microbial pathways and prediction of functional interactions within communities (Mishra *et al.*, 2022). These integrative datasets support systems biology approaches that improve predictive models of plant productivity and guide microbiome manipulation towards enhanced plant health.

Omics-based studies also contribute to ecological understanding by simultaneously revealing microbial identity, metabolic capability, and interaction networks. For instance, metagenomic and metabolomic profiling of rhizosphere communities has provided insights into nutrient cycling, chemical communication, and cooperative behavior among microorganisms and plant hosts (Li *et al.*, 2021).

Both HTS and multi-omics approaches have shifted microbiome research from descriptive cataloguing to mechanistic interpretation. These tools now form the foundation of functional genomics, microbial ecology, and systems biology, enabling the development of microbiome engineering strategies and innovative applications in agricultural biotechnology and environmental management.

5.2 Overview of High-throughput Sequencing Technologies

High-throughput sequencing technologies enable rapid and large-scale characterization of microbial genomes, functional potential, and evolutionary relationships. The advent of next-generation sequencing (NGS) platforms has transformed genomic research by permitting massively parallel sequencing, thereby improving resolution in microbial ecology, plant–microbe interactions, and functional genomics.

5.2.1 Next-generation sequencing platforms

Next-generation sequencing platforms allow simultaneous sequencing of millions to billions of DNA fragments in a single run, greatly increasing analytical speed and depth. These technologies are broadly grouped into short-read and long-read sequencing systems (Fig. 5.2), each contributing distinct advantages to genomic investigations (Mandlik *et al.*, 2024).

Short-read sequencing platforms such as Illumina® and Ion Torrent® generate highly accurate reads with very high throughput. These systems are cost-effective and widely applied in genome resequencing, transcriptome profiling (RNA-seq), metagenomics, and single-nucleotide variant detection. However, short read lengths limit their ability to resolve repetitive genomic regions, structural variants, and complex genome architectures. Ion Torrent systems may also display reduced precision in homopolymer stretches due to signal interpretation constraints (Mandlik *et al.*, 2024).

Long-read sequencing technologies, including PacBio SMRT® and Oxford Nanopore® sequencing, analyze single DNA molecules in real time and produce reads extending from several kilobases to ultra-long fragments (Satam *et al.*, 2023). PacBio SMRT sequencing now provides high-fidelity reads typically ranging from 10 to 25 kb with improved accuracy, facilitating detection of structural variation, repeat regions, and epigenetic modifications. Oxford Nanopore sequencing detects nucleotide sequences by monitoring ionic current disruptions as DNA or RNA strands pass through nanopores,

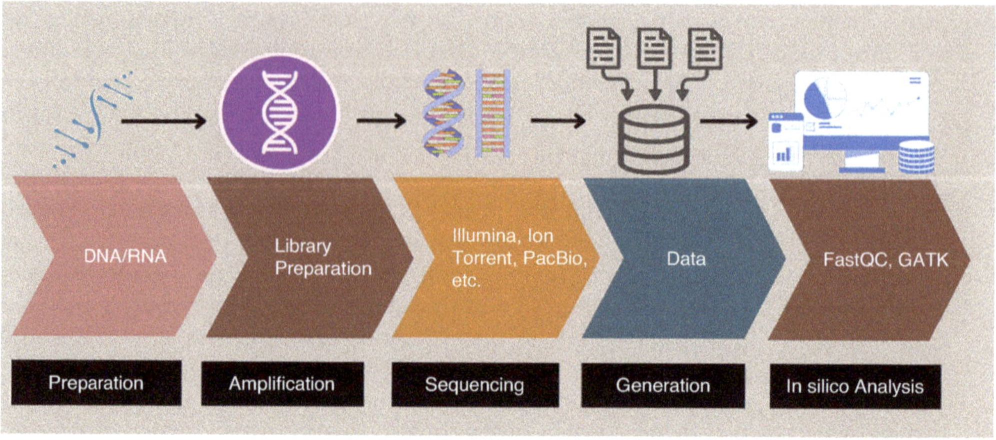

Fig. 5.2. Correlation of short-read and long-read next-generation sequencing platforms. Overview of the integrated workflow combining short-read and long-read sequencing technologies for comprehensive genomic analysis. The process begins with DNA/RNA extraction followed by library preparation and amplification. Sequencing is performed using platforms such as Illumina (short-read) and PacBio or Oxford Nanopore (long-read), generating complementary datasets. Short reads provide high accuracy and depth, while long reads enable improved genome assembly, structural variant detection, and resolution of repetitive regions. Subsequent data processing, quality control, and bioinformatic analysis using tools such as FastQC and GATK® enable accurate genome reconstruction and downstream functional interpretation.

enabling direct methylation detection and direct RNA sequencing with minimal laboratory infrastructure.

Long-read approaches are particularly useful for identifying large structural changes, resolving repetitive genomic regions, reconstructing transcript isoforms, and supporting de novo genome assembly. Their application often requires deeper sequencing coverage, advanced computational pipelines, and hybrid assembly strategies to compensate for historically higher raw error rates compared with short-read systems (Satam *et al.*, 2023).

5.2.2 Third-generation sequencing (long-read sequencing)

Third-generation sequencing (TGS), commonly referred to as long-read sequencing, produces read lengths typically ranging from ~10 kb to several megabases, enabling more comprehensive genome reconstruction than short-read technologies (Tafazoli *et al.*, 2025). As illustrated in Fig. 5.2, these platforms facilitate contiguous genome assembly and improved detection of structural variation, thereby enhancing genomic interpretation in complex biological systems.

5.2.2.1 *Advantages*

- Long read lengths enable more accurate de novo genome assembly by spanning repetitive and structurally complex regions, improving genome continuity and scaffolding (Amarasinghe *et al.*, 2020).
- Sequencing can directly detect epigenetic modifications, such as DNA methylation, without requiring chemical conversion or enrichment steps (Amarasinghe *et al.*, 2020).
- Individual reads can capture full-length genes, transcripts, and large structural variants, supporting detailed genome architecture analysis (Tafazoli *et al.*, 2025).

5.2.2.2 *Limitations*

- Raw sequencing error rates remain higher than in short-read platforms, historically ranging from approximately 10% to 20% in SMRT sequencing and 5% to 40% in Nanopore sequencing, although improvements in base-calling algorithms and consensus polishing continue to reduce these values (Marx, 2023).
- Sequencing costs remain comparatively higher, with reported estimates roughly US$22–90 per gigabase for PacBio and US$65–200 per gigabase for Nanopore sequencing (Xiao and Zhou, 2020).
- Long-read analysis requires high molecular weight DNA and more advanced computational pipelines to manage error correction, assembly, and structural variant detection (Amarasinghe *et al.*, 2020).
- Hybrid sequencing strategies that combine long- and short-read data substantially improve assembly accuracy and reduce error rates, increasing the reliability of downstream genomic analyses (Luan *et al.*, 2024).

5.2.3 Hybrid sequencing strategies

Hybrid sequencing strategies integrate short-read and long-read sequencing technologies to exploit the complementary strengths of both approaches (Antipov *et al.*, 2015). Short reads provide high accuracy and depth at relatively low cost, whereas long reads enable improved assembly continuity, structural variant detection, and resolution of repetitive regions. Their combined use enhances genome completeness, annotation quality, and transcriptome reconstruction.

In hybrid workflows, as illustrated in Fig. 5.3, long reads are typically used to generate a scaffold or backbone assembly, while short reads are employed for polishing, error correction, and fine-scale variant detection. This integration substantially improves consensus accuracy and reduces assembly fragmentation, particularly in complex plant genomes and metagenomic datasets.

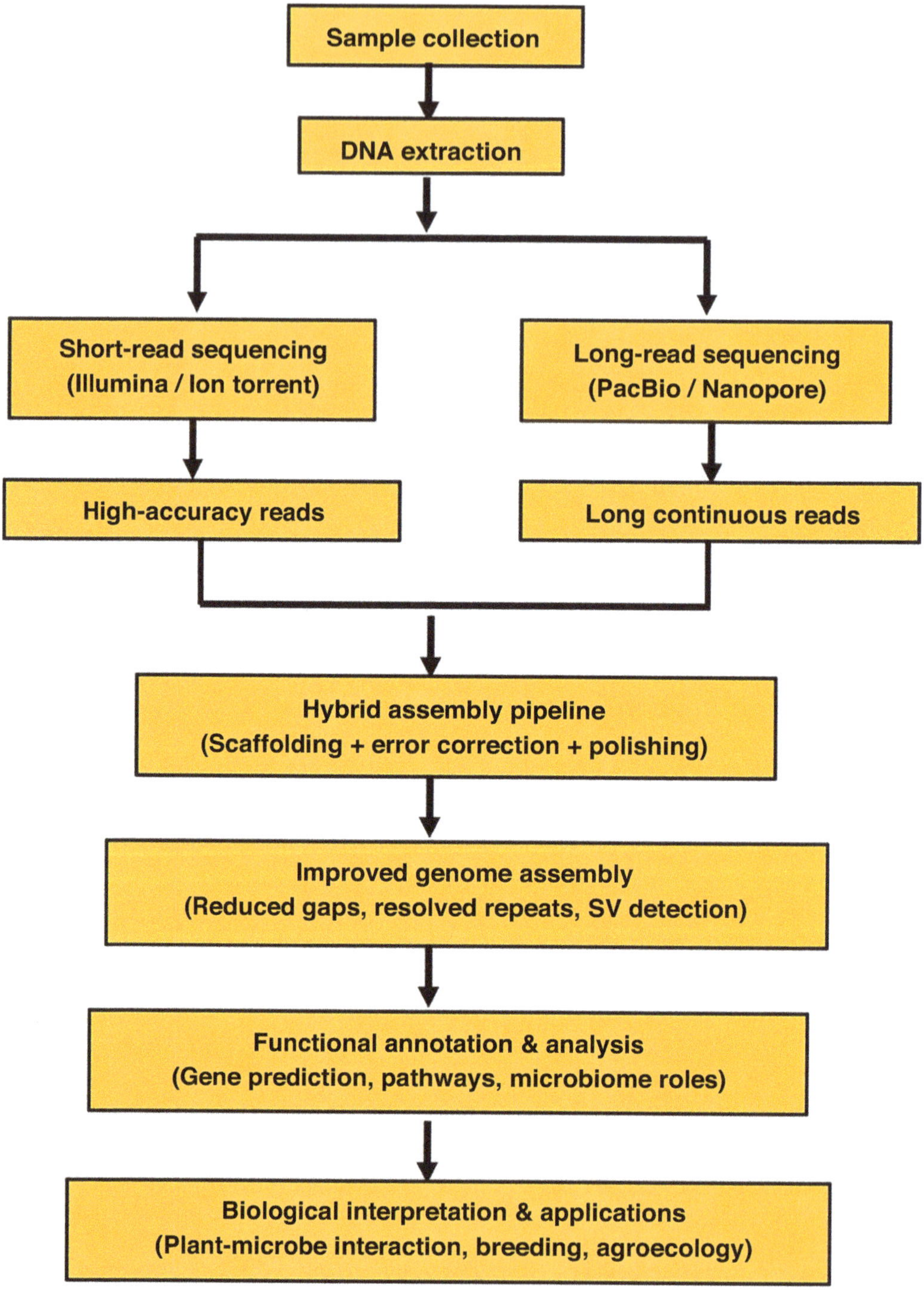

Fig. 5.3. General workflow of hybrid sequencing integrating short-read and long-read technologies for improved genome assembly, functional annotation, and microbiome analysis.

Hybrid sequencing has become increasingly valuable in rhizosphere microbiome studies, where microbial communities often contain closely related taxa and repetitive genomic regions (Jin *et al.*, 2022). The approach supports more reliable reconstruction of metagenome-assembled genomes (MAGs), improved functional annotation, and more accurate identification of biosynthetic pathways (Jin *et al.*, 2022; Mirete *et al.*, 2025). Furthermore, hybrid approaches facilitate multi-omics integration by linking genomic structure with transcriptomic and metabolomic profiles. This enables better interpretation of microbial functionality, plant–microbe interactions, and nutrient cycling processes in agroecosystems.

Hybrid sequencing pipelines typically require careful data balancing, computational resources, and optimized assembly algorithms; however, their ability to produce near-complete genome assemblies and high-confidence annotations makes them increasingly central to modern plant microbiome research (Antipov *et al.*, 2015).

5.2.4 Advances in metagenomics and metatranscriptomics

5.2.4.1 Shotgun metagenomics for gene catalogue construction

Shotgun metagenomic sequencing has become a central approach for generating comprehensive gene catalogues in complex microbial communities. Unlike marker-gene surveys such as 16S rRNA profiling, shotgun sequencing captures all DNA present in a sample, allowing simultaneous recovery of taxonomic diversity and functional potential. Pudake *et al.* (2021) demonstrated that this approach enables the assembly and annotation of biosynthetic gene clusters from uncultivated rhizosphere microorganisms, revealing previously inaccessible metabolic capacities. Similarly, Terrón-Camero *et al.* (2022) constructed reference gene catalogues from shotgun datasets and subsequently mapped metatranscriptomic reads onto these catalogues, facilitating integrated functional analysis. Such strategies support the identification of metabolic pathways and functional genes in poorly characterized ecosystems such as the rhizosphere (Fig. 5.4).

5.2.4.2 Metatranscriptomics for active gene expression profiling

Metatranscriptomics complements DNA-based metagenomics by identifying genes that are actively transcribed under specific environmental conditions. Kumari *et al.* (2022) showed that metatranscriptomic profiling of rhizosphere communities can reveal microbial responses to environmental cues, including nutrient limitation and root-derived stress signals.

Standard workflows typically involve total RNA extraction, removal of ribosomal RNA, enrichment of messenger RNA, conversion to cDNA, and high-throughput sequencing. Downstream analyses often employ differential expression tools such as DESeq2, edgeR or Cufflinks, with pipeline selection depending on dataset complexity and study objectives (Terrón-Camero *et al.*, 2022).

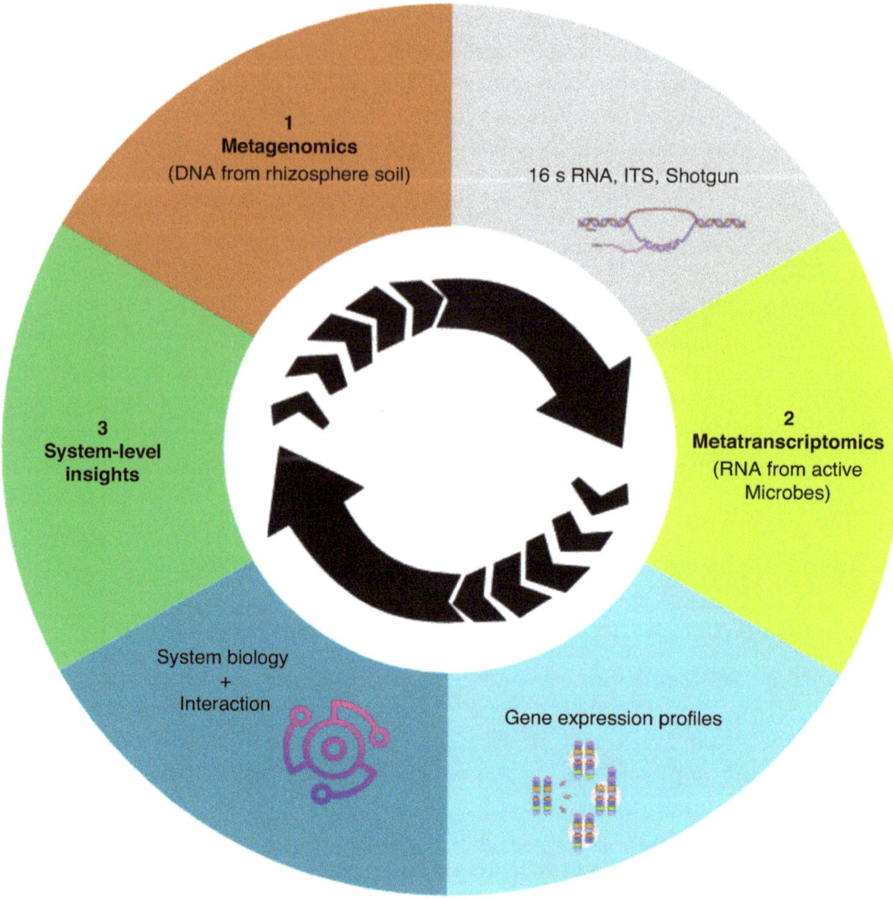

Fig. 5.4. Integrated framework of metagenomics and metatranscriptomics for rhizosphere consortia analysis. Schematic representation of the complementary roles of metagenomics and metatranscriptomics in decoding rhizosphere microbial communities. Metagenomics (DNA-based analysis) reveals taxonomic composition and genetic potential of soil microorganisms, including functional genes detected through approaches such as 16S rRNA sequencing and shotgun metagenomics. Metatranscriptomics (RNA-based profiling) captures actively expressed genes and metabolic pathways of microbial populations under prevailing environmental conditions. Integration of these datasets enables system-level insights into microbial interactions, gene expression dynamics, and functional networks within the rhizosphere, supporting systems biology approaches for understanding plant–microbe interactions and soil ecosystem functioning.

5.2.4.3 Challenges in integrative data analysis

The combined interpretation of metagenomic and metatranscriptomic datasets provides powerful functional insight but also presents technical challenges. In rhizosphere studies, contamination from plant DNA and uneven microbial biomass can bias sequencing depth and downstream analysis, necessitating host DNA depletion and optimized extraction procedures

(Pudake *et al.*, 2021). A major limitation in transcriptomic workflows is the dominance of ribosomal RNA, which can constitute over 90% of total RNA. Efficient rRNA depletion or mRNA enrichment is therefore essential for meaningful expression analysis (Kumari *et al.*, 2022).

Accurate mapping of reads to reference gene catalogues is another critical step. Terrón-Camero *et al.* (2022) recommend constructing sample-specific assemblies or curated reference databases to reduce alignment bias. Reliable comparisons across samples further require normalization procedures that account for sequencing depth, gene copy number variation, and differential microbial abundance. Recent advances in bioinformatics, including machine learning-assisted noise reduction and improved normalization algorithms, are helping to enhance signal detection in complex microbiomes, thereby strengthening integrative multi-omics analyses (Fig. 5.4).

5.3 Multi-omics Approaches for Rhizosphere Microbiome Analysis

Multi-omics approaches have emerged as powerful tools for deciphering the complexity of the rhizosphere microbiome by providing integrated insights into microbial diversity, functional potential, and ecological interactions. Metagenomics has substantially advanced our understanding of microbial community structure by enabling the direct recovery of environmental DNA and the identification of taxonomic composition, gene content, and metabolic capacity (Pang *et al.*, 2021). Studies have demonstrated that metagenomic profiling can reveal microbial groups associated with plant health, disease suppression, and nutrient acquisition (Wang *et al.*, 2023), while comparative analyses across soil types help explain community resilience and ecological adaptation (Pramanik *et al.*, 2020). Early work also highlighted the ability of metagenomics to uncover previously unrecognized taxa shaping rhizosphere microbial assemblages (Carvalhais *et al.*, 2013).

Metatranscriptomics complements these findings by examining actively expressed genes, thereby enabling real-time assessment of microbial metabolic processes and stress responses. Recent studies indicate that transcript-level analyses provide insight into how microbial genes respond to environmental fluctuations and contribute directly to plant growth and adaptive capacity (Law *et al.*, 2022; Liu *et al.*, 2023).

Beyond individual omics layers, integrated multi-omics frameworks allow system-level interpretation of plant–microbe interactions. Such integrative analyses reveal coordinated metabolic pathways, signaling processes, and community dynamics, strengthening the application of microbiome research in sustainable agriculture and ecosystem management (Vidal *et al.*, 2022).

5.3.1 Metagenomics: Elucidating community composition and functional potential

Metagenomics enables the comprehensive characterization of microbial communities by directly sequencing total DNA extracted from environmental samples such as rhizosphere soil. This cultivation-independent approach overcomes the limitations of classic culture-based microbiology, which historically detected only a small fraction of microbial diversity, a phenomenon described as the "great plate count anomaly" (Staley and Konopka, 1985).

High-throughput sequencing of environmental DNA generates detailed taxonomic profiles of microbial assemblages, encompassing bacteria, archaea, fungi ,and viruses across multiple phylogenetic levels (Quast *et al.*, 2012). In addition to community composition, metagenomic analysis reveals the collective functional repertoire of the microbiome by identifying genes involved in key ecological processes such as nutrient cycling, stress tolerance, pathogen suppression, and biosynthesis of plant growth-promoting compounds (Tringe *et al.*, 2005; Fierer, 2017).

These datasets provide a genetic blueprint of microbial metabolic capacity, allowing researchers to infer how microbiomes contribute to plant fitness and ecosystem productivity (Handelsman, 2004; Shreshtha *et al.*, 2024). Comparative metagenomic studies across host genotypes, soil types, and environmental conditions further help to identify ecological drivers shaping microbial community structure and genomic potential (Shade *et al.*, 2013).

Sequential advances in omics technologies, summarized in Table 5.1, illustrate the progression from early genomic surveys to integrated multi-omics frameworks now applied in rhizosphere research.

5.3.2 Metatranscriptomics: Quantifying *in situ* functional gene expression

While metagenomics characterizes the collective genetic potential of microbial communities, metatranscriptomics provides a direct measure of actively expressed genes, enabling assessment of functional activity under *in situ* conditions (Carvalhais *et al.*, 2012). This approach relies on HTS of the total RNA pool—the metatranscriptome, which represents the complete set of transcripts produced by community members at a specific spatial and temporal point (Frias-Lopez *et al.*, 2008).

Because transcriptional regulation responds rapidly to environmental fluctuations, metatranscriptomic datasets capture immediate microbial responses to nutrient availability, host-derived signals, and interspecies competition. For instance, transcriptional profiling can reveal enhanced expression of nitrogen fixation genes under nitrogen limitation or the activation of resistance determinants during microbial antagonistic interactions (Forsberg *et al.*, 2012).

Quantitative analysis of mRNA abundance allows identification of metabolically active taxa and their associated biochemical pathways (Hultman *et al.*, 2015). Such transcript-level insights extend beyond predictions based

Table 5.1. Chronological development of multi-omics approaches in plant research.

Year	Milestone	Reference
2008–2014	Launch of the 1000 Plant Transcriptomes Project (1KP), representing one of the first large-scale plant omics initiatives. It applied high-throughput transcriptomics across more than 1000 species, paving the way for comprehensive comparative omics research	(Kassim *et al.*, 2025)
2015	Emergence of integrated multi-omics frameworks in plant research, enabling deeper understanding of metabolic networks and regulatory pathways	(Scossa *et al.*, 2021)
2019–2022	Development of single-cell and spatial omics approaches in plant science, allowing simultaneous molecular profiling at cellular and tissue resolution	(Yang *et al.*, 2023; Yu *et al.*, 2023)
2020	Introduction of DBiT-seq technology enabling spatially resolved transcriptomic and proteomic analysis at near-single-cell resolution *in situ*	(Su *et al.*, 2021)
2022–2025	Rapid expansion of integrated multi-omics strategies incorporating machine learning tools for biomarker discovery, functional prediction, and precision agriculture applications	(Agamah *et al.*, 2022; Jiang *et al.*, 2025)
2024	Growth of AI-driven multi-omics modelling integrating central dogma layers (genome–transcriptome–proteome–metabolome) for phenotype prediction and systems biology analysis	(Xin *et al.*, 2024; Wu and Xie, 2025)
2025	Application of multi-omics in reproductive biology and developmental regulation, revealing pathways such as amino acid metabolism and reproductive signaling networks	(Kassim *et al.*, 2025)

This table outlines the progressive evolution of multi-omics strategies in plant science, highlighting key milestones from early transcriptional analyses to the integration of advanced computational biology frameworks. It emphasizes how the convergence of genomics, transcriptomics, proteomics, metabolomics, and bioinformatics has enabled deeper systems-level insights into complex plant biological processes and previously unexplored molecular mechanisms.

solely on genomic inventories, linking potential metabolic functions to their realized ecological activity. By documenting active microbial processes within plant-associated environments, metatranscriptomics contributes mechanistic understanding of plant–microbe interactions and ecosystem functioning (Knight *et al.*, 2018).

5.3.3 Integrated multi-omics analysis: Towards a systems-level understanding

A comprehensive interpretation of rhizosphere ecology requires integration of multiple molecular datasets to resolve interactions across biological scales. Multi-omics frameworks combine metagenomics, metatranscriptomics, metaproteomics, and metabolomics to generate a layered representation of microbiome structure and activity (Hettich *et al.*, 2013).

Within this hierarchy, metagenomics defines the genetic repertoire of the microbial community, whereas metatranscriptomics identifies actively transcribed genes. Metaproteomics extends this information by cataloguing translated proteins that represent the enzymatic and structural machinery of microbial cells (Wilmes and Bond, 2006). Complementing these molecular layers, metabolomics characterizes the small molecule products and substrates generated by microbial metabolism, providing a direct biochemical readout of community activity (Patti *et al.*, 2012).

Integration of these datasets enables reconstruction of functional cascades linking genes to transcripts, proteins, and metabolites, thereby clarifying cause–effect relationships within microbial ecosystems (He *et al.*, 2010). For example, multi-omics analyses may connect the genomic presence of nitrogen fixation operons to their transcriptional activation, subsequent synthesis of nitrogenase proteins, and measurable increases in bioavailable ammonium within the rhizosphere (López-Mondéjar *et al.*, 2020).

Such integrative strategies facilitate identification of regulatory networks, metabolic interdependencies, and functional redundancies that remain obscured when single-omics approaches are used in isolation (Segata *et al.*, 2013). The resulting datasets support development of predictive ecological models describing microbial consortia behavior, nutrient flux dynamics, and downstream impacts on plant phenotype and productivity (Yu *et al.*, 2024).

5.4 Experimental Design and Data Collection

Reliable microbiome profiling requires standardized soil sampling, appropriate controls, and careful nucleic acid extraction coupled with stringent quality assessment. These methodological considerations ensure that downstream sequencing data reflect biological variation rather than technical artefacts.

5.4.1 Sampling strategies for rhizosphere studies

Accurate characterization of rhizosphere microbial communities depends on systematic collection of compartment-specific soil fractions. Standardized protocols allow separation of bulk soil, rhizosphere soil, and rhizoplane fractions from plant roots, each representing distinct microbial habitats (Barillot *et al.*, 2013).

Plants are excavated carefully to preserve root integrity, and loosely attached bulk soil is removed by gentle shaking. Soil that remains tightly associated with the root surface is then detached by controlled agitation in saline solution (0.9% NaCl), allowing recovery of rhizosphere material while minimizing contamination from root tissues. Measurement of soil mass before and after washing enables quantitative comparison between fractions.

High-throughput sequencing studies require replication, randomization, and sampling across spatial and temporal gradients to capture heterogeneity within root systems. Sampling multiple plants and locations reduces site-specific bias, while temporal sampling across developmental stages or growing seasons captures dynamic shifts in microbial composition (Barillot *et al.*, 2013).

Bulk soil serves as a baseline reference representing ambient microbial assemblages, whereas rhizosphere and rhizoplane compartments reveal selective enrichment or depletion of taxa driven by plant roots. Gradients in microbial abundance—typically highest in the rhizosphere and lowest in bulk soil—demonstrate the strong filtering effect of root exudates on microbial colonization.

Compartment-specific sampling is further supported by independent studies showing that even within identical soil and host contexts, rhizosphere and rhizoplane communities display distinct microbial signatures requiring separate analysis. Spatial replication across sites and within-plot random sampling improves robustness of diversity metrics, differential abundance analyses, and cooccurrence network inference (Lakshmanan *et al.*, 2017).

5.4.2 Nucleic acid extraction and library preparation

Commercial extraction kits remain widely used in plant microbiome studies due to their reproducibility and compatibility with high-throughput workflows. Comparative assessments across diverse soil types indicate that automated bead-based platforms can produce optimal balances of DNA yield, purity, fragment size, and community representation (Jensen *et al.*, 2024).

The QIAGEN DNeasy® platform and similar kits frequently provide consistent results across heterogeneous samples, while magnetic bead-based approaches allow automated processing of large sample sets with minimal handling. Direct extraction methods using bead-purification systems have also been shown to generate DNA quality comparable to conventional protocols in plant root microbiome studies (Kumaishi *et al.*, 2022).

Postextraction quality assessment is essential prior to sequencing. DNA purity is typically evaluated using absorbance ratios near 260/280 ≈ 1.8, while electrophoretic methods such as agarose gel analysis or Bioanalyzer® profiling verify fragment integrity compatible with Illumina sequencing platforms (Haro *et al.*, 2021).

Library preparation workflows increasingly employ miniaturized and automated strategies to reduce cost and processing time. Miniaturized metagenomic and amplicon library protocols using nanoliter dispensing systems can reduce reagent volumes by 5–10-fold without altering microbial community profiles (Jensen *et al.*, 2024).

Two-step PCR approaches incorporating exonuclease treatment and magnetic bead clean-up improve representation of low-abundance taxa while reducing size selection bias, although slight reductions in overall library yield may occur. Sequencing depth and diversity patterns produced by such workflows remain comparable to those generated by standard Illumina TruSeq protocols (Kumaishi *et al.*, 2022).

5.4.3 Experimental challenges and limitations

Environmental samples often contain low microbial biomass relative to host DNA, which can limit sequencing sensitivity and reduce taxonomic resolution. Shotgun metagenomic sequencing typically requires tens of nanograms of target DNA, a threshold difficult to achieve in low-biomass systems (Afridi *et al.*, 2022).

Reduced-representation approaches such as 2bRAD-M sequencing have demonstrated accurate microbial profiling from picogram-level DNA inputs, even in samples dominated by host contamination. However, host DNA still consumes sequencing capacity and increases analytical cost.

Host DNA depletion strategies can enrich microbial DNA by 10–50-fold, improving detection of rare taxa, though aggressive depletion may introduce compositional bias or remove genuine microbial sequences (Marín *et al.*, 2021). Amplicon-based sequencing approaches remain susceptible to PCR bias arising from primer mismatches, amplification efficiency differences, and chimera formation. Fragment length variation and GC content bias in restriction-based methods can further distort community profiles, particularly in low-template samples where stochastic amplification effects become pronounced (Afridi *et al.*, 2022).

Low-biomass studies are also highly vulnerable to contamination originating from reagents, laboratory environments, or cross-sample transfer. Rigorous contamination control measures, including negative controls, batch-specific reagent testing, and *in silico* contaminant filtering, are therefore essential for distinguishing biological signal from artefacts (Sergaki *et al.*, 2018).

Mock microbial communities are frequently employed to evaluate extraction efficiency, PCR uniformity, and sequencing reproducibility. Even with these controls, minor procedural variations—such as PCR cycle number, storage conditions, or extraction chemistry—can significantly alter community composition estimates, emphasizing the importance of methodological standardization across studies (Marín *et al.*, 2021).

5.5 Bioinformatics and Data Analysis

High-throughput sequencing of microbial communities produces extremely large volumes of raw sequence data that require standardized, reproducible analytical workflows for biological interpretation. A robust *in silico* pipeline typically includes sequence preprocessing, taxonomic and functional annotation, and downstream statistical modeling. Such workflows commonly integrate specialized analytical platforms together with clustering strategies such as operational taxonomic units (OTUs), amplicon sequence variants (ASVs), and MAGs.

5.5.1 Data processing pipelines

Initial quality control is a critical step in ensuring data reliability and removing sequencing artefacts. Raw reads are subjected to trimming and filtering to eliminate low-quality bases and adapter contamination. For amplicon datasets, chimera removal is essential to prevent artificial inflation of diversity estimates. Pipelines based on QIIME2 workflows have demonstrated that preprocessing steps such as primer trimming and region-specific read separation significantly influence downstream taxonomic assignment accuracy (Maki *et al.*, 2023).

Widely used amplicon analysis frameworks integrate trimming, denoising, chimera filtering, and taxonomic assignment against curated databases such as SILVA or Greengenes. Marker gene-based profiling approaches allow efficient classification of microbial taxa, while shotgun metagenomic tools extend analysis to functional inference. Marker-based strategies provide species-level resolution, whereas pathway-mapping tools enable reconstruction of enzyme families and metabolic networks from metagenomic reads.

Recent modular meta-omics workflows allow both assembly-free profiling and genome-resolved approaches using de novo assembly and binning to reconstruct MAGs. Such strategies enable linkage of gene abundance and transcriptional activity to specific genomes, improving ecological interpretation of community function (Saenz *et al.*, 2022).

Sequence clustering approaches strongly influence biological interpretation. Traditional OTU clustering groups sequences at fixed similarity thresholds (commonly 97%), which may mask ecologically meaningful variation. In contrast, ASV-based approaches employ statistical error correction algorithms to resolve exact biological sequences, providing improved taxonomic resolution and reproducibility. Genome-resolved metagenomics further advances analysis by linking genes and pathways to reconstructed microbial genomes, facilitating ecological and evolutionary interpretation of community structure.

Benchmarking studies indicate that trimming strategies, reference database selection, and choice of variable regions in 16S rRNA datasets can significantly influence observed community composition. Modular pipelines must therefore be adapted to dataset characteristics and research objectives, particularly when analyzing transcriptomic shifts or temporal turnover in microbial communities (Saenz *et al.*, 2022).

5.5.2 Network analysis and microbial interactions

Network analysis has become an important tool for exploring microbial associations and ecological interactions within complex communities. Correlation-based approaches are frequently used to infer cooccurrence relationships; however, compositional constraints, zero-inflated data structures, and indirect associations can produce misleading connections if not carefully interpreted (Floćh *et al.*, 2020).

Methods specifically designed for compositional microbiome data, such as log-ratio correlation frameworks, improve robustness of inferred associations. Multimetric approaches that combine several similarity measures further enhance detection of meaningful cooccurrence patterns (Jamil *et al.*, 2022). Network visualization platforms (Cytoscape®) allow graphical exploration of these associations, enabling identification of clusters, hubs, and potential functional guilds.

Within microbial networks, taxa displaying disproportionately high connectivity or centrality are often described as keystone species. Network metrics such as degree, betweenness centrality, and clustering coefficients are commonly used to identify such influential taxa. Studies of plant-associated microbiomes have repeatedly detected members of *Rhizobiales* and *Ktedonobacterales* as highly connected taxa contributing to community stability across root and soil compartments (Floćh *et al.*, 2020).

Although network inference can guide ecological interpretation and identify candidates for microbiome engineering, inferred keystone roles require validation through perturbation experiments or longitudinal studies to distinguish direct ecological effects from statistical artefacts.

5.5.3 Machine learning and predictive models

Machine learning (ML) methods have become increasingly important for analyzing high-dimensional sequencing datasets, particularly for predicting plant health status, disease susceptibility, and ecosystem function. Unsupervised techniques such as principal component analysis and clustering algorithms are widely used for exploratory data analysis, dimensionality reduction, and identification of latent community structure.

Supervised approaches, including random forest and support vector machine models, are frequently applied for classification tasks when labeled datasets are available. Ensemble tree-based models are particularly suitable for microbiome datasets because they handle nonlinear relationships, tolerate noisy features, and resist overfitting in high-dimensional data (Papoutsoglou *et al.*, 2023).

Studies using microbiome features derived from sequencing data have demonstrated strong predictive performance in disease classification models, with validation metrics approaching high diagnostic accuracy (Mathieu *et al.*, 2022). Support vector machines also perform effectively when appropriate normalization and feature scaling are applied.

Deep learning architectures have further expanded predictive capabilities by capturing complex nonlinear interactions among microbial taxa and host traits. Neural network-based approaches have shown promise in modeling plant–microbiome interactions and predicting phenotypic outcomes from sequencing-derived features (Deng *et al.*, 2021).

Effective ML analysis of sequencing data requires careful feature selection to reduce dimensionality, remove redundant variables, and improve interpretability. Filtering strategies based on variance, compositional transformations, and recursive feature elimination are commonly employed prior to model fitting. Rigorous validation procedures, including nested cross-validation and bias-corrected resampling, are essential to prevent overfitting and ensure model generalizability (Papoutsoglou *et al.*, 2023).

Microbiome datasets pose specific analytical challenges, including compositionality, sparsity, batch effects, and high dimensionality. Methodological frameworks for ecological ML applications therefore emphasize preprocessing, model tuning, interpretability assessment, and alignment of performance metrics with research objectives (Walsh *et al.*, 2024).

5.6 Applications and Insights from Rhizosphere Microbiome Studies

Research on the rhizosphere microbiome has revealed the central role of soil microorganisms in enhancing plant growth, improving resilience to biotic and abiotic stresses, and enabling sustainable agroecological interventions through bioinoculants and microbial biotechnology.

5.6.1 Plant growth promotion and stress mitigation

Plant health and adaptability are strongly influenced by the diverse microbial communities inhabiting the rhizosphere, the narrow soil zone shaped by root exudates and intense microbial activity. Among these, plant growth-promoting rhizobacteria (PGPR) play a central role in stimulating plant productivity and improving tolerance to environmental stress. Members of PGPR communities contribute to plant growth through both direct and indirect mechanisms (Tsotetsi *et al.*, 2022).

Direct growth-promoting mechanisms include biological nitrogen fixation, solubilization of mineral phosphates, and synthesis of phytohormones such as indole-3-acetic acid, cytokinins, and gibberellins. Certain rhizobacteria enhance iron acquisition through siderophore production, thereby improving micronutrient availability in the rhizosphere. Others produce antimicrobial compounds and cell wall-degrading enzymes that suppress phytopathogens and reduce disease incidence.

Indirect mechanisms involve stimulation of plant defense responses and modification of rhizosphere structure. Induced systemic resistance and biofilm formation strengthen plant immunity and improve root colonization efficiency, thereby enhancing protection against pathogens and environmental stressors (Tsotetsi *et al.*, 2022).

Advances in HTS have substantially expanded understanding of how rhizosphere microbial communities respond to environmental pressures such as drought, salinity, and heavy metal contamination. Stress-adapted microbial populations often increase in abundance under harsh conditions, suggesting that natural selection favors microorganisms capable of supporting plant adaptation.

Under saline conditions, certain rhizobacteria enhance plant tolerance through production of exopolysaccharides and osmoprotective compounds that stabilize ionic balance and improve soil water retention. Similarly, drought-adapted microbial strains influence expression of plant stress response genes and promote deeper or more extensive root systems, thereby improving water acquisition (Pantigoso *et al.*, 2022).

In metal-contaminated soils, rhizobacteria contribute to phytoremediation by transforming toxic metal ions into less bioavailable forms or sequestering them within microbial cells. Some strains release organic acids, chelating compounds, and metallothionein-like proteins that reduce metal toxicity and facilitate plant survival in polluted environments (Pantigoso *et al.*, 2022).

As illustrated in Fig. 5.5, these plant–microbe interactions collectively demonstrate how rhizosphere microorganisms enhance plant stress tolerance while contributing to soil restoration and sustainable agricultural productivity. Integration of metagenomics and other omics approaches has further enabled identification of functional genes associated with stress resilience, supporting the design of targeted microbial consortia tailored to specific agroecological conditions.

5.6.2 Biocontrol of plant pathogens

Rhizosphere microbiome investigations have revealed a diverse community of microorganisms possessing strong antifungal and antibacterial properties. These beneficial microbes suppress plant pathogens through multiple mechanisms, including production of hydrolytic enzymes, siderophores,

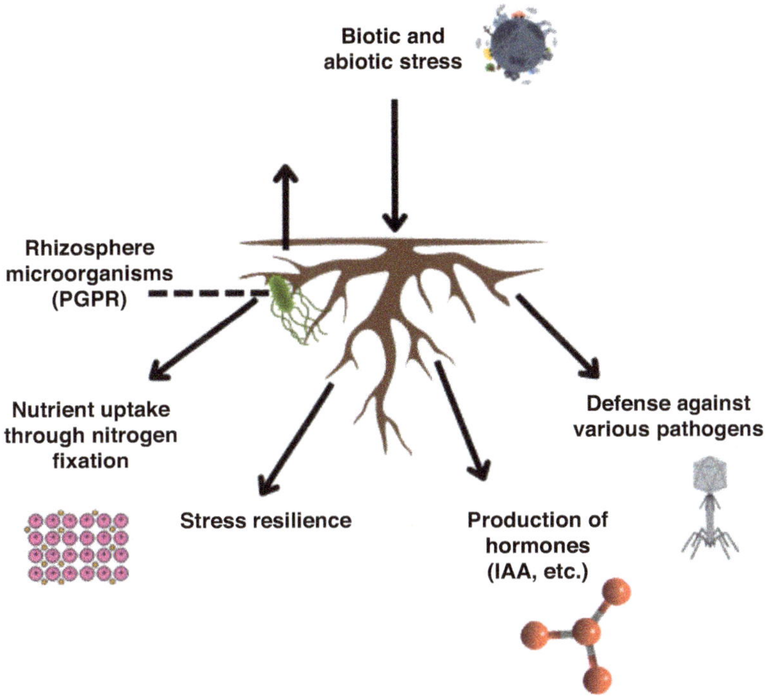

Fig. 5.5. Role of rhizosphere microbial consortia in plant growth and stress adaptation. Diagram illustrating how rhizosphere microorganisms, particularly PGPR, enhance plant performance by improving nutrient acquisition (e.g. nitrogen fixation), producing phytohormones such as indole-3-acetic acid, strengthening tolerance to abiotic stress, and protecting plants against pathogens through induced resistance and antimicrobial activity.

antimicrobial metabolites, and volatile organic compounds. PGPR employ strategies such as nutrient competition, antibiosis, enzymatic degradation of pathogens, and induction of systemic resistance in plants (Egamberdieva *et al.*, 2023).

Species belonging to the genus *Bacillus* are widely reported as effective biocontrol agents. Rhizosphere isolates produce siderophores, antifungal metabolites, and hydrolytic enzymes that inhibit important soilborne pathogens such as *Fusarium oxysporum*. These compounds interfere with pathogen growth by limiting iron availability, degrading fungal cell walls, and disrupting pathogen metabolism (Boulahouat *et al.*, 2023).

Fungal biocontrol agents are also prominent in rhizosphere systems. *Trichoderma* species are extensively documented for their antagonistic activity against *Fusarium* through mycoparasitism, secretion of antifungal metabolites, and rapid colonization of root surfaces, which prevents pathogen establishment (Petrucci *et al.*, 2023). Similarly, members of the genus *Streptomyces* produce volatile organic compounds with antifungal properties that suppress pathogen development while stimulating host resistance responses.

Siderophore production represents an important competitive mechanism in rhizosphere biocontrol. By chelating iron from the surrounding environment, beneficial microbes restrict iron availability to pathogens and thereby limit their proliferation (Egamberdieva *et al.*, 2023). In addition, enzymes such as chitinases, glucanases, and proteases directly degrade fungal cell walls, further strengthening plant protection.

High-throughput sequencing studies demonstrate that rhizosphere microbial diversity contributes significantly to disease suppression. Beneficial microbial communities often exhibit overlapping and synergistic traits, allowing coordinated pathogen inhibition through combined enzymatic activity, nutrient competition, volatile production, and activation of plant immunity. Such multifunctional microbial consortia are increasingly recognized as key components of sustainable disease management strategies that reduce reliance on chemical pesticides.

5.6.3 Engineering microbiomes for sustainable agriculture

Advances in sequencing technologies and systems biology have enabled the development of synthetic microbial communities (SynComs), which consist of functionally selected microorganisms assembled to perform specific ecological roles. Rhizosphere microbiomes are shaped largely by environmental recruitment from soil, with only a minor proportion determined directly by host genotype (Xun *et al.*, 2024). SynCom design therefore focuses on identifying beneficial taxa that consistently enhance plant performance across environmental conditions and reconstructing them into optimized microbial consortia.

The advancement of microbiome engineering strategies has been driven by HTS, synthetic biology, and multi-omics integration, enabling the rational design of plant-associated microbial communities for sustainable agriculture. These approaches range from the construction of synthetic microbial consortia and targeted probiotic applications to genome editing and microbiome-assisted breeding. Collectively, such strategies aim to enhance nutrient cycling, improve stress tolerance, and suppress plant pathogens while maintaining ecosystem stability. The principal microbiome engineering approaches, their methodologies, and their functional outcomes in agricultural systems are summarized in Table 5.2.

Rhizosphere engineering approaches often combine microbial inoculation strategies with plant-mediated recruitment processes. Probiotic approaches involve application of plant growth-promoting microorganisms, whereas prebiotic strategies manipulate root exudates or soil amendments to selectively stimulate beneficial microbial populations. Plant-derived compounds such as flavonoids, phenolics, indole-3-acetic acid, and quorum-sensing molecules influence microbial colonization, metabolic activity, and community stability (Jamil *et al.*, 2022). These interactions modulate nutrient cycling, pathogen suppression, and plant stress responses.

Table 5.2. Microbiome engineering strategies and their functional outcomes in sustainable agriculture.

Engineering strategy	Approach/methodology	Target function	Agricultural outcome
Synthetic microbial communities (SynComs)	Assembly of selected beneficial taxa with complementary traits	Nutrient mobilization, stress tolerance, pathogen suppression	Improved plant growth stability and yield under variable environments
Probiotic inoculation (PGPM/ PGPR application)	Introduction of plant growth-promoting microbes through seed, soil, or root treatments	Nitrogen fixation, phytohormone production, biocontrol activity	Enhanced plant vigor, nutrient uptake, and disease resistance
Prebiotic rhizosphere manipulation	Use of root exudate mimics, organic amendments, or signaling molecules to stimulate beneficial microbes	Selective recruitment of functional microbial groups	Increased microbial diversity and rhizosphere resilience
Microbiome-assisted plant breeding	Selection of plant genotypes that favor beneficial microbial recruitment	Stable plant–microbe interactions and improved stress tolerance	Development of microbiome-optimized crop varieties
Genome editing of microbial strains	CRISPR-based modification to enhance beneficial metabolic pathways or remove deleterious traits	Improved metabolite production, colonization efficiency, or stress protection	Tailored bio-inoculants with predictable performance
Host–microbiome cooptimization	Integration of plant genomics, GWAS, and microbiome profiling to design compatible plant–microbe systems	Improved functional compatibility between host and microbiome	Greater ecosystem stability and productivity
Multi-omics-guided microbiome design	Use of metagenomics, transcriptomics, proteomics, and metabolomics to identify key functional taxa	Evidence-based selection of microbial consortia	Precision agriculture approaches with improved soil health and sustainability

GWAS, genome-wide association study; PGPM, plant growth-promoting microorganism; PGPR, plant growth-promoting rhizobacteria.

Applications of plant growth-promoting microorganisms extend to improvement of soil fertility, activation of stress tolerance pathways, and restoration of degraded soils. Microbial inoculants designed for salinity tolerance, nutrient mobilization, or pathogen suppression are increasingly incorporated into sustainable crop management programs (Hakim *et al.*, 2021). Such interventions support ecological intensification of agriculture while reducing dependence on chemical inputs.

The emergence of genome editing and synthetic biology has further expanded the potential of microbiome engineering. CRISPR-based modification of microbial strains enables targeted enhancement of beneficial metabolic pathways while reducing undesirable traits. Integration of host genetic information derived from approaches such as genome-wide association studies (GWAS) facilitates cooptimization of plant genotypes and microbial consortia (Xun *et al.*, 2024). These developments support the creation of stable bio-inoculant formulations tailored to specific agroecosystems.

5.7 Future Perspectives and Challenges

Research on microbial consortia is being transformed by rapid developments in multi-omics approaches, HTS, and data-intensive analytics. These advances open new opportunities for ecological interpretation and agricultural innovation, while also raising challenges related to environmental monitoring, biosafety, and data management.

5.7.1 Advances in sequencing and omics technologies

Modern multi-omics strategies now provide unprecedented depth and resolution for analyzing plant–microbe interactions (Bhimani *et al.*, 2024). The integration of multiple omics layers with next-generation sequencing enables comprehensive mapping of molecular interactions between plants and their associated microbiota. Such integrated datasets facilitate the construction of communication networks that decode biochemical signaling between soil microbes and plant hosts (Pantigoso *et al.*, 2022).

High-throughput sequencing has significantly improved our ability to characterize rhizosphere microbiomes and to understand how root exudates recruit and shape microbial consortia that influence plant nutrition and health. Recent developments in single-cell transcriptomics and epigenomics further allow functional links to be established between individual taxa and their biological roles, representing a major advance in soil microbiology.

Portable sequencing platforms, such as those developed by Oxford Nanopore Technologies, now permit real-time DNA and RNA sequencing in field conditions, producing ultra-long reads suitable for *in situ* metagenomic and metatranscriptomic analyses. Emerging spatial transcriptomics and metaepigenomics approaches add further resolution by revealing not only which microbial species are present, but also where and under what environmental conditions they are active.

Integration of epigenomic and metabolomic signals has been emphasized as essential for deciphering plant–microbe signaling networks in soil (Bhimani *et al.*, 2024). Meanwhile, local rhizosphere structure appears to be strongly shaped by microenvironmental gradients and root exudate heterogeneity (Pantigoso *et al.*, 2022). Metaepigenomics can reveal microbial adaptive responses to plant-derived compounds by mapping DNA methylation and chromatin accessibility patterns, whereas spatial transcriptomics clarifies how gene expression varies across root zones. Together, these tools enable the development of functionally annotated and spatially resolved soil maps that can support ecological modeling and targeted biotechnological interventions.

5.7.2 Ethical and environmental considerations

Although HTS has enabled the identification and engineering of microbial strains with beneficial traits, such as improved nutrient cycling or pest resistance, ecological risks remain. The release of genetically modified microbes into the environment may result in unintended transmission to nontarget organisms or horizontal gene transfer events (Jeganathan and Holmes, 2021). Ethical oversight therefore requires rigorous environmental risk assessment, controlled field trials and robust containment strategies. Postrelease monitoring, including sequencing-based traceability of engineered strains, is essential to ensure biosafety.

Soil microbial diversity plays a fundamental role in plant health, ecosystem stability, and biogeochemical cycling. Multi-omics approaches allow comprehensive investigation of microbial consortia and enable targeted manipulation strategies (Arikan and Muth, 2023). When combined with multi-omics data, HTS has substantially improved our understanding of rhizosphere ecology (White III *et al.*, 2017). Taxonomic surveys can now be complemented with transcriptomic and metabolomic analyses to link microbial taxa with functional traits relevant to plant fitness (Jain *et al.*, 2024).

Root exudates contain complex mixtures of phytochemicals, signaling molecules, and protective metabolites that regulate microbial communities. Multi-omics studies have revealed how these compounds influence both the structure and activity of rhizosphere microbiota (Jain *et al.*, 2024). Integrating taxonomic data from sequencing with functional omics evidence enables the identification of microbes responsible for hormone production, siderophore release, and nutrient metabolism. This level of understanding supports microbiome engineering strategies tailored to specific crop genotypes and environmental conditions (White III *et al.*, 2017).

5.8 Conclusion

Research on rhizosphere microbiomes has been fundamentally advanced by HTS and multi-omics approaches. These technologies have expanded ecological insight and enabled the development of data-driven agronomic strategies, supporting more precise soil management and crop production systems.

5.8.1 Key insights from high-throughput sequencing and multi-omics

The complexity of rhizosphere systems necessitates advanced analytical tools. Integration of modern analytical chemistry with computational modeling has facilitated the discovery of previously undetected root exudates (Pantigoso *et al.*, 2022). Studies have also demonstrated links between plant root traits such as root diameter and microbial community structure (Herms *et al.*, 2022). These interdisciplinary insights can be strengthened further through artificial intelligence and remote sensing technologies.

Machine learning approaches are increasingly used to predict plant–microbe interactions and link microbial dynamics with plant phenotypes *in situ* (Chien and Larsen, 2017). Similarly, UAV-based multispectral imaging combined with spectral modelling enables real-time mapping of soil nutrient variability (Hossen *et al.*, 2021). Translating rhizosphere knowledge into agricultural practice requires collaboration among researchers, policymakers, and farmers, alongside standardized functional testing and exudate analysis protocols (Pantigoso *et al.*, 2022).

Breeding strategies may also benefit from selecting root traits that promote beneficial microbiomes, such as fine root architectures or enhanced exudation patterns (Herms *et al.*, 2022). Integration of metagenomic sequencing with soil sensor data and environmental monitoring could enable predictive soil health dashboards, supporting improved agronomic decision making, targeted biostimulant application, and cultivar selection.

5.8.2 Future directions in rhizosphere microbiome research

Future progress in rhizosphere microbiome research will depend on the integration of advanced analytical tools, supportive regulatory frameworks, and personalized agricultural approaches. Artificial intelligence combined with remote sensing represents a major pathway for development, enabling predictive modeling and real-time monitoring of plant health and soil microbial dynamics (Herms *et al.*, 2022; Tyagi *et al.*, 2022).

The development of microbiome-based agricultural inputs is another promising direction. Institutional support and regulatory systems recognizing microbial consortia as agricultural inputs will be necessary for scaling biofertilizer and biopesticide technologies (Pantigoso *et al.*, 2022).

Standardized management strategies could improve consistency, efficacy, and farmer adoption.

High-throughput sequencing also enables personalized soil health profiling. Identifying the specific microbial composition of individual soils allows tailored microbial amendments to be developed for local crop varieties and environmental conditions (Pantigoso *et al.*, 2022). Such precision approaches can enhance crop productivity, improve soil sustainability, and reduce reliance on chemical inputs. By linking microbial taxa to functional roles in soil processes, sequencing platforms provide a foundation for predictive microbiome-based agricultural systems (Herms *et al.*, 2022).

References

Afridi, M.S., Ali, S., Salam, A., Cesar Terra, W., Hafeez, A. *et al.* (2022) Plant microbiome engineering: Hopes or hypes. *Biology* 11(12), 1782.

Agamah, F.E., Bayjanov, J.R., Niehues, A., Njoku, K.F., Skelton, M. *et al.* (2022) Computational approaches for network-based integrative multi-omics analysis. *Frontiers in Molecular Biosciences* 9, 967205.

Amarasinghe, S.L., Su, S., Dong, X., Zappia, L., Ritchie, M.E. *et al.* (2020) Opportunities and challenges in long-read sequencing data analysis. *Genome Biology* 21(1), 30.

Antipov, D., Korobeynikov, A., McLean, J.S. and Pevzner, P.A. (2015) HybridSPAdes: An algorithm for hybrid assembly of short and long reads. *Bioinformatics* 32(7), 1009–1015. DOI: 10.1093/bioinformatics/btv688.

Arikan, M. and Muth, T. (2023) Integrated multi-omics analyses of microbial communities: A review of the current state and future directions. *Molecular Omics* 19(8), 607–623.

Barillot, C.D.C., Sarde, C.O., Bert, V., Tarnaud, E. and Cochet, N. (2013) A standardized method for the sampling of rhizosphere and rhizoplan soil bacteria associated to a herbaceous root system. *Annals of Microbiology* 63(2), 471–476.

Bhimani, P., Mahavar, P., Rajguru, B., Bhatt, V.D., Nathani, N.M. *et al.* (2024) Unveiling the green dialogue: Advancements in omics technologies for deciphering plant–microbe interactions in soil. *Discover Plants* 1(1), 4.

Boulahouat, S., Cherif-Silini, H., Silini, A., Bouket, A.C., Luptakova, L. *et al.* (2023) Biocontrol efficiency of rhizospheric *Bacillus* against the plant pathogen *Fusarium oxysporum*: A promising approach for sustainable agriculture. *Microbiology Research* 14(3), 892–908.

Carvalhais, L.C., Dennis, P.G., Tyson, G.W. and Schenk, P.M. (2012) Application of metatranscriptomics to soil environments. *Journal of Microbiological Methods* 91(2), 246–251.

Carvalhais, L.C., Dennis, P.G., Tyson, G.W. and Schenk, P.M. (2013) Rhizosphere metatranscriptomics: Challenges and opportunities. *Molecular Microbial Ecology of the Rhizosphere* 1, 1137–1144.

Chien, J. and Larsen, P. (2017) Predicting the plant root-associated ecological niche of 21 pseudomonas species using machine learning and metabolic modelling. Available at: https://arxiv.org/abs/1701.03220 (accessed 9 May 2026).

Deng, Z., Zhang, J., Li, J. and Zhang, X. (2021) Application of deep learning in plant–microbiota association analysis. *Frontiers in Genetics* 12, 697090.

Egamberdieva, D., Eshboev, F., Shukurov, O., Alaylar, B. and Arora, N.K. (2023) Bacterial bioprotectants: Biocontrol traits and induced resistance to phytopathogens. *Microbiology Research* 14(2), 689–703.

Fierer, N. (2017) Embracing the unknown: Disentangling the complexities of the soil microbiome. *Nature Reviews Microbiology* 15(10), 579–590.

Floćh, J.-B., Hamel, C., Lupwayi, N., Harker, K.N., Hijri, M. *et al.* (2020) Bacterial communities of the canola rhizosphere: Network analysis reveals a core bacterium shaping microbial interactions. *Frontiers in Microbiology* 11, 1587.

Forsberg, K.J., Reyes, A., Wang, B., Selleck, E.M., Sommer, M.O. *et al.* (2012) The shared antibiotic resistome of soil bacteria and human pathogens. *Science* 337(6098), 1107–1111.

Frias-Lopez, J., Shi, Y., Tyson, G.W., Coleman, M.L., Schuster, S.C. *et al.* (2008) Microbial community gene expression in ocean surface waters. *Proceedings of the National Academy of Sciences USA* 105(10), 3805–3810.

Hakim, S., Naqqash, T., Nawaz, M.S., Laraib, I., Siddique, M.J. *et al.* (2021) Rhizosphere engineering with plant growth-promoting microorganisms for agriculture and ecological sustainability. *Frontiers in Sustainable Food Systems* 5, 617157.

Handelsman, J. (2004) Metagenomics: Application of genomics to uncultured microorganisms. *Microbiology and Molecular Biology Reviews* 68(4), 669–685.

Haro, C., Anguita-Maeso, M., Metsis, M., Navas-Cortés, J.A. and Landa, B.B. (2021) Evaluation of established methods for DNA extraction and primer pairs targeting 16S rRNA gene for bacterial microbiota profiling of olive xylem sap. *Frontiers in Plant Science* 12, 640829.

He, Z., Deng, Y., Van Nostrand, J.D., Tu, Q., Xu, M. *et al.* (2010) GeoChip 3.0 as a high-throughput tool for analyzing microbial community composition, structure and functional activity. *ISME Journal* 4(9), 1167–1179.

Herms, C.H., Hennessy, R.C., Bak, F., Dresbøll, D.B. and Nicolaisen, M.H. (2022) Back to our roots: Exploring the role of root morphology as a mediator of beneficial plant–microbe interactions. *Environmental Microbiology* 24(8), 3264–3272.

Hettich, R.L., Pan, C., Chourey, K. and Giannone, R.J. (2013) Metaproteomics: Harnessing the power of high performance mass spectrometry to identify the suite of proteins that control metabolic activities in microbial communities. *Analytical Chemistry* 85(9), 4203–4214.

Hossen, M.A., Diwakar, P.K. and Ragi, S. (2021) Total nitrogen estimation in agricultural soils via aerial multispectral imaging and LIBS. *Scientific Reports* 11(1), 12693.

Hultman, J., Waldrop, M.P., Mackelprang, R., David, M.M., McFarland, J. *et al.* (2015) Multi-omics of permafrost, active layer and thermokarst bog soil microbiomes. *Nature* 521(7551), 208–212.

Jain, A., Sarsaiya, S., Singh, R., Gong, Q., Wu, Q. *et al.* (2024) Omics approaches in understanding the benefits of plant–microbe interactions. *Frontiers in Microbiology* 15, 1391059.

Jamil, F., Mukhtar, H., Fouillaud, M. and Dufossé, L. (2022) Rhizosphere signaling: Insights into plant–rhizomicrobiome interactions for sustainable agronomy. *Microorganisms* 10(5), 899.

Jeganathan, P. and Holmes, S.P. (2021) A statistical perspective on the challenges in molecular microbial biology. *Journal of Agricultural, Biological and Environmental Statistics* 26(2), 131–160.

Jensen, T.B.N., Dall, S.M., Knutsson, S., Karst, S.M. and Albertsen, M. (2024) High-throughput DNA extraction and cost-effective miniaturized metagenome and amplicon library preparation of soil samples for DNA sequencing. *PLOS ONE* 19(4), e0301446.

Jiang, W., Ye, W., Tan, X. and Bao, Y.J. (2025) Network-based multi-omics integrative analysis methods in drug discovery: A systematic review. *BioData Mining* 18(1), 27.

Jin, H., You, L., Zhao, F., Li, S., Ma, T. *et al.* (2022) Hybrid, ultra-deep metagenomic sequencing enables genomic and functional characterization of low-abundance species in the human gut microbiome. *Gut Microbes* 14(1), 2021790. DOI: 10.1080/19490976.2021.2021790.

Kassim, Y., Sheng, H., Xu, G., Jin, H., Iqbal, T. *et al.* (2025) Integrated multi-omics analysis reveals key regulators of bovine oocyte maturation. *International Journal of Molecular Sciences* 26(9), 3973.

Kimotho, R.N. and Maina, S. (2024) Unraveling plant–microbe interactions: Can integrated omics approaches offer concrete answers? *Journal of Experimental Botany* 75(5), 1289–1313.

Knight, R., Vrbanac, A., Taylor, B.C., Aksenov, A., Callewaert, C. *et al.* (2018) Best practices for analysing microbiomes. *Nature Reviews Microbiology* 16(7), 410–422.

Kumaishi, K., Usui, E., Suzuki, K., Kobori, S., Sato, T. *et al.* (2022) High throughput method of 16S rRNA gene sequencing library preparation for plant root microbial community profiling. *Scientific Reports* 12(1), 19289.

Kumari, N., Iquebal, M.A., Jaiswal, S., Rai, A. and Kumar, D. (2022) Metatranscriptomic approach to understand the role of the rhizosphere microbiome: A review. *Bhartiya Krishi Anusandhan Patrika* 37(3), 197–201.

Lakshmanan, V., Ray, P. and Craven, K.D. (2017) Rhizosphere sampling protocols for microbiome (16S/18S/ITS rRNA) library preparation and enrichment for the isolation of drought tolerance-promoting microbes. In: Sunkar, R. (ed.) *Plant Stress Tolerance: Methods and Protocols*. Springer, Dordrecht, The Netherlands, pp. 349–362.

Law, S.R., Serrano, A.R., Daguerre, Y., Sundh, J., Schneider, A.N. *et al.* (2022) Metatranscriptomics captures dynamic shifts in mycorrhizal coordination in boreal forests. *Proceedings of the National Academy of Sciences USA* 119(26), e2118852119.

Li, J., Wang, C., Liang, W. and Liu, S. (2021) Rhizosphere microbiome: The emerging barrier in plant-pathogen interactions. *Frontiers in Microbiology* 12, 772420.

Ling, N., Wang, T. and Kuzyakov, Y. (2022) Rhizosphere bacteriome structure and functions. *Nature Communications* 13(1), 836.

Liu, Q., Cheng, L., Nian, H., Jin, J. and Lian, T. (2023) Linking plant functional genes to rhizosphere microbes: A review. *Plant Biotechnology Journal* 21(5), 902–917.

López-Mondéjar, R., Tláskal, V., Větrovský, T., Štursová, M., Toscan, R. *et al.* (2020) Metagenomics and stable isotope probing reveal the complementary contribution of fungal and bacterial communities in the recycling of dead biomass in forest soil. *Soil Biology and Biochemistry* 148, 107875.

Luan, T., Commichaux, S., Hoffmann, M., Jayeola, V., Jang, J.H. *et al.* (2024) Benchmarking short and long read polishing tools for nanopore assemblies: Achieving near-perfect genomes for outbreak isolates. *BMC Genomics* 25(1), 679.

Maki, K.A., Wolff, B., Varuzza, L., Green, S.J. and Barb, J.J. (2023) Multi-amplicon microbiome data analysis pipelines for mixed orientation sequences using QIIME2: Assessing reference database, variable region and pre-processing bias in classification of mock bacterial community samples. *PLOS ONE* 18(1), e0280293.

Mandlik, J.S., Patil, A.S. and Singh, S. (2024) Next-generation sequencing (NGS): Platforms and applications. *Journal of Pharmacy and Bioallied Sciences* 16(Suppl 1), S41–S45.

Marx, V. (2023) Method of the year: Long-read sequencing. *Nature Methods* 20(1), 6–11.

Marín, O., González, B. and Poupin, M.J. (2021) From microbial dynamics to functionality in the rhizosphere: A systematic review of the opportunities with synthetic microbial communities. *Frontiers in Plant Science* 12, 650609.

Mathieu, A., Leclercq, M., Sanabria, M., Perin, O. and Droit, A. (2022) Machine learning and deep learning applications in metagenomic taxonomy and functional annotation. *Frontiers in Microbiology* 13, 811495.

Mendes, R., Garbeva, P. and Raaijmakers, J.M. (2013) The rhizosphere microbiome: Significance of plant beneficial, plant pathogenic, and human pathogenic microorganisms. *FEMS Microbiology Reviews* 37(5), 634–663.

Mirete, S., Sánchez-Costa, M., Díaz-Rullo, J., González de Figueras, C., Martínez-Rodríguez, P. *et al.* (2025) Metagenome-assembled genomes (MAGS): Advances, challenges, and ecological insights. *Microorganisms* 13(5), 985. DOI: 10.3390/microorganisms13050985.

Mishra, A.K., Sudalaimuthuasari, N., Hazzouri, K.M., Saeed, E.E., Shah, I. *et al.* (2022) Tapping into plant–microbiome interactions through the lens of multi-omics techniques. *Cells* 11(20), 3254.

Pang, Z., Dong, F., Liu, Q., Lin, W., Hu, C. *et al.* (2021) Soil metagenomics reveals effects of continuous sugarcane cropping on the structure and functional pathway of rhizospheric microbial community. *Frontiers in Microbiology* 12, 627569.

Pantigoso, H.A., Newberger, D. and Vivanco, J.M. (2022) The rhizosphere microbiome: Plant–microbial interactions for resource acquisition. *Journal of Applied Microbiology* 133(5), 2864–2876.

Papoutsoglou, G., Tarazona, S., Lopes, M.B., Klammsteiner, T., Ibrahimi, E. *et al.* (2023) Machine learning approaches in microbiome research: Challenges and best practices. *Frontiers in Microbiology* 14, 1261889.

Patti, G.J., Yanes, O. and Siuzdak, G. (2012) Metabolomics: The apogee of the omics trilogy. *Nature Reviews Molecular Cell Biology* 13(4), 263–269.

Petrucci, A., Khairullina, A., Sarrocco, S., Jensen, D.F., Jensen, B. *et al.* (2023) Understanding the mechanisms underlying biological control of Fusarium diseases in cereals. *European Journal of Plant Pathology* 167(4), 453–476.

Pramanik, K., Khairullina, A., Sarrocco, S., Jensen, D.F., Jensen, B. *et al.* (2020) Metagenomic insights into rhizospheric microbiome profiling in lentil cultivars unveils differential microbial nitrogen and phosphorus metabolism under rice-fallow ecology. *International Journal of Molecular Sciences* 21(23), 8895.

Pudake, R.N., Sahu, B.B., Kumari, M. and Sharma, A.K. (eds) (2021) *Omics Science for Rhizosphere Biology*. Springer, Cham, Switzerland.

Quast, C., Pruesse, E., Yilmaz, P., Gerken, J., Schweer, T. *et al.* (2012) The SILVA ribosomal RNA gene database project: Improved data processing and web-based tools. *Nucleic Acids Research* 41(D1), D590–D596.

Saenz, C., Nigro, E., Gunalan, V. and Arumugam, M. (2022) MIntO: A modular and scalable pipeline for microbiome metagenomic and metatranscriptomic data integration. *Frontiers in Bioinformatics* 2, 846922.

Satam, H., Joshi, K., Mangrolia, U., Waghoo, S., Zaidi, G. *et al.* (2023) Next-generation sequencing technology: Current trends and advancements. *Biology* 12(7), 997.

Scossa, F., Alseekh, S. and Fernie, A.R. (2021) Integrating multi-omics data for crop improvement. *Journal of Plant Physiology* 257, 153352.

Segata, N., Boernigen, D., Tickle, T.L., Morgan, X.C., Garrett, W.S. *et al.* (2013) Computational meta-omics for microbial community studies. *Molecular Systems Biology* 9(1), 666.

Sergaki, C., Lagunas, B., Lidbury, I., Gifford, M.L. and Schäfer, P. (2018) Challenges and approaches in microbiome research: From fundamental to applied. *Frontiers in Plant Science* 9, 1205.

Shade, A., Gregory Caporaso, J., Handelsman, J., Knight, R. and Fierer, N. (2013) A meta-analysis of changes in bacterial and archaeal communities with time. *ISME Journal* 7(8), 1493–1506.

Shreshtha, K., Raj, S., Pal, A.K., Tripathi, P., Choudhary, K.K. *et al.* (2024) Isolation and identification of rhizospheric and endophytic bacteria from cucumber plants irrigated with wastewater: Exploring their roles in plant growth promotion and disease suppression. *Current Research in Microbial Sciences* 7, 100256.

Staley, J.T. and Konopka, A. (1985) Measurement of in situ activities of nonphotosynthetic microorganisms in aquatic and terrestrial habitats. *Annual Review of Microbiology* 39(1), 321–346.

Su, G., Qin, X., Enninful, A., Bai, Z., Deng, Y. *et al.* (2021) Spatial multi-omics sequencing for fixed tissue via DBiT-seq. *STAR Protocols* 2(2), 100532.

Tafazoli, A., Hemmati, M., Rafigh, M., Alimardani, M., Khaghani, F. *et al.* (2025) Leveraging long-read sequencing technologies for pharmacogenomic testing: Applications, analytical strategies, challenges, and future perspectives. *Frontiers in Genetics* 16, 1435416.

Terrón-Camero, L.C., Gordillo-González, F., Salas-Espejo, E. and Andrés-León, E. (2022) Comparison of metagenomics and metatranscriptomics tools: A guide to making the right choice. *Genes* 13(12), 2280.

Tringe, S.G., Von Mering, C., Kobayashi, A., Salamov, A.A., Chen, K. *et al.* (2005) Comparative metagenomics of microbial communities. *Science* 308(5721), 554–557.

Tsotetsi, T., Nephali, L., Malebe, M. and Tugizimana, F. (2022) Bacillus for plant growth promotion and stress resilience: What have we learned? *Plants* 11(19), 2482.

Tyagi, R., Pradhan, S., Bhattacharjee, A., Dubey, S. and Sharma, S. (2022) Management of abiotic stresses by microbiome-based engineering of the rhizosphere. *Journal of Applied Microbiology* 133(2), 254–272.

Vidal, C., González, F., Santander, C., Pérez, R., Gallardo, V. *et al.* (2022) Management of rhizosphere microbiota and plant production under drought stress: A comprehensive review. *Plants* 11(18), 2437.

Walsh, C., Stallard-Olivera, E. and Fierer, N. (2024) Nine (not so simple) steps: A practical guide to using machine learning in microbial ecology. *MBio* 15(2), e02050-23.

Wang, B., Wang, X., Wang, Z., Zhu, K. and Wu, W. (2023) Comparative metagenomic analysis reveals rhizosphere microbial community composition and functions help protect grapevines against salt stress. *Frontiers in Microbiology* 14, 1102547.

White III, R.A., Rivas-Ubach, A., Borkum, M.I., Köberl, M., Bilbao, A. *et al.* (2017) The state of rhizospheric science in the era of multi-omics: A practical guide to omics technologies. *Rhizosphere* 3, 212–221.

Wilmes, P. and Bond, P.L. (2006) Metaproteomics: Studying functional gene expression in microbial ecosystems. *Trends in Microbiology* 14(2), 92–97.

Wu, Y. and Xie, L. (2025) AI-driven multi-omics integration for multi-scale predictive modeling of genotype-environment-phenotype relationships. *Computational and Structural Biotechnology Journal* 27, 267–277.

Xiao, T. and Zhou, W. (2020) The third generation sequencing: The advanced approach to genetic diseases. *Translational Pediatrics* 9(2), 163.

Xin, L., Huang, C., Li, H., Huang, S., Feng, Y. *et al.* (2024) Artificial intelligence for central dogma-centric multi-omics: Challenges and breakthroughs. Available at: https://arxiv.org/abs/2412.12668 (accessed 9 May 2026).

Xun, W., Liu, Y., Ma, A., Yan, H., Miao, Y. *et al.* (2024) Dissection of rhizosphere microbiome and exploiting strategies for sustainable agriculture. *New Phytologist* 242(6), 2401–2410.

Yang, L., Yang, Y., Huang, L., Cui, X. and Liu, Y. (2023) From single-to multi-omics: Future research trends in medicinal plants. *Briefings in Bioinformatics* 24(1), bbac485.

Yu, X., Liu, Z. and Sun, X. (2023) Single-cell and spatial multi-omics in the plant sciences: Technical advances, applications, and perspectives. *Plant Communications* 4(3), 100508.

Yu, Z., Gan, Z., Tawfik, A. and Meng, F. (2024) Exploring interspecific interaction variability in microbiota: A review. *Engineering Microbiology* 4(4), 100178.

6 Biogeochemical Processes Occurring at the Plant Root–Soil Interface, Influenced by Root Activities and Chemicals Released by the Roots

Vandana Anand and Udit Yadav*

Department of Botany, Institute of science BHU, Varanasi

6.1 Introduction

The plant root–soil interface, often termed the rhizosphere, is a dynamic zone where complex biogeochemical processes take place, strongly influenced by root activities and the diverse chemical compounds released by plants. This microenvironment plays a pivotal role in plant growth, soil fertility, and ecosystem functioning, as shown in Fig. 6.1. Root exudates, comprising sugars, amino acids, organic acids, phenolics, and secondary metabolites, mediate crucial interactions between plants, soil, and microorganisms. These compounds not only influence nutrient mobilization and availability but also regulate microbial community composition and activity (Jones *et al.*, 2009; Sun *et al.*, 2025).

Biogeochemical processes at the rhizosphere are central to the cycling of carbon, nitrogen, phosphorus, and other essential nutrients. Roots actively modify the soil pH, redox potential, and organic matter dynamics, thereby creating microhabitats with unique physicochemical properties. Such modifications enhance nutrient uptake, support beneficial microbial consortia, and influence soil structural stability (Philippot *et al.*, 2013). Additionally, root–microbe interactions foster processes like nitrogen fixation, mineral weathering, and organic matter decomposition, linking below-ground activity with above-ground productivity.

Understanding these processes is essential for sustainable agriculture, soil management, and climate change mitigation. Insights into rhizosphere

*Corresponding author: udityadav48@gmail.com

DOI: 10.1079/9781836990475.0006

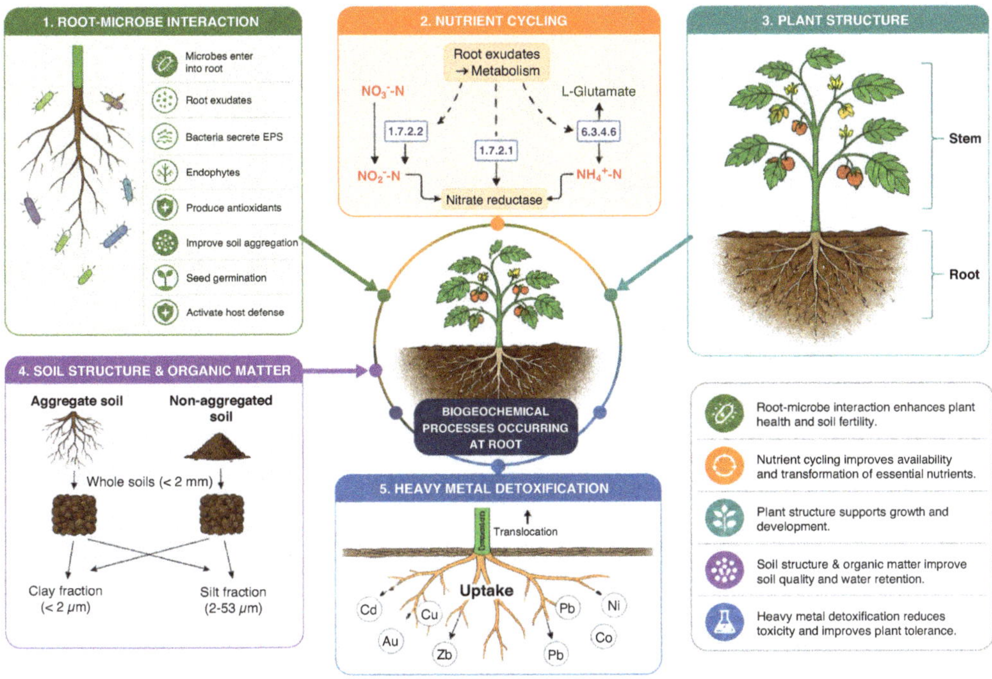

Fig. 6.1. Schematic diagram showing the biogeochemical processes occurring at the plant root–soil interface.

biogeochemistry provide a foundation for developing strategies to optimize nutrient use efficiency and harness plant–microbe partnerships for improved crop resilience.

6.2 Root Exudates and Their Role in Biogeochemical Processes

Plant roots are not passive organs limited to nutrient and water uptake; instead, they act as active mediators of complex chemical dialogues with their surrounding soil environment.

Central to this communication are root exudates, a diverse array of compounds that plants release into the rhizosphere. These exudates include primary metabolites such as sugars, amino acids, and organic acids, as well as secondary metabolites like flavonoids, phenolics, terpenoids, and signaling molecules (Bais *et al.*, 2006). Collectively, they shape the physicochemical and biological characteristics of the soil, thereby driving key biogeochemical processes.

One of the most critical functions of root exudates lies in nutrient mobilization. For example, organic acids secreted by roots can solubilize otherwise unavailable forms of phosphorus and micronutrients, while protons alter rhizosphere pH to facilitate mineral weathering (Dakora and Phillips, 2002). These processes not only enhance plant nutrient acquisition but also contribute to broader element cycling within ecosystems. Exudates also influence soil microbial communities by acting as carbon and energy sources, thereby stimulating microbial activity and altering community composition (Philippot *et al.*, 2013). Such interactions foster processes like nitrogen fixation, denitrification, and organic matter decomposition, which are crucial for maintaining soil fertility.

Beyond nutrient dynamics, root exudates regulate soil structural properties. Polysaccharides and mucilage enhance soil aggregation, improving water retention and aeration (Carminati and Vetterlein, 2013). Additionally, signaling molecules in exudates mediate symbiotic relationships with mycorrhizal fungi and nitrogen-fixing bacteria, reinforcing nutrient exchange networks and stabilizing biogeochemical cycles.

Taken together, root exudates function as powerful drivers of rhizosphere chemistry and biology. They integrate plant metabolism with soil processes, creating feedback loops that sustain nutrient availability, microbial diversity, and ecosystem resilience. In this way, root exudation serves as a pivotal mechanism through which plants actively shape biogeochemical processes at the soil interface.

6.3 Nutrient Cycling at the Root–Soil Interface

The root–soil interface is a hotspot of nutrient cycling, where plant roots, soil minerals, and microorganisms interact to regulate the transformation and availability of essential elements. These cycles are tightly interconnected, forming the foundation of plant productivity and soil fertility. Root exudates, microbial processes, and physicochemical modifications by roots are key drivers of carbon (C), nitrogen (N), phosphorus (P), and sulfur (S) cycling in the rhizosphere.

6.3.1 Carbon cycling

Carbon cycling in the rhizosphere begins with photosynthetically fixed carbon transported below ground as root exudates, mucilage, and sloughed-off cells. These carbon inputs serve as substrates for microbial respiration and biomass production, creating a highly active zone of decomposition and nutrient mineralization (Sun *et al.*, 2025). Microbes transform root-derived carbon into humic substances, influencing soil organic matter formation and stabilization. Through priming effects, root exudates also accelerate the breakdown of native soil organic matter, thereby linking plant metabolism with long-term carbon sequestration and atmospheric CO_2 fluxes (Jones *et al.*, 2009).

6.3.2 Nitrogen cycling

Nitrogen cycling at the root–soil interface is governed by processes such as biological nitrogen fixation, nitrification, and denitrification. Symbiotic associations with diazotrophic bacteria, like *Rhizobium* and free-living *Azotobacter*, enrich soil nitrogen through fixation, providing plants with ammonium (NH_4^+) (Richardson *et al.*, 2009). Root exudates stimulate microbial activity that drives ammonification and nitrification, while anaerobic microsites in the rhizosphere can foster denitrification, releasing N_2 and N_2O gases. These interactions balance plant nitrogen availability and gaseous losses, tightly linking plant nutrition with environmental sustainability.

6.3.3 Phosphorus cycling

Phosphorus (P) is often immobile in soils due to its strong binding with minerals, but root activities enhance its availability. Organic acids exuded by roots, such as citrate and malate, solubilize mineral-bound phosphates, while phosphatase enzymes released by roots and microbes mineralize organic P compounds (Richardson and Simpson, 2011). Mycorrhizal fungi further extend root absorptive capacity, accessing phosphorus beyond the root depletion zone. Thus, plant–microbe–soil interactions create dynamic phosphorus hotspots essential for sustaining growth under nutrient-limited conditions.

6.3.4 Sulfur cycling

Sulfur (S) cycling at the root–soil interface involves mineralization of organic S compounds, oxidation of elemental sulfur, and reduction processes driven by microbial communities. Root exudates stimulate sulfur-oxidizing and sulfur-reducing bacteria, which regulate the transformation of sulfate (SO_4^{2-}), the primary form taken up by plants (Gahan and Schmalenberger, 2014). Additionally, sulfur availability influences root exudate composition and plant defense metabolites, linking sulfur dynamics to both nutrition and stress resilience.

6.4 Soil pH and Redox Reactions in the Rhizosphere

The rhizosphere is characterized by steep chemical gradients, strongly influenced by root activity. Among the most important factors regulating nutrient availability and microbial processes at the root–soil interface are soil pH and redox reactions. Roots actively modify these parameters through ion exchange, exudation, and respiration, thereby creating localized microenvironments that determine nutrient dynamics and element cycling.

6.4.1 Soil pH in the rhizosphere

Soil pH in the rhizosphere often differs significantly from bulk soil due to root-induced processes. The uptake of cations such as NH_4^+, K^+, Ca^{2+}, and Mg^{2+} is typically balanced by the release of protons (H^+), leading to rhizosphere acidification, whereas the uptake of anions such as NO_3^- can result in proton consumption and local alkalinization (Hinsinger *et al.*, 2003). Organic acids exuded by roots further modulate pH, enhancing the solubility of otherwise unavailable nutrients, especially phosphorus and micronutrients like iron, zinc, and manganese (Dakora and Phillips, 2002). These localized pH shifts play a vital role in nutrient mobilization, microbial activity, and metal detoxification.

6.4.2 Redox reactions in the rhizosphere

In addition to pH regulation, plant roots influence the redox status of the rhizosphere. Oxygen release from aerenchyma tissues in wetland species, such as rice, creates oxidized microzones around roots, favoring aerobic microbial processes, including nitrification and iron oxidation (Colmer, 2003). Conversely, oxygen depletion through root and microbial respiration generates reducing conditions that promote denitrification, sulfate reduction, and iron/manganese reduction (Neumann and Römheld, 2007). These redox transformations are central to nutrient cycling, influencing the availability of nitrogen, phosphorus, and trace elements, while also determining the fate of pollutants and greenhouse gas emissions.

6.4.3 Integrated role of pH and redox dynamics

Soil pH and redox potential interact dynamically in the rhizosphere, jointly shaping nutrient transformations. For example, low pH and reducing conditions can enhance the solubility of toxic metals, while moderate acidification under oxidized conditions favors phosphorus mobilization. Through these mechanisms, plant roots act as active regulators of soil chemistry, influencing biogeochemical cycles and sustaining plant–microbe interactions that underpin ecosystem productivity.

6.5 Microbial Interactions at the Root–Soil Interface

The root–soil interface is one of the most biologically active zones in terrestrial ecosystems, where plant roots interact intimately with an enormous diversity of microorganisms. These interactions range from beneficial associations that enhance nutrient uptake and stress tolerance, to antagonistic relationships that can impair plant growth. Root exudates act as chemical signals and nutrient sources that shape microbial community structure, driving multiple forms of microbial interactions.

6.5.1 Symbiotic interactions

Symbiotic associations between plants and microorganisms are fundamental to nutrient acquisition. The most well-studied examples include mycorrhizal fungi and nitrogen-fixing bacteria. Mycorrhizal fungi extend the absorptive surface area of roots, improving phosphorus and micronutrient uptake while receiving carbohydrates from the host plant (Smith and Read, 2008). Similarly, rhizobia establish nodulation in leguminous plants, converting atmospheric nitrogen into plant-available ammonium, thereby reducing reliance on external nitrogen sources (Oldroyd *et al.*, 2011). These interactions are mediated by root exudates such as flavonoids, which trigger microbial signaling pathways.

6.5.2 Commensal interactions

A large proportion of rhizosphere microbes establish commensal relationships, in which they utilize root exudates without directly affecting plant health. These microbes help maintain rhizosphere biodiversity and can modulate the activity of pathogenic or beneficial organisms indirectly by competing for space and nutrients (Philippot *et al.*, 2013). Commensals often act as ecological buffers, stabilizing community dynamics under changing environmental conditions.

6.5.3 Antagonistic interactions

Not all microbial associations are beneficial. Pathogenic fungi, bacteria, and nematodes exploit root tissues, causing diseases and reducing crop yields. However, antagonistic microbes also exist that suppress pathogens. For instance, *Trichoderma* spp. and *Pseudomonas* spp. release antibiotics, lytic enzymes, and siderophores that inhibit harmful organisms (Compant *et al.*, 2010). These antagonistic interactions are a cornerstone of biological control strategies in sustainable agriculture.

6.5.4 Microbial competition and cooperation

Microbes at the root–soil interface engage in both competition and cooperation. Competition arises through resource limitation, allelopathic compounds, or niche exclusion. At the same time, cooperative interactions such as crossfeeding, quorum sensing, and biofilm formation enhance microbial survival and stability in the rhizosphere (Berendsen *et al.*, 2012). Such cooperative networks support multifunctionality in nutrient cycling, organic matter decomposition, and plant health promotion.

6.5.5 Integrated role in plant health and soil functioning

The interplay of symbiotic, commensal, and antagonistic microbes creates a dynamic rhizosphere ecosystem that directly influences plant productivity and soil function. By shaping nutrient dynamics, suppressing pathogens, and enhancing stress resilience, microbial interactions form the biological backbone of biogeochemical processes at the root–soil interface.

6.6 Soil Structure and Organic Matter Dynamics

The physical structure of soil and the dynamics of organic matter are central to the functioning of the root–soil interface. Plant roots, microbial activity, and organic inputs collectively shape soil porosity, aggregation, and carbon turnover, which in turn regulate nutrient cycling, water retention, and root penetration.

6.6.1 Soil aggregation and stability

Soil aggregation refers to the binding of mineral particles into larger units, which determines soil porosity and aeration. Roots contribute to aggregation by exerting physical pressure and by releasing mucilage and polysaccharides that act as binding agents (Bronick and Lal, 2005). Fungal hyphae, especially those of mycorrhizae, further stabilize aggregates by enmeshing soil particles and secreting glomalin-related proteins. Stable aggregates not only protect organic matter from decomposition but also enhance water infiltration and erosion resistance.

6.6.2 Role of root exudates in soil structure

Root exudates, particularly mucilage and low molecular weight organic acids, modify soil hydraulic properties by increasing water retention around roots (Carminati and Vetterlein, 2013). Mucilage acts as a temporary adhesive, binding soil particles into microaggregates and reducing friction during root growth. Additionally, exudates influence soil wettability, allowing roots to maintain contact with soil under drying conditions, which is critical for water and nutrient uptake.

6.6.3 Organic matter decomposition and turnover

Soil organic matter (SOM) is continuously decomposed and transformed at the root–soil interface. Root-derived carbon inputs, including exudates, sloughed cells, and decaying root tissues, provide substrates for microbial communities that drive decomposition (Kuzyakov and Domanski, 2000). Microbial activity converts labile organic matter into stable humic substances, influencing long-term carbon sequestration. This decomposition not only recycles nutrients but also controls greenhouse gas fluxes from soils.

6.6.4 Protection and stabilization of soil organic matter

The stabilization of SOM occurs through physical, chemical, and biological mechanisms. Physical protection occurs when organic matter is occluded within aggregates, reducing microbial accessibility. Chemically, SOM interacts with clay minerals and metal oxides, forming organo-mineral complexes (Six *et al.*, 2002). Biologically, microbial residues contribute to the formation of persistent organic compounds, often more stable than plant-derived inputs. These stabilization pathways are essential for maintaining soil fertility and mitigating carbon loss under environmental stress.

6.6.5 Integrated role in soil function and plant productivity

The interplay between soil structure and SOM dynamics creates a self-sustaining system that underpins plant productivity. Stable aggregates improve root growth, while SOM provides energy for microbial communities that cycle nutrients. In turn, plants reinforce these processes by contributing organic inputs and modifying soil structure through their roots. This tight feedback loop illustrates the fundamental importance of soil structure and organic matter dynamics in sustaining biogeochemical processes at the root–soil interface.

6.7 Biogeochemical Processes in Heavy Metal Detoxification

Heavy metals such as cadmium (Cd), lead (Pb), arsenic (As), and chromium (Cr) often accumulate in soils due to natural weathering, industrial emissions, mining, and excessive agrochemical use. At high concentrations, these elements can impair plant growth, reduce microbial activity, and disrupt soil biogeochemistry. However, the rhizosphere is not a passive sink; instead, it hosts a suite of biogeochemical processes that detoxify or immobilize heavy metals, reducing their bioavailability. These processes are mediated by root exudates, microbial transformations, and soil physicochemical reactions.

6.7.1 Root-induced pH and redox modifications

Roots influence heavy metal solubility by altering rhizosphere pH and redox conditions. Acidification through proton release increases the mobility of metals such as Zn and Cu, whereas alkalinization by nitrate uptake can precipitate metals in less soluble forms (Hinsinger *et al.*, 2003). Similarly, root-driven redox changes regulate the speciation of metals like Fe, Mn, and As. For instance, oxygen release from wetland roots promotes Fe (III) precipitation, which can coprecipitate toxic As, thereby reducing its availability (Colmer, 2003).

6.7.2 Chelation and complexation by root exudates

Organic acids, phenolics, and amino acids secreted by roots play a crucial role in metal detoxification. These exudates chelate heavy metals, forming stable complexes that lower their toxicity to plants (Dakora and Phillips, 2002). In some cases, chelation enhances phytoextraction by increasing metal solubility, while in others it promotes immobilization by binding metals within the rhizosphere matrix. Such dual roles highlight the adaptive strategies of plants in regulating heavy metal stress.

6.7.3 Microbial transformations of heavy metals

Microorganisms at the root–soil interface contribute significantly to metal detoxification. Some bacteria and fungi reduce toxic metal ions into less harmful forms, such as the reduction of Cr(VI) to Cr(III) (Gadd, 2010). Others immobilize metals through biosorption, bioaccumulation, or the production of extracellular polymers. Mycorrhizal fungi also act as biofilters, sequestering heavy metals within their hyphae and reducing translocation to plant tissues (Rajkumar *et al.*, 2012). These microbial processes enhance the resilience of plants in contaminated environments.

6.7.4 Precipitation and immobilization reactions

Heavy metals can also be detoxified through rhizosphere-mediated precipitation and immobilization. Roots and microbes release phosphates, carbonates, and sulfides that react with metals to form insoluble precipitates (e.g. Pb-phosphates or Cd-sulfides). These geochemical reactions reduce metal solubility, effectively locking contaminants into less bioavailable forms (Park *et al.*, 2011). Immobilization is particularly important for stabilizing polluted soils and preventing leaching into groundwater.

6.7.5 Integrated role in plant and soil health

Together, root exudates, microbial processes, and geochemical transformations form an integrated detoxification network at the rhizosphere. By reducing metal bioavailability, enhancing sequestration, or facilitating controlled phytoextraction, these processes protect plant health and contribute to soil remediation. Understanding these mechanisms provides a foundation for developing phytoremediation and bioremediation strategies to manage heavy metal pollution in agroecosystems.

6.8 Conclusion

The root–soil interface is a dynamic microenvironment where plants, soil, and microorganisms interact through a complex web of biogeochemical processes. Far from being a passive zone of nutrient uptake, the rhizosphere represents an active biochemical frontier, shaped by root exudates, microbial communities, and soil physicochemical properties. These interactions regulate fundamental processes such as nutrient cycling, pH and redox dynamics, organic matter turnover, soil structural development, and even the detoxification of heavy metals. By modifying soil pH and redox potential, roots create microsites with unique chemical gradients that control nutrient availability and microbial metabolism. Through the release of exudates, plants not only mobilize essential nutrients but also establish signaling pathways that structure microbial communities, fostering symbiosis, competition, and antagonism. Organic matter inputs from roots further sustain soil carbon pools and aggregation, thereby linking below-ground processes with long-term soil fertility and ecosystem stability. Importantly, these same mechanisms also underpin the detoxification of harmful heavy metals, highlighting the multifunctional role of rhizosphere processes in both productivity and environmental remediation.

Understanding the biogeochemical processes at the root–soil interface is therefore critical for advancing sustainable agriculture, soil health management, and climate resilience. By harnessing plant–microbe–soil interactions, it is possible to improve nutrient use efficiency, enhance crop performance under stress, and mitigate the ecological impacts of pollutants. The rhizosphere, in essence, is not only the foundation of plant growth but also a central driver of ecosystem functioning and sustainability.

References

Bais, H.P., Weir, T.L., Perry, L.G., Gilroy, S. and Vivanco, J.M. (2006) The role of root exudates in rhizosphere interactions with plants and other organisms. *Annual Review of Plant Biology* 57, 233–266. DOI: 10.1146/annurev.arplant.57.032905.105159.

Berendsen, R.L., Pieterse, C.M.J. and Bakker, P.A.H.M. (2012) The rhizosphere microbiome and plant health. *Trends in Plant Science* 17(8), 478–486. DOI: 10.1016/j.tplants.2012.04.001.

Bronick, C.J. and Lal, R. (2005) Soil structure and management: A review. *Geoderma* 124(1–2), 3–22. DOI: 10.1016/j.geoderma.2004.03.005.

Carminati, A. and Vetterlein, D. (2013) Plasticity of rhizosphere hydraulic properties as a key for efficient utilization of scarce resources. *Annals of Botany* 112(2), 277–290. DOI: 10.1093/aob/mcs262.

Colmer, T.D. (2003) Long-distance transport of gases in plants: A perspective on internal aeration and radial oxygen loss from roots. *Plant, Cell and Environment* 26(1), 17–36. DOI: 10.1046/j.1365-3040.2003.00846.x.

Compant, S., Clément, C. and Sessitsch, A. (2010) Plant growth-promoting bacteria in the rhizo- and endosphere of plants: Their role, colonization, mechanisms involved and prospects for utilization. *Soil Biology and Biochemistry* 42(5), 669–678. DOI: 10.1016/j.soilbio.2009.11.024.

Dakora, F.D. and Phillips, D.A. (2002) Root exudates as mediators of mineral acquisition in low-nutrient environments. *Plant and Soil* 245(1), 35–47. DOI: 10.1023/A:1020809400075.

Gadd, G.M. (2010) Metals, minerals and microbes: Geomicrobiology and bioremediation. *Microbiology* 156(Pt 3), 609–643. DOI: 10.1099/mic.0.037143-0.

Gahan, J. and Schmalenberger, A. (2014) The role of bacteria and mycorrhiza in plant sulfur supply. *Frontiers in Plant Science* 5, 723. DOI: 10.3389/fpls.2014.00723.

Hinsinger, P., Plassard, C., Tang, C. and Jaillard, B. (2003) Origins of root-mediated pH changes in the rhizosphere and their responses to environmental constraints: A review. *Plant and Soil* 248(1–2), 43–59. DOI: 10.1023/A:1022371130939.

Jones, D.L., Nguyen, C. and Finlay, R.D. (2009) Carbon flow in the rhizosphere: Carbon trading at the soil–root interface. *Plant and Soil* 321(1–2), 5–33. DOI: 10.1007/s11104-009-9925-0.

Kuzyakov, Y. and Domanski, G. (2000) Carbon input by plants into the soil. Review. *Journal of Plant Nutrition and Soil Science* 163(4), 421–431. DOI: 10.1002/1522-2624(200008)163:4<421::AID-JPLN421>3.0.CO;2-R.

Neumann, G. and Römheld, V. (2007) The release of root exudates as affected by the plant physiological status. In: Pinton, R., Varanini, Z. and Nannipieri, P. (eds) *The Rhizosphere: Biochemistry and Organic Substances at the Soil–Plant Interface*. CRC Press, Boca Raton, USA, pp. 23–72. DOI: 10.1201/9781420005585.ch2.

Oldroyd, G.E.D., Murray, J.D., Poole, P.S. and Downie, J.A. (2011) The rules of engagement in the legume-rhizobial symbiosis. *Annual Review of Genetics* 45, 119–144. DOI: 10.1146/annurev-genet-110410-132549.

Park, J.H., Bolan, N., Megharaj, M. and Naidu, R. (2011) Isolation of phosphate solubilizing bacteria and their potential for lead immobilization in soil. *Journal of Hazardous Materials* 185(2–3), 829–836. DOI: 10.1016/j.jhazmat.2010.09.095.

Philippot, L., Raaijmakers, J.M., Lemanceau, P. and Van Der Putten, W.H. (2013) Going back to the roots: The microbial ecology of the rhizosphere. *Nature Reviews Microbiology* 11(11), 789–799. DOI: 10.1038/nrmicro3109.

Rajkumar, M., Sandhya, S., Prasad, M.N.V. and Freitas, H. (2012) Perspectives of plant-associated microbes in heavy metal phytoremediation. *Biotechnology Advances* 30(6), 1562–1574. DOI: 10.1016/j.biotechadv.2012.04.011.

Richardson, A.E. and Simpson, R.J. (2011) Soil microorganisms mediating phosphorus availability. *Plant Physiology* 156(3), 989–996. DOI: 10.1104/pp.111.175448.

Richardson, A.E., Barea, J.M., McNeill, A.M. and Prigent-Combaret, C. (2009) Acquisition of phosphorus and nitrogen in the rhizosphere and plant growth promotion by microorganisms. *Plant and Soil* 321(1–2), 305–339. DOI: 10.1007/s11104-009-9895-2.

Six, J., Conant, R.T., Paul, E.A. and Paustian, K. (2002) Stabilization mechanisms of soil organic matter: Implications for C-saturation of soils. *Plant and Soil* 241(2), 155–176. DOI: 10.1023/A:1016125726789.

Smith, S.E. and Read, D.J. (2008) *Mycorrhizal Symbiosis*, 3rd edn. Academic Press, New York, USA.

Sun, H., Wang, L., Kumar, A., Auwal, M., Zwieten, L.V. *et al.* (2025) Nutrient availability mediates organic carbon turnover in paddy soils through regulating microbial metabolism. *Geoderma* 458, 117313. DOI: 10.1016/j.geoderma.2025.117313.

7 Chemically Diverse Primary and Secondary Metabolites Released by Plant Roots and Their Bioactive Effects on Microorganisms

Ankita Chaubey[1], Manoj Kumar Mishra[2] and Neha Verma[3*]

[1]iGlobal Research and Publishing Foundation, Uttam Nagar, New Delhi, India
[2]Department of Biotechnology, S.R. Institute of Management and Technology, Lucknow, India
[3]Department of Biotechnology, Meerut Institute of Engineering and Technology, Meerut, India

7.1 Introduction

All plants have a natural tendency to release different compounds through their roots, leaves, and shoots in the surrounding atmosphere. These compounds may be solid, liquid or gas. Phytochemicals are naturally occurring compounds produced by plants. Many phytochemicals have shown some therapeutic activity, thus making them useful for humans (Hasler and Blumberg, 1999). These metabolites are not only used for flavor, colour, and fragrance but also provide protection from injury and disease by neutralizing the effect of stressed conditions or pathogenic infections (Gibson *et al.*, 1998). Direct extraction of these compounds from plants is a lengthy and costly process but they can be chemically synthesized or produced by plant cell culture, which is a more promising way to procure metabolites that are difficult to obtain by plant extraction or chemical synthesis (Tiwari and Rana, 2015).

The compounds secreted from roots are also known as root exudates. Root exudates are a mixture of low and high molecular weight compounds

*Corresponding author: neha.emo@gmail.com

 Medicinal Plants and the Rhizosphere (eds M.K. Mishra and S. Tiwari)
DOI: 10.1079/9781836990475.0007

that can be categorized as primary and secondary metabolites. Because of their complex structure and diversity, the classification of phytochemicals is based on different parameters. Primary metabolites (PMs) are directly involved in the growth and development of the plant. Secondary metabolites (SMs) are not involved directly in plant growth but play crucial roles in the survival of plants as they contribute to plant defense, stress tolerance, and other ecological complications (Balunas and Kinghorn, 2005). PMs also act as precursors for the formation of secondary metabolites. In some reports, three categories of plant metabolites are suggested as plant hormones are placed in separate categories from primary and secondary metabolites (Erb and Kliebenstein, 2020). Plant hormones can be explained as small molecules that interact with receptor proteins to regulate living mechanisms along with metabolite formation (Davies, 2004). These metabolites have been extensively studied in commonly used plant species such as rice and beans (Strehmel *et al.*, 2014; Tawaraya *et al.*, 2014a, b). However, most of the identification and quantification of metabolites from roots has been done with herbaceous plants and shrubs. The number of metabolites obtained from tree roots is muchlower as very few species of trees have been studied for metabolites (Zhang *et al.*, 2007).

Root exudates have so many direct and indirect impacts on plant growth and development because they contain compounds that facilitate several interactions within the rhizosphere. These interactions exhibit modification in soil properties, nutrient availability and its absorption. Exudates are also responsible for the circulation of macronutrients such as carbon and nitrogen from root to soil (Jones *et al.*, 2009). Therefore, a comprehensive approach to the exploration of these metabolites and combining this with modern techniques could be very useful for agricultural studies to promote improved crops and sustainable agricultural products.

7.2 Classification of Plant Metabolites

Plant metabolites are classified based on their contribution to plant metabolism. Phytochemicals such as proteins, nucleic acids, sugars, amino acids, and chlorophylls, which are required by plants for their growth and development, are known as primary metabolites. Secondary metabolites are compounds that are not directly involved in plant metabolism but are required for their survival, so they are also known as specialized plant metabolites. Alkaloids, phenolics, flavonoids, terpenoids, steroids, and glucosides are in the secondary metabolite category (Ramawat *et al.*, 2009). Details of each category are discussed in the following sections.

7.2.1 Primary metabolites

Primary metabolites are the main ingredients of metabolic processes that facilitate different functions in plants including growth, reproduction, and other cellular functions. PMs are also required for energy production, respiratory and photosynthetic process. They also act as signaling molecules to trigger responses in stress conditions (Wang *et al.*, 2012). Carbohydrates, proteins, lipids, organic acids, and nucleic acids are the most common and crucial classes of PMs (Rogers, 2010). PMs are mostly produced during the growth phase, i.e. trophophase, depending on the requirement of cellular functions.

Plants produce carbohydrates through the process of photosynthesis which is the main source of energy (Saxena *et al.*, 2009). On the basis of the number of monomer units, carbohydrates can be classified as monosaccharides, disaccharides, oligosaccharides, and polysaccharides. Carbohydrates can also be classified as structural or functional. Amino acids are another important class of PMs as they help in the formation of proteins and nucleic acids. As they are precursors for various protein molecules that form growth factors, vitamins, and enzymes, they are essential for signaling and regulating processes (Heinemann and Hildebrandt, 2021).

Lipids are the main building blocks of biological membranes such as plasma (Hou *et al.*, 2016). Lipids can be further classified based on their chemical structure and nature into fatty acids, glycerides, phosphoglycerides, sphingolipids, steroids, isoprenoids, glycolipids, and polyketides (Fahy *et al.*, 2005).

7.2.2 Secondary metabolites

Secondary metabolites of plants have gained much attention in recent decades for their wide applications in various industries. Scientists are exploiting many species of plants to obtain bioactive compounds, studying their isolation, structural and functional characterization along with toxicity. Secondary metabolites include various classes of compounds but they can mainly be categorized in three divisions: terpenoids, phenolic compounds, and nitrogen-containing compounds (Samuni-Blank *et al.*, 2012).

7.2.2.1 Terpenoids

This is the largest class of plant SMs which contain 5-carbon isoprene units. Terpenes consist only of hydrocarbons while terpenoids are derivatives of terpenes, containing other atoms such as oxygen, nitrogen, and many more. The classification of terpenes is based on number of isoprene units as described in Table 7.1.

Another class, called norisoprenoids, is derived from isoprenoids by substitution or removal of methylene groups. Monoterpenes, sesquiterpenes, and diterpenes are further divided into groups as shown in Table 7.2.

Monoterpenes are key components of essential oils obtained from plants of various families including Limoaceae, Apiaceae, and Rutaceae (Elshafie *et al.*, 2023). Diterpenes are widely used for their pharmacological properties as they have anticancerous activity (Eksi *et al.*, 2020). Commercially available

Table 7.1. Description of types of terpenes.

Class	Number of isoprene units	Molecular formula
Monoterpenes	2	$C_{10}H_{16}$
Sesquiterpenes	3	$C_{15}H_{24}$
Diterpenes	4	$C_{20}H_{32}$
Sesterterpenes	5	$C_{25}H_{32}$
Triterpenes	6	$C_{30}H_{40}$
Sesquarterpenes	7	$C_{35}H_{48}$
Tetraterpenes	8	$C_{40}H_{56}$
Polyterpenes	Chain of isoprene units	$(C_5H_8)_n$

Table 7.2. Description of terpene subclasses.

Class	Subclass	Example	Reference
Monoterpenes	Unsaturated hydrocarbon	Limonene	(Breitmaier, 2006)
	Alcohol	Linalool	
	Alcohol esters	Linalyl acetate	
	Aldehydes	Citronellal	
	Ketones	Carvone	
Sesquiterpenes	Acyclic	Farnesol	(Zárybnický *et al.*, 2018)
	Monocyclic	Bisabolol	
	Bicyclic	Caryophyllene	
Diterpenes	Acyclic	Phytol	(Almeida *et al.*, 2025)
	Monocyclic	Afpolines	
	Bicyclic	Clerodanes, halimanes, and labdanes	
	Tricyclic	levopimaric acid	
	Tetracyclic	Cauranoic acids	
	Macrocyclic	Cembrane	

anticancerous drugs like paclitaxel (classified as a diterpenoid), oridonin, and andrographolide are derived from diterpenes (Jing *et al.*, 2019). Vitamin K1 and A are diterpenoids that are found in many medicinal plants (Davis and Croteau, 2000).

The most crucial metabolites of this group are phytohormones including gibberellins, which are categorized as diterpenes, and abscisic acid classified as a sesquiterpene, which are required at different developmental stages (Xu *et al.*, 1998). Terpenoids not only play a very important role in growth and development of plants, but they also provide protection in various plant species like lemon, basil, cotton, tobacco, maize, and peppermint (Gershenzon and Croteau, 1991).

7.2.2.2 Phenolic compounds

This is the most important and widespread class of SM, characterized by the presence of one or more phenol groups that provide colour and flavor to the plant species (Davin and Lewis, 1995). Phenolic compounds have demonstrated antioxidant, antiinflammatory, and antihepatotoxic activity. This class can be categorized into three groups based on their structure as simple phenolics, tannins, and flavonoids (Muniyandi *et al.*, 2019). Flavonoids are the most abundant in nature and mostly found in higher plants (Evans, 2009). Flavonoids play an important role in processes including nitrogen fixation, flower pigmentation, and symbiotic association, when secreted by roots (Samanta *et al.*, 2011; Liu and Murray, 2016). This is the most explored class of phenolic compounds, as more than 5000 flavonoid compounds have been identified and isolated from plants. Flavonoids are further classified based on their chemical structure that differs due to oxidation of the central ring. The main classes are flavonols, flavones, flavanones, anthocyanidins, and isoflavones as shown in Table 7.3 (Hättenschwiler and Vitousek, 2000). There are also a few other groups which are derived from flavonoids like coumarines, chalcones, dihydroflavonols, and aurones (de Souza Farias *et al.*, 2021).

Table 7.3. Description of flavonoid classes.

Class	Structure	Example	Reference
Flavonols	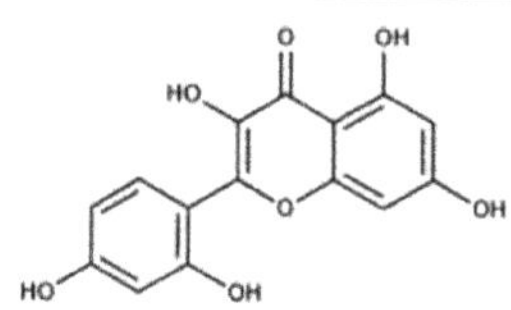	Quercetin, kaempferol, myricetin, galangin	(Edenharder *et al.*, 1993; Stewart *et al.*, 2000; Cushnie and Lamb, 2005)
Flavones	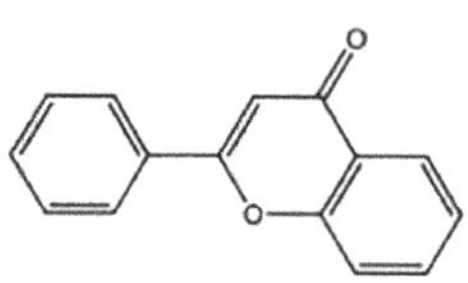	Luteolin, apigenin, tangeritin	(Edenharder *et al.*, 1993; Cushnie and Lamb, 2005)
Flavanones	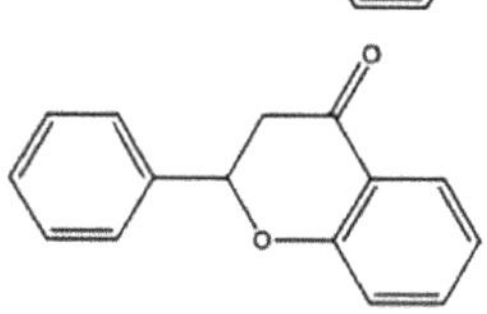	Hesperetin, naringenin, eriodictyol	(Edenharder *et al.*, 1993; Miyake *et al.*, 2000; Cushnie and Lamb, 2005)
Anthocyanidins	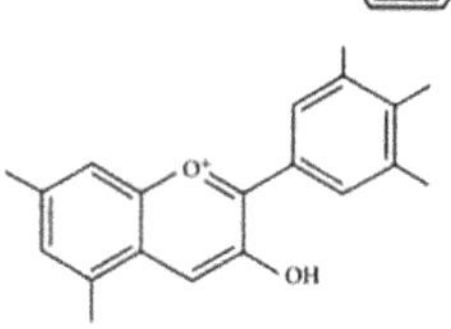	Cyanidin, delphinidin, peonidin	(Stewart *et al.*, 2000; Kong *et al.*, 2003)
Isoflavones	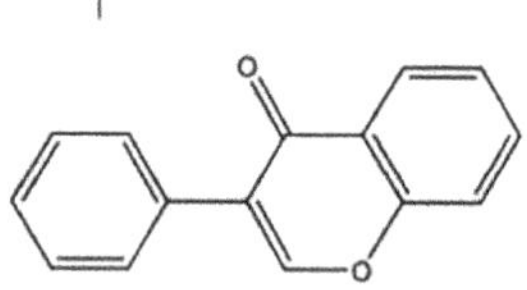	Daidzein, genistein, glycitein	(Edenharder *et al.*, 1993; Cushnie and Lamb, 2005)

7.2.2.3 Nitrogen-containing compounds

Plant SMs that contain nitrogen atoms are categorized as alkaloids, cyanogenic glucoside, and glucosinate (Taiz and Zeiger, 2005). Alkaloids are an important class of nitrogen-containing compounds as they have antimicrobial and antiviral activity along with cytotoxic effects on vertebrates (Evans, 2009). Many compounds of this class, such as quinine, ephedrine, and homoharringtonine, have shown remarkable medicinal value in treating diseases including malaria, cancer, cardiac dysfunction, diabetes, etc. (Elshafie *et al.*, 2023). Alkaloids are mostly found in plant species that belong to the orders Caryophyllales, Magnoliales, Rosales, and Gentiales (Evans, 2009). There is structural diversity in alkaloids, as they can be heterocyclic and nonheterocyclic. Alkaloids serve as reservoirs of nitrogen and act as protective agents for plants against external attacks and aid in growth regulation.

Cyanogenic glucosides are the next most important class of nitrogen-containing compounds, found in more than 2500 plant species. They provide protection to plants from herbivores by inducing release of hydrogen cyanide on tissue disruption (Elshafie *et al.*, 2023). Also, they are responsible for the bitter taste of almond kernels and other fruits like apricots, peaches, and apples.

There is a wide spectrum of compounds present in plants which have different functional and structural aspects, and are categorized based on structural similarities. Plant root exudates have been explored for the presence of metabolites and have shown multiple bioactive compounds that are involved in their growth and development (Table 7.4).

7.3 Factors Affecting Root Metabolite Secretion

Plants release their metabolites into the rhizosphere through various processes that can be broadly classified as passive or active.

Passive transport is the most common method of metabolite secretion, facilitated by diffusion, ionic channels, and vesicle transport (Baetz and Martinoia, 2014). The process of secretion depends on the structural and functional properties of metabolites. Secretion of low molecular weight compounds such as sugars, carboxylic acids, phenolics, and amino acids takes place through diffusion, which is driven by the concentration gradient between root cell and rhizosphere. Factors affecting this process include permeability of root membrane, polarity of compound to be secreted, and structure of root cells (Badri and Vivanco, 2009). The second pathway of passive metabolite release is through ionic channels. This method mainly facilitates the secretion of carboxylates such as malate and oxalate, and some carbohydrates (Vives-Peris *et al.*, 2020). In this process, the compounds are transported with the help of proteins. Another method of passive transport is vesicle-mediated transport, also known as exocytosis. High molecular weight metabolites are also secreted from roots (Badri and Vivanco, 2009);

Table 7.4. List of metabolites released by plant roots. Modified from Vives-Peris *et al.* (2020) and Bhatt *et al.* (2020).

Category	Class	Compounds
Primary	Amino acid	α-Alanine, β-alanine, γ-aminobutyric, α-aminoadipic, arginine, asparagine, aspartic, citrulline, cystathionine, cysteine, cystine, 3-epihydroxymugineic, glutamine, glutamic, glycine, histidine, homoserine, isoleucine, leucine, lysine, methionine, ornithine, phenylalanine, proline, serine, threonine, tryptophan, tyrosine, valine
	Carbohydrates	Arabinose, fructose, galactose, glucose, maltose, mannose, mucilages of various compositions, oligosaccharides, rafnose, rhamnose, ribose, sucrose, xylose, deoxyribose
	Lipids	Linoleic, linolenic, oleic, palmitic, stearic
	Growth factors and vitamins	p-Amino benzoic acid, biotin, choline, inositol, N-methyl nicotinic acid, niacin, pantothenic, pantothenate, pyridoxine, riboflavin, strigolactones, thiamine
	Enzymes	Amylase, invertase, peroxidase, phenolase, acid/alkaline phosphatase, polygalacturonase, protease
	Organic acids	Acetic, aconitic, ascorbic, benzoic, butyric, citric, formic, glutaric, glycolic, lactic, malic, malonic, oxaloacetic, oxalic, propionic, pyruvic, succinic
	Nucleotides	Adenine, guanine, uridine/cytidine
Secondary	Sterols	Campesterol, cholesterol, sitosterol, stigmasterol
	Flavonoids	Chalcone, coumarine, licochalcone A, naringenin, kaempferol, quercitin, myricetin, naringin, rutin, genistein, strigolactone and their substitutes with sugar
	Glucosinolates	Cyclobrassinone, desulfogluconapin, desulfoprogoitrin, desulfonapoleiferin, desulfoglucoalyssin
	Aurones	Benzyl aurones, synapates, sinapoyl choline
	Anthocyanins	Cyanidin, delphinidin, pelargonidin and their substitutes with sugar molecules
	Coumarins	Umbelliferone
	Alkaloids	Betalains, camptothecin, tropane alkaloids
	Terpenes	Artemisinin
	Phenolic compounds	Rosmarinic acid, aryltetralin lignans, stilbenoids
	Phenolic acids	Vanillic, cis-ferulic, trans-ferulic, cis-coumaric, trans-coumaric, syringic, p-hydroxybenzoic, salicylic

the metabolites travel through the endoplasmic reticulum and Golgi apparatus and act against pathogens to protect the plant (Weston *et al.*, 2012).

Active transportation of metabolites also takes place with the help of proteins present in the root plasma membrane (Baetz and Martinoia, 2014). This pathway is mainly facilitated by two families of membrane transporters:

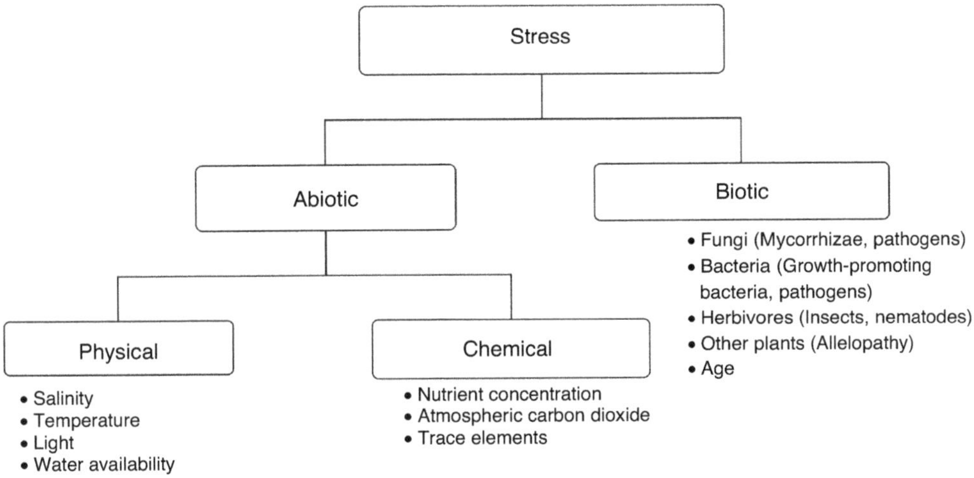

Fig. 7.1. Factors affecting the composition and amount of root exudates.

the ATP-binding cassette (ABC) and multidrug and toxic compound extrusion (MATE) (Kang *et al.*, 2011).

Secretion of metabolites from roots is affected by various factors that can be categorized as abiotic or biotic (Fig. 7.1).

Nutrient availability and atmospheric carbon dioxide are the most explored abiotic factors that can cause alteration in root secretions. Deficient or excess nutrients both affect the composition of plant root exudates. For example, in sugarbeet plants, there is an increase in the amount of salicylic acid which is induced by a deficiency of phosphorus. High concentration of organic acids contributes to acidification of soil, resulting in solubilization of soil phosphorus (Khorassani *et al.*, 2011). The second important factor affecting root exudates is carbon dioxide concentration (Jones *et al.*, 2009). Studies on grass species have shown that carbon rhizodeposition is directly related to plant root and leaf biomass so as plant growth takes place, rhizodeposition also increases, affecting the total fixed carbon dioxide concentration (Baptist *et al.*, 2015).

Water availability also affects the composition of root exudates. Scarcity or excess of water both have the same impact on root exudates, which is to increase organic carbon in root secretions. A study by Henry *et al.* (2007) reported that *Agropyron cristatum* plants demonstrated a 71% increase in organic carbon followed by elevated production of succinic, fumaric, and malic acids, when drought conditions were imposed. In flood conditions, this increase of organic carbon was 45% while the organic acids found were oxalic and malonic acids.

Salinity has an impact on the composition of root exudates. Dardanelli *et al.* (2010, 2012) studied the effect of salt stress on two plant species: *Glycine max* (soybean) and *Phaseolus vulgaris* (common bean). Both species demonstrated an increased quantity of flavonoids under salt stress. A similar effect was shown in the excised roots of almonds which demonstrated increased production of proline when treated with salt stress under *in vitro* conditions (Marin *et al.*, 2010).

Temperature and light have important effects on the composition and quantity of compounds in root exudates. A study conducted by Pramanik *et al.* (2000) reported that an increase in temperature up to 5°C promotes concentration of benzoic, 4-hydroxybenzoic, phthalic, and palmitic acids in root exudates of *Cucumis sativus* plants, grown in hydroponic conditions. Similarly, the availability of light, type of light, its wavelength and photoperiod can all influence the quality and quantity of compounds in exudates. Martin *et al.* (2018) demonstrated the effect of low light which induced changes in exudates, resulting in a decrease in beneficial microorganisms in the rhizosphere of seagrass. Pomilio *et al.* (2000) revealed the effect of light on root exudates of *Avena fatua*, which demonstrated altered composition when grown under far-infrared radiation. Moreover, studies have shown that longer photoperiod enhances production of benzoic and 4-hydroxybenzoic acids in *Cucumis sativus* plants (Pramanik *et al.*, 2000).

Other abiotic factors that affect root exudation include trace elements, presence of heavy metal ions and other toxic ions in plant root surroundings. Biotic factors that impact root exudate have ecological implications which are discussed in the following section.

7.4 Ecological Implications of Root Exudate Metabolites

The metabolites present in root secretions widely affect surrounding plant species as well as microorganisms, exhibiting modification in the local environment leading to changes in plant and microbial communities. The impact of these exudates is described in detail in the following section.

7.4.1 Impact of root exudates on other plant species

The chemicals present in root exudates facilitate interactions among plants of the same or different species. The way these exudates act depends on their composition and the mechanism of the relationship. There can be negative as well as positive interactions.

In negative interactions, several types of mechanisms can modulate these relationships, including allelopathy, which is the release of phytotoxins to the rhizosphere to create an advantage over other competitors. Most of these phytotoxins have a similar chemical structure, including aromatic components and hydroxyl or ketone groups, with metabolites such as flavonoids, quinones, quinolines, and hydroxamic acids (Bais *et al.*, 2006). Allelochemicals can negatively affect other plants in several ways, causing changes in cell structures and preventing cell division as well as cell elongation. They also destabilize the antioxidant system, increase membrane permeability, affect plant growth regulators and enzymes, and influence respiration, photosynthesis, metabolism, and water and nutrient uptake (Cheng and Cheng, 2015).

However, affected plants continue to adapt to secreted allelochemicals, leading to evolutionary reduction in the adverse effects of these phytotoxins.

If the host plant does not adapt to these phytotoxins, these compounds can be used as herbicides against invasive plants. However, the specificity and efficacy of these compounds used as herbicides are usually limited (Bhadoria, 2011). There are examples of secondary metabolites such as methoxy-1,4-naphthoquinone, obtained from root exudates of *Impatiens glandulifera*, that inhibit seed germination of other plant species like *Hieracium murorum* and *Scrophularia nodosa* (Ruckli *et al.*, 2014). Another example is momilactone B, which is present in rice root exudates, that also exhibits inhibitory effects on the growth of nearby plants (Kato-Noguchi, 2004).

However, there are other root exudate compounds that have a positive impact on neighboring plants by introducing resistance against herbivores. For example, the root exudates of *Elytrigia repens* (containing carboline) protect another plant, *Hordeum vulgare*, from aphids (Glinwood *et al.*, 2003). Reports have shown that some compounds in root exudates attract the natural enemies of herbivores to defend the plant. For instance, an aphid-affected plant produces volatile compounds that attract parasitoids to protect other plants from aphids. This alteration of root exudates happened because of stressful conditions like aphid infection. For example, *Vicia faba* plants infected by *Acyrthosiphon pisum* aphids produce compounds in their root exudates that induce volatile compound production in unaffected plants to attract the parasitoid *Aphidius ervi* that can protect them from the aphids (Guerrieri *et al.*, 2002).

Similarly, primary metabolites present in plant root exudates have shown positive effects on the growth of nearby plants by secreting growth regulators such as auxins and strigolactones that are crucial for the establishment of root morphology and facilitate growth of root and shoot (Haichar *et al.*, 2014).

7.4.2 Impact of root metabolites on microorganisms

The composition of plant root exudates directly affects the nature of the surrounding soil and therefore the microorganisms present in the soil. In stressful conditions, the composition of root exudates alters so that the soil pH changes, affecting beneficial as well as pathogenic bacteria, fungi, and nematodes (Lareen *et al.*, 2016). For example, any alteration in phenolic compounds present in *Arabidopsis thaliana* root exudates causes changes in the microbiome of the rhizosphere, as shown by pyrosequencing of bacterial communities. The sensitivity found in the analysis towards phenolic compounds demonstrates that even small changes can have a wide impact on microbiomes that could be positive or negative for different bacterial communities (Badri *et al.*, 2013). The positive impact of metabolites is also found in symbiotic relationships between nitrogen-fixing microbes and legume plants, as flavonoids present in legume root exudates attract bacterial communities that facilitate nitrogen fixation (Hassan and Mathesius, 2012). Similar effects are shown by citrus plant root exudates that attract beneficial rhizobacteria such as *Pseudomonas putida* and *Novosphingobium* sp. (Vives-Peris *et al.*, 2018). In adverse conditions, this positive association helps the bacteria to survive by providing them with useful metabolites like proline as a nutrient source (Vílchez *et al.*, 2000; Vives-Peris *et al.*, 2018).

Several reports have shown that root exudates also inhibit growth of some fungal species. This kind of effect is found in *Citrullus lanatus* (watermelon) which shows resistance to infection with *Fusarium oxysporum* when cultivated with *Triticum aestivum* (wheat), as the root exudates of wheat help to inhibit fungal growth, leading to a decrease in damage to the watermelon plants (Xu *et al.*, 2015). However, some reports demonstrate the positive effect of root exudates on beneficial fungi. Secondary metabolites, like flavonoids and strigolactones, attract mycorrhizal fungi and help in their colonization, leading to establishment of mutualism (Steinkellner *et al.*, 2007).

Thus, plant root exudates have significant effects on adjacent plant and microbial communities. It is also evident from various reports that plants can adapt to adverse conditions and accordingly can alter the composition of their root exudates, leading to modification in the rhizosphere microbiome by either attracting beneficial microbes or inhibiting the growth of pathogenic microbes (Vives-Peris *et al.*, 2020).

7.5 Application of Plant Root Exudates

Comprehensive study of root exudates can provide multifaceted applications in soil and crop improvement. Studies on wheat and rice have revealed that knowledge of root exudates has real-time utility in improving agricultural production and weed management by encouraging sustainable farming practices and use of plants' allelopathic nature instead of synthetic herbicides (Zambelli *et al.*, 2025). Metabolomics provides an insight into the metabolic processes and response to environmental conditions, which could be used to engineer root secretions, rhizospheric microbiome, and soil conditions to improve the growth and development of plants.

The composition of root exudates is plant and species specific. The exploration of root exudates and their interaction with microbiomes is essential to exploit the potential of the rhizosphere. Rhizospheric engineering provides a promising approach for sustainable agriculture ecosystems (Kudjordjie *et al.*, 2019). As the plant is exposed to stress conditions, the composition of its root exudates changes, positively affecting microbial signaling compounds and causing changes in the rhizosphere microflora. It also affects the physical properties of soil such as pH, nutrient availability, etc. Studies have revealed that this alteration of microflora facilitates enrichment of the rhizosphere, which promotes sustainable agriculture.

Exploration of root exudates can also help with rhizoremediation, in which decontamination of pollutants in the soil is performed by the rhizospheric microbiome. Plant root exudates alter soil conditions followed by attracting beneficial microbes that help to create a favorable habitat for better plant growth and development (Singh *et al.*, 2017). So, the efficiency of rhizoremediation by plants depends on the rhizomicrobiome, which is affected by the root exudates.

Metabolomics studies of root secretions may offer a broad spectrum of applications by combining it with advanced omics tools such as genomics, transcriptomics, proteomics, etc. (Scossa *et al.*, 2021). Through these tools, gene mapping is possible for specific metabolites along with their metabolic pathways that could help in establishing the relationship between genes and metabolites. Therefore, metabolite editing or targeted gene engineering could provide crop improvement and sustainable agricultural ecosystems (Razzaq *et al.*, 2022).

7.6 Conclusion

Root secretions are a complex mixture of primary and secondary metabolites that facilitate biological processes in plants. This chapter highlights the classification of compounds present in root secretions, factors affecting their secretions and their effect on surrounding microorganisms and soil conditions. A detailed exploration of root exudates may help in developing sustainable agricultural practices. The metabolites of root exudates have multiple functions to perform for growth and development of plants, including modification of soil, recruitment of beneficial microbes, and protection of plants from weeds. All these properties can be exploited to develop a more robust and productive agricultural system. The power of root exudate metabolites can lead to better weed control methods and advancements in agroecology to establish a healthier interaction between agriculture and environment.

References

Almeida, J.E., de Oliveira, A.C., de Castro Alves, C.E., Filho, S.M.C., de Oliveira, E.C.P. *et al.* (2025) Diterpenes: Nature's hidden gems of immunomodulation. *International Journal of Molecular Sciences* 26(5), 2250. DOI: 10.3390/ijms26052250.

Badri, D.V. and Vivanco, J.M. (2009) Regulation and function of root exudates. *Plant, Cell and Environment* 32(6), 666–681. DOI: 10.1111/j.1365-3040.2008.01926.x.

Badri, D.V., Chaparro, J.M., Zhang, R., Shen, Q. and Vivanco, J.M. (2013) Application of natural blends of phytochemicals derived from the root exudates of Arabidopsis to the soil reveal that phenolic-related compounds predominantly modulate the soil microbiome. *Journal of Biological Chemistry* 288(7), 4502–4512. DOI: 10.1074/jbc.M112.433300.

Baetz, U. and Martinoia, E. (2014) Root exudates: The hidden part of plant defense. *Trends in Plant Science* 19(2), 90–98. DOI: 10.1016/j.tplants.2013.11.006.

Bais, H.P., Weir, T.L., Perry, L.G., Gilroy, S. and Vivanco, J.M. (2006) The role of root exudates in rhizosphere interactions with plants and other organisms. *Annual Review of Plant Biology* 57, 233–266. DOI: 10.1146/annurev.arplant.57.032905.105159.

Balunas, M.J. and Kinghorn, A.D. (2005) Drug discovery from medicinal plants. *Life Sciences* 78(5), 431–441. DOI: 10.1016/j.lfs.2005.09.012.

Baptist, F., Aranjuelo, I., Legay, N., Lopez-Sangil, L., Molero, G. *et al.* (2015) Rhizodeposition of organic carbon by plants with contrasting traits for resource acquisition: Responses to different fertility regimes. *Plant and Soil* 394(1–2), 391–406. DOI: 10.1007/s11104-015-2531-4.

Bhadoria, P. (2011) Allelopathy: A natural way towards weed management. *American Journal of Experimental Agriculture* 1(1), 7–20. DOI: 10.9734/AJEA/2011/002.

Bhatt, P., Verma, A., Verma, S., Anwar, M.S., Prasher, P. *et al.* (2020) Understanding phytomicrobiome: A potential reservoir for better crop management. *Sustainability* 12(13), 5446. DOI: 10.3390/su12135446.

Breitmaier, E. (2006) Hemi- and monoterpenes. In: *Terpenes: Flavors, Fragrances, Pharmaca, Pheromones*. John Wiley and Sons, Hoboken, USA, pp. 10–23. DOI: 10.1002/9783527609949.

Cheng, F. and Cheng, Z. (2015) Research progress on the use of plant allelopathy in agriculture and the physiological and ecological mechanisms of allelopathy. *Frontiers in Plant Science* 6, 1020. DOI: 10.3389/fpls.2015.01020.

Cushnie, T.P.T. and Lamb, A.J. (2005) Antimicrobial activity of flavonoids. *International Journal of Antimicrobial Agents* 26(5), 343–356. DOI: 10.1016/j.ijantimicag.2005.09.002.

Dardanelli, M.S., Manyani, H., González-Barroso, S., Rodríguez-Carvajal, M.A., Gil-Serrano, A.M. *et al.* (2010) Effect of the presence of the plant growth promoting rhizobacterium (PGPR) *Chryseobacterium balustinum* Aur9 and salt stress in the pattern of flavonoids exuded by soybean roots. *Plant and Soil* 328(1–2), 483–493. DOI: 10.1007/s11104-009-0127-6.

Dardanelli, M.S., de Córdoba, F.J.F., Estévez, J., Contreras, R., Cubo, M.T. *et al.* (2012) Changes in flavonoids secreted by *Phaseolus vulgaris* roots in the presence of salt and the plant growth-promoting rhizobacterium *Chryseobacterium balustinum*. *Applied Soil Ecology* 57, 31–38. DOI: 10.1016/j.apsoil.2012.01.005.

Davies, P.J. (2004) *Plant Hormones: Biosythesis, Signal Transduction, Action*, 3rd edn. Kluwer Academic, Dordrecht, The Netherlands.

Davin, L.B. and Lewis, N.G. (1995) Lignin and lignan biochemical pathways in plants: An unprecedented discovery in phenolic coupling. *Anais da Academia Brasileira de Ciencias* 63(3), 363–378.

Davis, E.M. and Croteau, R. (2000) Cyclization enzymes in the biosynthesis of monoterpenes, sesquiterpenes, and diterpenes. *Topics in Current Chemistry* 209, 53–95. DOI: 10.1007/3-540-48146-X_2.

de Souza Farias, S.A., da Costa, K.S. and Martins, J.B.L. (2021) Analysis of conformational, structural, magnetic, and electronic properties related to antioxidant activity: Revisiting flavan, anthocyanidin, flavanone, flavonol, isoflavone, flavone, and flavan-3-ol. *ACS Omega* 6(13), 8908–8918. DOI: 10.1021/acsomega.0c06156.

Edenharder, R., von Petersdorff, I. and Rauscher, R. (1993) Antimutagenic effects of flavonoids, chalcones and structurally related compounds on the activity of 2-amino-3-methylimidazo[4,5-f]quinoline (IQ) and other heterocyclic amine mutagens from cooked food. *Mutation Research* 287(2), 261–274. DOI: 10.1016/0027-5107(93)90019-c.

Eksi, G., Kurbanoglu, S. and Erdem, S.A. (2020) Analysis of diterpenes and diterpenoids. In: Sanches-Silva, A., Nabavi, S.F., Saeedi, M. and S.M, N. (eds) *Recent Advances in Natural Products Analysis*. Elsevier, Amsterdam, The Netherlands, pp. 313–345. DOI: 10.1016/B978-0-12-816455-6.00009-3.

Elshafie, H.S., Camele, I. and Mohamed, A.A. (2023) A comprehensive review on the biological, agricultural and pharmaceutical properties of secondary metabolites based-plant origin. *International Journal of Molecular Sciences* 24(4), 3266. DOI: 10.3390/ijms24043266.

Erb, M. and Kliebenstein, D.J. (2020) Plant secondary metabolites as defenses, regulators, and primary metabolites: The blurred functional trichotomy. *Plant Physiology* 184(1), 39–52. DOI: 10.1104/pp.20.00433.

Evans, W.C. (2009) *Trease and Evans Pharmacognosy*, 16th edn. Elsevier Health Sciences, Amsterdam, The Netherlands.

Fahy, E., Subramaniam, S., Brown, H.A., Glass, C.K., Merrill Jr, A.H. *et al.* (2005) A comprehensive classification system for lipids. *Journal of Lipid Research* 46(5), 839–861. DOI: 10.1194/jlr.E400004-JLR200.

Gershenzon, J. and Croteau, R. (1991) Terpenoids. In: Rosenthal, G.A. and Berenbaum, M.R. (eds) *Herbivores: Their Interactions with Secondary Plant Metabolites*, 2nd edn. Academic Press, Orlando, Florida, pp. 165–219. DOI: 10.1016/B978-0-12-597183-6.50010-3.

Gibson, E.L., Wardle, J. and Watts, C.J. (1998) Fruit and vegetable consumption, nutritional knowledge and beliefs in mothers and children. *Appetite* 31(2), 205–228. DOI: 10.1006/appe.1998.0180.

Glinwood, R., Pettersson, J., Ahmed, E., Ninkovic, V., Birkett, M. *et al.* (2003) Change in acceptability of barley plants to aphids after exposure to allelochemicals from couch-grass (*Elytrigia repens*). *Journal of Chemical Ecology* 29(2), 261–274. DOI: 10.1023/a:1022687025416.

Guerrieri, E., Poppy, G.M., Powell, W., Rao, R. and Pennacchio, F. (2002) Plant-to-plant communication mediating in-flight orientation of *Aphidius ervi*. *Journal of Chemical Ecology* 28(9), 1703–1715. DOI: 10.1023/a:1020553531658.

Haichar, F.Z., Santaella, C., Heulin, T. and Achouak, W. (2014) Root exudates mediated interactions belowground. *Soil Biology and Biochemistry* 77, 69–80. DOI: 10.1016/j.soilbio.2014.06.017.

Hasler, C.M. and Blumberg, J.B. (1999) Phytochemicals: Biochemistry and physiology. Introduction. *Journal of Nutrition* 129(3), 756S–757S. DOI: 10.1093/jn/129.3.756S.

Hassan, S. and Mathesius, U. (2012) The role of flavonoids in root-rhizosphere signalling: Opportunities and challenges for improving plant-microbe interactions. *Journal of Experimental Botany* 63(9), 3429–3444. DOI: 10.1093/jxb/err430.

Hättenschwiler, S. and Vitousek, P.M. (2000) The role of polyphenols in terrestrial ecosystem nutrient cycling. *Trends in Ecology and Evolution* 15(6), 238–243. DOI: 10.1016/S0169-5347(00)01861-9.

Heinemann, B. and Hildebrandt, T.M. (2021) The role of amino acid metabolism in signaling and metabolic adaptation to stress-induced energy deficiency in plants. *Journal of Experimental Botany* 72(13), 4634–4645. DOI: 10.1093/jxb/erab182.

Henry, A., Doucette, W., Norton, J. and Bugbee, B. (2007) Changes in crested wheatgrass root exudation caused by food, drought, and nutrient stress. *Journal of Environmental Quality* 36, 904–912. DOI: 10.2134/jeq2006.0425sc.

Hou, Q., Ufer, G. and Bartels, D. (2016) Lipid signalling in plant responses to abiotic stress. *Plant, Cell and Environment* 39(5), 1029–1048. DOI: 10.1111/pce.12666.

Jing, W., Zhang, X., Zhou, H., Wang, Y., Yang, M. *et al.* (2019) Naturally occurring cassane diterpenoids (CAs) of Caesalpinia: A systematic review of its biosynthesis, chemistry and pharmacology. *Fitoterapia* 134, 226–249. DOI: 10.1016/j.fitote.2019.02.023.

Jones, D.L., Nguyen, C. and Finlay, R.D. (2009) Carbon flow in the rhizosphere: Carbon trading at the soil–root interface. *Plant and Soil* 321(1–2), 5–33. DOI: 10.1007/s11104-009-9925-0.

Kang, J., Park, J., Choi, H., Burla, B., Kretzschmar, T. *et al.* (2011) Plant ABC transporters. *Arabidopsis Book* 9, e0153. DOI: 10.1199/tab.0153.

Kato-Noguchi, H. (2004) Allelopathic substance in rice root exudates: Rediscovery of momilactone B as an allelochemical. *Journal of Plant Physiology* 161(3), 271–276. DOI: 10.1078/0176-1617-01188.

Khorassani, R., Hettwer, U., Ratzinger, A., Steingrobe, B., Karlovsky, P. *et al.* (2011) Citramalic acid and salicylic acid in sugar beet root exudates solubilize soil phosphorus. *BMC Plant Biology* 11, 121. DOI: 10.1186/1471-2229-11-121.

Kong, J.M., Chia, L.S., Goh, N.K., Chia, T.F. and Brouillard, R. (2003) Analysis and biological activities of anthocyanins. *Phytochemistry* 64(5), 923–933. DOI: 10.1016/s0031-9422(03)00438-2.

Kudjordjie, E.N., Sapkota, R., Steffensen, S.K., Fomsgaard, I.S. and Nicolaisen, M. (2019) Maize synthesized benzoxazinoids affect the host associated microbiome. *Microbiome* 7(1), 59. DOI: 10.1186/s40168-019-0677-7.

Lareen, A., Burton, F. and Schäfer, P. (2016) Plant root-microbe communication in shaping root microbiomes. *Plant Molecular Biology* 90(6), 575–587. DOI: 10.1007/s11103-015-0417-8.

Liu, C.W. and Murray, J.D. (2016) The role of flavonoids in nodulation host-range specificity: An update. *Plants* 5(3), 33. DOI: 10.3390/plants5030033.
Marin, J.A., Andreu, P., Carrasco, A. and Arbeloa, A. (2010) Determination of proline concentration, an abiotic stress marker, in root exudates of excised root cultures of fruit tree rootstocks under salt stress. *Revue des Régions Arides* 24, 722–727.
Martin, B.C., Gleeson, D., Statton, J., Siebers, A.R., Grierson, P. *et al.* (2018) Low light availability alters root exudation and reduces putative beneficial microorganisms in seagrass roots. *Frontiers in Microbiology* 8, 2667. DOI: 10.3389/fmicb.2017.02667.
Miyake, Y., Shimoi, K., Kumazawa, S., Yamamoto, K., Kinae, N. *et al.* (2000) Identification and antioxidant activity of flavonoid metabolites in plasma and urine of eriocitrin-treated rats. *Journal of Agricultural and Food Chemistry* 48(8), 3217–3224. DOI: 10.1021/jf990994g.
Muniyandi, K., George, E., Sathyanarayanan, S., George, B.P., Abrahamse, H. *et al.* (2019) Phenolics, tannins, flavonoids and anthocyanins contents influenced antioxidant and anticancer activities of Rubus fruits from Western Ghats, India. *Food Science and Human Wellness* 8(1), 73–81. DOI: 10.1016/j.fshw.2019.03.005.
Pomilio, A.B., Leicach, S.R., Grass, M.Y., Ghersa, C.M., Santoro, M. *et al.* (2000) Constituents of the root exudate of *Avena fatua* grown under far-infrared-enriched light. *Phytochemical Analysis* 11(5), 304–308. DOI: 10.1002/1099-1565(200009/10)11:5<304::AID-PCA531>3.0.CO;2-G.
Pramanik, M.H.R., Nagai, M., Asao, T. and Matsui, Y. (2000) Effects of temperature and photoperiod on phytotoxic root exudates of cucumber (*Cucumis sativus*) in hydroponic culture. *Journal of Chemical Ecology* 26(8), 1953–1967. DOI: 10.1023/A:1005509110317.
Ramawat, K.G., Dass, S. and Mathur, M. (2009) The chemical diversity of bioactive molecules and therapeutic potential of medicinal plants. In: Ramawat, K.G. (ed.) *Herbal Drugs: Ethnomedicine to Modern Medicine*. Springer-Verlag, Berlin, Germany.
Razzaq, A., Wishart, D.S., Wani, S.H., Hameed, M.K., Mubin, M. *et al.* (2022) Advances in metabolomics-driven diagnostic breeding and crop improvement. *Metabolites* 12(6), 511. DOI: 10.3390/metabo12060511.
Rogers, K. (2010) *The Components of Life: From Nucleic Acids to Carbohydrates*. Britannica Educational Publishing, New York, USA.
Ruckli, R., Hesse, K., Glauser, G., Rusterholz, H.P. and Baur, B. (2014) Inhibitory potential of naphthoquinones leached from leaves and exuded from roots of the invasive plant *Impatiens glandulifera*. *Journal of Chemical Ecology* 40(4), 371–378. DOI: 10.1007/s10886-014-0421-5.
Samanta, A., Das, G. and Das, S.K. (2011) Roles of flavonoids in plants. *International Journal of Pharmaceutical Sciences and Technology* 6, 12–35.
Samuni-Blank, M., Izhaki, I., Dearing, M.D., Gerchman, Y., Trabelcy, B. *et al.* (2012) Intraspecific directed deterrence by the mustard oil bomb in a desert plant. *Current Biology* 22(13), 1218–1220. DOI: 10.1016/j.cub.2012.04.051.
Saxena, R.C., Adhikari, D.K. and Goyal, H.B. (2009) Biomass-based energy fuel through biochemical routes: A review. *Renewable and Sustainable Energy Reviews* 13(1), 167–178. DOI: 10.1016/j.rser.2007.07.011.
Scossa, F., Alseekh, S. and Fernie, A.R. (2021) Integrating multi-omics data for crop improvement. *Journal of Plant Physiology* 257, 153352. DOI: 10.1016/j.jplph.2020.153352.
Singh, H., Verma, A., Kumar, M., Sharma, R., Gupta, R. *et al.* (2017) Phytoremediation: A green technology to clean up the sites with low and moderate level of heavy metals. *Austin Biochemistry* 2(2), 1–8.
Steinkellner, S., Lendzemo, V., Langer, I., Schweiger, P., Khaosaad, T. *et al.* (2007) Flavonoids and strigolactones in root exudates as signals in symbiotic and pathogenic plant-fungus interactions. *Molecules* 12(7), 1290–1306. DOI: 10.3390/12071290.
Stewart, A.J., Bozonnet, S., Mullen, W., Jenkins, G.I., Lean, M.E. *et al.* (2000) Occurrence of flavonols in tomatoes and tomato-based products. *Journal of Agricultural and Food Chemistry* 48(7), 2663–2669. DOI: 10.1021/jf000070p.

Strehmel, N., Böttcher, C., Schmidt, S. and Scheel, D. (2014) Profiling of secondary metabolites in root exudates of *Arabidopsis thaliana*. *Phytochemistry* 108, 35–46. DOI: 10.1016/j.phytochem.2014.10.003.

Taiz, L. and Zeiger, E. (2005) *Plant Physiology*, 3rd edn. Sinauer Association Inc, Sunderland, USA.

Tawaraya, K., Horie, R., Saito, S., Wagatsuma, T., Saito, K. *et al.* (2014a) Metabolite profiling of root exudates of common bean under phosphorus deficiency. *Metabolites* 4(3), 599–611. DOI: 10.3390/metabo4030599.

Tawaraya, K., Horie, R., Shinano, T., Wagatsuma, T., Saito, K.A. *et al.* (2014b) Metabolite profiling of soybean root exudates under phosphorus deficiency. *Soil Science and Plant Nutrition* 60(5), 679–694. DOI: 10.1080/00380768.2014.945390.

Tiwari, R. and Rana, C.S. (2015) Plant secondary metabolites: A review. *International Journal of Engineering Research General Science* 3, 661–670. DOI: 10.18488/journal.aefr/2015.5.4/102.4.661.670.

Vives-Peris, V., Molina, L., Segura, A., Gómez-Cadenas, A. and Pérez-Clemente, R.M. (2018) Root exudates from citrus plants subjected to abiotic stress conditions have a positive effect on rhizobacteria. *Journal of Plant Physiology* 228, 208–217. DOI: 10.1016/j.jplph.2018.06.003.

Vives-Peris, V., de Ollas, C., Gómez-Cadenas, A. and Pérez-Clemente, R.M. (2020) Root exudates: From plant to rhizosphere and beyond. *Plant Cell Reports* 39(1), 3–17. DOI: 10.1007/s00299-019-02447-5.

Vílchez, S., Molina, L., Ramos, C. and Ramos, J.L. (2000) Proline catabolism by *Pseudomonas putida*: Cloning, characterization, and expression of the put genes in the presence of root exudates. *Journal of Bacteriology* 182(1), 91–99. DOI: 10.1128/JB.182.1.91-99.2000.

Wang, W., Wang, S.X. and Guan, H.S. (2012) The antiviral activities and mechanisms of marine polysaccharides: An overview. *Marine Drugs* 10(12), 2795–2816. DOI: 10.3390/md10122795.

Weston, L.A., Ryan, P.R. and Watt, M. (2012) Mechanisms for cellular transport and release of allelochemicals from plant roots into the rhizosphere. *Journal of Experimental Botany* 63(9), 3445–3454. DOI: 10.1093/jxb/ers054.

Xu, W., Liu, D., Wu, F. and Liu, S. (2015) Root exudates of wheat are involved in suppression of *Fusarium wilt* in watermelon in watermelon-wheat companion cropping. *European Journal of Plant Pathology* 141(1), 209–216. DOI: 10.1007/s10658-014-0528-0.

Xu, X., van Lammeren, A.A., Vermeer, E. and Vreugdenhil, D. (1998) The role of gibberellin, abscisic acid, and sucrose in the regulation of potato tuber formation *in vitro*. *Plant Physiology* 117(2), 575–584. DOI: 10.1104/pp.117.2.575.

Zambelli, A., Nocito, F.F. and Araniti, F. (2025) Unveiling the multifaceted roles of root exudates: Chemical interactions, allelopathy, and agricultural applications. *Agronomy* 15(4), 845. DOI: 10.3390/agronomy15040845.

Zárybnický, T., Boušová, I., Ambrož, M. and Skálová, L. (2018) Hepatotoxicity of monoterpenes and sesquiterpenes. *Archives of Toxicology* 92(1), 1–13. DOI: 10.1007/s00204-017-2062-2.

Zhang, J.H., Mao, Z.Q., Wang, L.Q. and Shu, H.R. (2007) Bioassay and identification of root exudates of three fruit tree species. *Journal of Integrative Plant Biology* 49(3), 257–261. DOI: 10.1111/j.1744-7909.2007.00307.x.

8 Major Organic Compounds Derived from Plant Root Exudates: Sugar, Amino Acids, Organic Acids, Phenolic Compounds, and Secondary Metabolites

Samreen Abbas[1], Saniya Shakir[1], Pranesh Prabhu[1], Manoj Kumar Mishra[2] and Neha Verma[3*]

[1]Department of Biotechnology, Institute of Engineering and Technology, Lucknow, India
[2]Department of Biotechnology, S.R. Institute of Management and Technology, Lucknow, India
[3]Department of Biotechnology, Meerut Institute of Engineering and Technology, Meerut, India

8.1 Introduction

Plants interact continuously with their surrounding environment through a sophisticated network of biochemical and physical processes, one of the most pivotal being the release of root exudates. The physicochemical and biological characteristics of the rhizosphere are greatly influenced by root exudates, which are described as a complex mixture of organic and inorganic compounds secreted into the soil by living plant roots (Dakora and Phillips, 2002). These molecules include a range of high molecular weight substances like enzymes and mucilage, as well as low molecular weight substances including sugars, amino acids, organic acids, phenolics, flavonoids, and secondary metabolites. Root exudates are not released at random; rather, they have an evolutionary adapted function that influences nutrient availability, soil microbial communities, and even interactions with nearby plants.

*Corresponding author: neha.emo@gmail.com

DOI: 10.1079/9781836990475.0008

8.1.1 Mechanisms of root exudation

The process of root exudation is governed by both passive and active transport mechanisms, which are closely regulated by plant physiology, developmental stage, and environmental factors (Fig. 8.1). These processes regulate the type, quantity, and timing of substances that plant roots release into the rhizosphere. Root exudation is a highly selective and frequently energy-dependent activity that is essential for microbial contact, environmental reaction, and nutrition acquisition. It is not a random loss of cellular contents.

8.1.1.1 Passive diffusion

One of the most simple and basic processes of root exudation is passive diffusion. This involves the movement of small, water-soluble, and nonionic compounds such as sugars, amino acids, and organic acids down their concentration gradient through the plasma membrane of root cells (Jones *et al.*, 2004). The permeability of the membrane and the concentration of metabolites within root cells relative to the external rhizosphere both affect the rate of diffusion. Passive diffusion is restricted to substances that can freely move through tiny holes or lipid bilayers, even though it does not require metabolic energy.

8.1.1.2 Facilitated diffusion and ion channels

Certain substances need help to pass through the hydrophobic lipid bilayer found in root cell membranes. In these situations, certain channel proteins or carrier proteins embedded in the plasma membrane enable diffusion. For

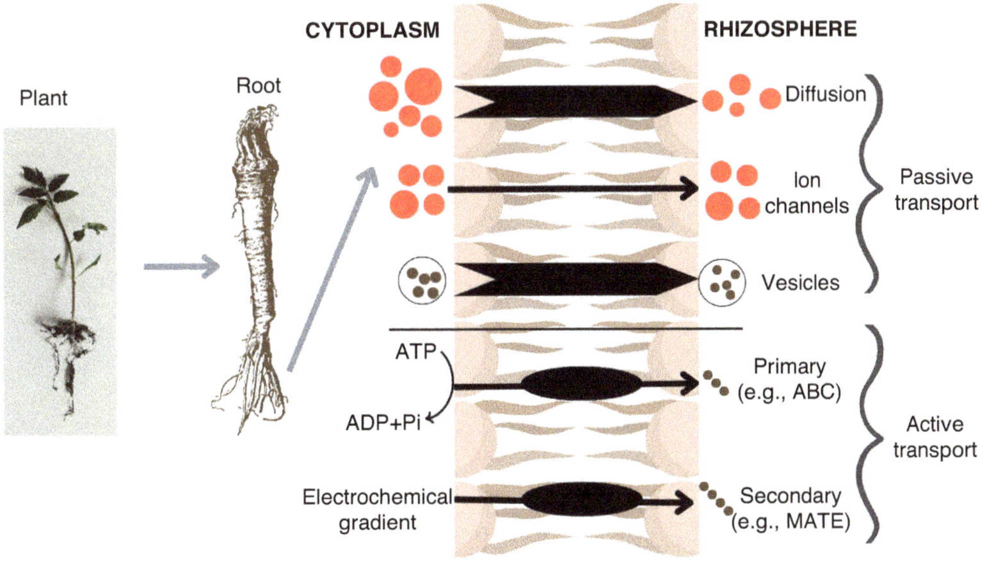

Fig. 8.1. The active and passive transport of root exudates from cytoplasm to rhizosphere.

example, in reaction to stress signals like aluminum poisoning, anion channels permit the production of negatively charged organic acids like citrate or malate (Ryan *et al.*, 2001). Likewise, it has been determined that sugar efflux transporters mediate the flow of sucrose and glucose into the rhizosphere (Chen *et al.*, 2010).

8.1.1.3 Active transport through membrane transporters

In contrast to diffusion, active transport requires energy, usually in the form of ATP, to move compounds against their concentration gradient. This is especially important for the selective exudation of metabolites that are present in low internal concentrations or for directional secretion under nutrient stress or pathogen attack. ATP-binding cassette (ABC) transporters, multidrug and toxic compound extrusion (MATE) transporters, and major facilitator superfamily (MFS) transporters are among the well-characterized protein families involved in this process (Taylor and Grotewold, 2005).

8.1.1.4 Vesicle-mediated exocytosis

For larger or hydrophobic compounds that cannot pass through membrane channels, plants utilize vesicle-mediated transport. In this mechanism, metabolites are packaged into secretory vesicles within the cytoplasm, which are then transported to and fuse with the plasma membrane, releasing their contents into the apoplast or directly into the soil (Bais *et al.*, 2006). This process is especially relevant for the exudation of phenolic compounds, phytoalexins, and signaling peptides. Vesicle transport is also linked to the Golgi apparatus and endoplasmic reticulum, implicating it in the secretion of cell wall-modifying enzymes and mucilage.

8.1.1.5 Programmed cell death and cell lysis

In root cap regions, particularly in the root border cells, controlled cell separation and programmed cell death (PCD) contribute to the release of a wide variety of exudates. These cells detach from the root tip and secrete mucilage and enzymes, creating a protective biochemical interface between the root and soil. Under environmental stress or mechanical injury, cell lysis may also lead to the release of intracellular contents, such as nucleotides and proteins, which enrich the rhizosphere and can act as nutrient sources or signaling cues for microbes (Hawes *et al.*, 2000).

8.1.1.6 Environmental and developmental regulation

The mechanism and rate of root exudation are not static but are dynamically influenced by plant developmental stage, diurnal cycles, and abiotic factors such as pH, salinity, nutrient availability, and heavy metal stress. For example, drought conditions can alter the composition of root exudates, favoring the release of osmoprotectants like proline, while iron deficiency often increases the exudation of siderophores or phenolic compounds that mobilize iron. Additionally, biotic factors such as colonization by arbuscular mycorrhizal fungi or rhizobia can upregulate specific exudation pathways. These feedback interactions make root exudation a central, responsive mechanism by which plants adapt to their environment (Walker *et al.*, 2003).

8.1.2 Functional significance of root exudation in rhizosphere ecology

One of the most dynamic areas of terrestrial ecosystems is the rhizosphere, which is the small area of soil that envelops and is impacted by plant roots. It acts as a vital interface for interactions between soil particles, microbes, and plants on a biological, chemical, and physical level. The process of root exudation, which allows plants to actively alter their immediate surroundings, is central to rhizosphere dynamics. Essential ecological processes, including microbial recruitment, pathogen defense, chemical signaling, nutrient mobilization, and soil structure regulation, are all mediated by root exudates.

8.1.2.1 Nutrient mobilization and solubilization

The enhancement of nutrient availability in the rhizosphere is one of the main ecological functions of root exudates. Important nutrients including phosphorus (P), iron (Fe), and zinc (Zn) are frequently poorly soluble in soils where plants thrive. Organic acids, particularly citric, malic, and oxalic acids, are released by roots in reaction, chelating metal ions and releasing phosphate from soil complexes. For instance, in phosphorus-deficient soils, it has been demonstrated that root production of malate and citrate increases phosphate availability (Fay and Knapp, 1995).

8.1.2.2 Recruitment and modulation of microbial communities

For soil microbes, root exudates serve as both a chemical signal and a selective source of nutrients. The composition and activity of the rhizosphere microbiome are significantly influenced by many kinds of exudates, including sugars, amino acids, and flavonoids. For instance, the symbiotic nitrogen fixation pathway is started by flavonoids released by leguminous plants, which act as chemoattractants and inducers of nod gene expression in rhizobia. In a similar vein, strigolactones generated by host roots attract arbuscular mycorrhizal fungus (AMF), resulting in symbiotic association and enhanced phosphorus acquisition (Akiyama *et al.*, 2005).

8.1.2.3 Defense against soil pathogens and pests

Additionally, root exudates aid in plant defense by releasing substances that inhibit the growth of harmful organisms. These include antimicrobial compounds such as phytoalexins, terpenoids, alkaloids, and phenolics. Apart from their direct antibacterial properties, root exudates have the potential to promote the production of antibiotics and siderophores by beneficial microorganisms, thereby establishing an antagonistic environment for pathogens. For instance, in iron-deficient environments, the coumarins released by

Arabidopsis help mobilize iron while also inhibiting the growth of fungi like *Fusarium* (Stringlis *et al.*, 2018).

8.1.2.4 Allelopathy and chemical communication

Certain root exudates have allelopathic qualities, which prevent nearby plant species from growing and germinating. Crop–weed interactions frequently involve chemical warfare, since crops like wheat and sorghum emit substances like benzoxazinoids or sorgoleone that reduce rival plants. Through the selective inhibition or promotion of particular microbial taxa, allelochemicals can also affect microbial activity. Apart from allelopathy, root exudates also play a role in communication between plants and microbes, allowing plants to identify their neighbors and modify their growth plans accordingly. Microbial collaborations, resource foraging, and root architecture are all facilitated by this kind of chemical signaling (Cumming, 2016).

8.1.2.5 Soil aggregation and physical modification

Root exudates, especially mucilage and polysaccharides, improve soil structure, porosity, and water-holding capacity by binding soil particles together and promoting soil aggregation. These physical alterations improve the flow of nutrients and encourage microbial colonization. Mucilage also facilitates root movement in compacted soils and serves as a lubricant for root growth (Compant *et al.*, 2010).

In the early 20th century, when scientists discovered that plants may affect soil microbial activity and even the growth of nearby plants, the idea of root exudation began to attract interest. However, specialized research did not start measuring the kinds and intensities of these exudates until the 1950s and 1960s (Watt, 2009). In recent decades, technological advancements in mass spectrometry and chromatography have made it possible to profile the chemistry of root exudates with extreme specificity and sensitivity. Root exudates are now understood to be important components of chemical communication, plant immunology, and environmental adaptability in addition to serving as a source of nutrients for rhizosphere bacteria in contemporary plant science and soil microbiology.

The objective of this chapter is to explore and critically examine the major organic compounds released as root exudates, particularly sugars, amino acids, organic acids, phenolic compounds, and secondary metabolites. Each class of compound will be discussed in terms of its chemical composition, the biological mechanisms that govern its release, and its role in shaping rhizosphere dynamics. The chapter also highlights the ecological and agricultural implications of root exudates and discusses emerging technologies and research frontiers in this rapidly evolving field. By gaining a better understanding of how root exudates function, we can unlock new possibilities for enhancing crop productivity, soil health, and ecological resilience (Badri and Vivanco, 2009).

8.2 Overview of Root Exudate Composition

Root exudates are classified into two classes on the basis of composition (Fig. 8.2).

8.2.1 Low molecular weight: Sugars, amino acids, organic acids, phenolics

Low molecular weight compounds which easily leak out from plants via cell junctions are called exudates. Low molecular weight (LMW) chemicals diffuse outward due to the wide concentration gradient of LMW solutes that typically exists between the cytosol (millimolar concentrations) and the rhizosphere (micromolar concentrations as a result of microbial degradation). Furthermore, LMW substances that are negatively charged at the cytosol's pH (7.0–7.5), such as organic acids and amino acids, are further encouraged to diffuse outward by the outward positive electrochemical potential gradient produced by proton extrusion via proton motive H+ ATPases (Ma *et al.*, 2022). Since they are naturally very soluble, exudation is a passive process that does not require plant energy. In the rhizosphere, these compounds may contribute to nutrient cycling as C and N sources for rhizosphere microorganisms.

Amino acids, sugars, organic acid anions, and different phenolics are the main components of LMW chemicals found in root exudates. Many LMW chemicals in the rhizosphere have a half-life of only 1–5 hours due to fast microbial degradation (Jones *et al.*, 2004). The root exudate fraction can be further separated into (i) LMW organic compounds that are permanently lost from root cells through diffusion (diffusates); (ii) root secretions that have unique roles in defense reactions, detoxification, and nutrient mobilization; (iii) root signals that are released through controlled mechanisms using transport proteins or membrane channels; and (iv) metabolic waste products that are released as root excretions.According to Jones *et al.* (2004), carbon input

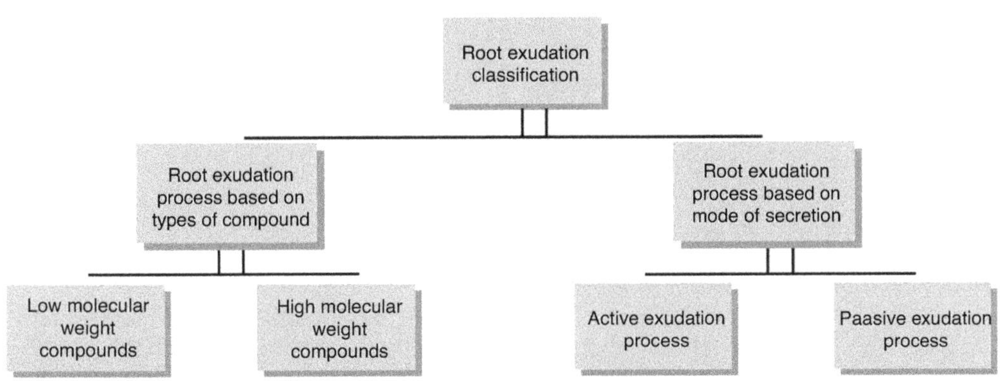

Fig. 8.2. Classification of root exudates.

into soils by root exudation may account for 5–10% of the net fixed carbon in plants cultivated in soil. Adsorption and complexation can immobilize exudates on the soil matrix, preventing biodegradation and significantly extending their residence duration in soils.

8.2.2 High molecular weight: Proteins, mucilage, enzymes

Similar to exudation, secretions are the result of active plant processes, meaning that the plants must use metabolic energy in order to produce high molecular weight (HMW) molecules. The primary way in which roots release polygalacturonic acid and polysaccharides is by secretion. Mucilage, a HMW gelatinous polygalacturonic acid polymer, is mostly released by the Golgi apparatus of hypersecretory cells of the root cap. It is then transmitted to older root zones during root elongation, although epidermal cells can also secrete mucilage. One of the few obvious indicators of organic C excretion from roots is the gelatinous coating of root mucilage that surrounds the terminals of the roots. The mucilage of maize (*Zea mays*) roots has lower concentrations of uronic acid, mannose, and glucuronic acid and higher concentrations of galactose, xylose, arabinose, rhamnose, and glucose. The range of mucilage generated *in vitro* is 11–47 μg MS/mg MS root development. Actually, the endoplasmic reticulum is where mucilage synthesis begins, and the Golgi saccules are where it finishes. The Golgi vesicles carry the mucilage to the plasmalemma, where exocytosis releases it between the plasmalemma and the cell wall. These are all energy-dependent processes (Nguyen, 2009).

Mucilage can build up at the inner wall surface and the flow through the cell wall is not systematic. The mucilage is thought to pass passively through the cell wall and form a droplet at the root tip if the mucilage's degree of hydration and the cell turgor are both adequate (Hale and Moore, 1980). The middle lamella's increased permeability is most likely what allows the passage through the cell wall. The droplet forms in a cycle of 3–4 hours under regulated conditions. The mucilage of the root cap has a large capacity for hydration but if the soil is unsaturated, the root cap mucilage appears as a dry coating over the apex and does not form a droplet, as is often observed *in vitro*. Mucilage and living border cells tightly adhering to the root form the rhizosheath.

8.2.2.1 Mucigel

Microorganisms perform additional transformations of mucilage. Mucigel is made up of mucilage, soil colloids, organic materials, microbes, and sloughed-off plant tissue cells. Mucilage in nonsterile media also contains materials made when the cell walls are broken down by microbes. Microorganisms typically infiltrate the mucilage of soil-grown plants, embedding both organic and inorganic soil particles. Mucigel is the name given to this mixture of soil particles, bacteria, and gelatinous substances.

8.2.2.2 Lysate

When the plasmalemma fails, mature cortical cells release lysate. Many plants are able to generate metabolically active root border cells that are designed to split apart and discharge into the surrounding environment from the root cap perimeter. The separated cells adhere to the root tip in the absence of free water. While release rates are heavily influenced by the current environmental conditions, the daily rate of border cell creation varies greatly among plant species, ranging from a dozen in tobacco to over 10,000 cells/day for cotton and pine. Numerous proteins, including different enzymes, are also released by plant roots as secretory proteins. Polysomes affixed to the endoplasmic reticulum (ER) produce secretory proteins, which are already separated into the ER lumen throughout the translation process.

Transfer vesicles containing the secretory proteins are differentiated from vesicles having a vacuolar destination during their passage through the Golgi apparatus. The proteins in the vesicles are exocytosed once they have passed through the plasma membrane and are discharged into the apoplasm. All exocytosis-related processes rely heavily on the availability of calcium ions.

The quantity and make-up of rhizodeposition are determined by a variety of internal and external variables. Increased mechanical impedance of the growing substrate, poisonous substances, low pH in the soil solution, nutrient limitation, intense light, high atmospheric CO_2 concentrations, temperature extremes, and microbes can all promote rhizodeposition. Rhizodeposition may be separated into two primary fractions based on origin and release mechanisms: (i) lysates of sloughed-off cells and tissues from root turnover, which can account for as much as 50% of the carbon translocation below ground; and (ii) organic compounds released from intact root cells as so-called root exudates. Five to ten percent of the net fixed carbon in plants grown in soil may come from carbon input into the soil through root exudation. A sizeable amount of carbon also enters the soil through mycorrhizal hyphae, which may be similar to fine roots. Hyphal carbon input may possibly be the primary method of carbon transfer into the soil for ectomycorrhizal fungi (Godbold *et al.*, 2006).

Table 8.1 Summarizes the different classes and subclasses of root exudates and rhizosphere roles.

8.3 Sugars

8.3.1 Common sugars in root exudates

Carbon flux to the rhizosphere microbiota is influenced by the relative amount and chemistry of a small spectrum of soluble sugars found in most

Table 8.1. Root exudate classes with their roles and examples.

Main class	Key subclasses (representative compounds)	Illustrative plant examples	Rhizosphere roles	References
Sugars	– Monosaccharides: glucose, fructose, xylose – Disaccharides: sucrose, maltose, trehalose – Sugar alcohols: ribitol, glycerol	Maize, tomato, *Arabidopsis*	Carbon source for microbes; chemotactic signals; driver of early rhizobacterial assembly	(Jiao *et al*., 2022; Lopes *et al*., 2022)
Amino acids	– Neutral α-AAs: alanine, serine, glycine – Basic AAs: arginine, lysine, histidine – Acidic AAs: aspartate, glutamate – Aromatic AAs: phenylalanine, tryptophan – Nonproteinogenic: γ-aminobutyric acid (GABA), β-alanine	Vegetable crops, maize	Nutrient source; signaling under biotic stress; modulation of microbial recruitment	(Ivarson and Sowden, 1969)
Organic acids	– Tricarboxylic: citric acid – Dicarboxylic: malic, succinic, oxalic acids – Monocarboxylic: acetic, lactic acids – Fumaric and tartaric acids	Alfalfa, barley, rice	Metal chelation, P mobilization, bacterial chemotaxis, alleviation of Al toxicity	(Lipton *et al*., 1987; Wang *et al*., 2015)
Phenolic compounds	– Simple phenolic acids: *p*-hydroxybenzoic, caffeic acids – Phenolic aldehydes: *p*-hydroxy-benzaldehyde – Flavonoids: quercetin, tricin, diosmetin, naringenin – Coumarins: scopoletin, esculetin	Johnson grass, alfalfa, *Triadica sebifera*	Allelopathy; recruitment of mycorrhiza/PGPR; pathogen inhibition	(Mandal *et al*., 2010; Li *et al*., 2013)
Secondary metabolites	– Benzoxazinoids: – DIMBOA, DIBOA, HDMBOA-glc – -trigolactones: 5-deoxystrigol, orobanchol – Glucosinolates: 2-phenylethyl-, indole-3-yl-glucosinolates – Terpenoids: mono- and sesquiterpenes, acylsugars	Maize, pea, *Brassica rapa*, sorghum	Signaling to AM fungi, neighbour detection, biofumigation, defense against pests and pathogens	(Neal *et al*., 2012; Canarini *et al*., 2019; Rasmann and Hiltpold, 2022)

DIBOA, 2,4-dihydroxy-1,4-benzoxazin-3-one; DIMBOA, (2,4-dihydroxy-7-methoxy-1,4-benzoxazin-3-one); HDMBOA, 4,7-dimethoxy-2-((3,4,5-trihydroxy-6-(hydroxymethyl)oxan-2-yl)oxy)–3,4-dihydro-2H-1,4-benzoxazin-3-one; PGPR, plant growth-promoting rhizobacteria.

root exudates. The most concentrated hexoses are often glucose and fructose, which offer easily absorbed energy that promotes chemotaxis and quick bacterial growth close to the root surface. SWEET transporters can export or passively leak sucrose, which is actively loaded into the phloem. While the trisaccharide raffinose, which is representative of raffinose-family oligosaccharides, accumulates under abiotic stress and may serve as an osmoprotectant signal, selectively enriching drought-tolerant microbes, maltose, a breakdown product derived from starch, is released in smaller quantities but provides brief bursts of carbon during diurnal starch remobilization (Lopes *et al.*, 2022).

8.3.1.1 Glucose

The simplest and often most abundant hexose in root exudates, glucose usually diffuses passively from cortical cells along steep cytosol-to-soil gradients. Typical concentrations in the rhizosphere solution of field crops reach 2–5 mg kg^{-1} soil, values that are rapidly scavenged by copiotrophic bacteria attracted by chemotactic receptors tuned to this sugar. Because microbes metabolize glucose within minutes, plants can steer microbial hotspots by modulating leakiness through changes in membrane integrity or suberin deposition (Eze and Amuji, 2024).

8.3.1.2 Fructose

A by-product of sucrose cleavage, fructose is typically found at somewhat higher concentrations (about 4 mg kg^{-1} in some rhizosphere samples) than glucose. Its increased chemical reactivity may stabilize organic carbon by encouraging Maillard browning on mineral surfaces. A high fructose to organic acid ratio can cause decomposer communities to shift toward sugar specialists, speeding up nitrogen mineralization in nutrient-poor soils since microorganisms preferentially oxidize organic acids.

8.3.1.3 Sucrose

Energy-dependent SWEET transporters, such as GmSWEET6 in soybeans, release the primary transport carbohydrate in plants more often than simple leakage. When these exporters are upregulated in mild or low-phosphate environments, there is a noticeable sucrose efflux, which benefits rapidly proliferating Proteobacteria and encourages the early production of biofilms along the elongation zone. Genetic inhibition of SWEET-mediated export highlights its signaling and nutritional significance by significantly lowering sugar in exudates and delaying microbial colonization (Zheng *et al.*, 2024).

8.3.1.4 Maltose

Maltose is a by-product of nocturnal starch remobilization and is detected at micromolar quantities. It provides a brief carbon burst that synchronizes microbial respiration with the plant's diurnal rhythm when it is pulsed into the rhizosphere at dawn. Maltose, less prevalent than mono- and trisaccharides, can trigger the synthesis of enzymes in specific strains of *Bacillus* and *Pseudomonas*, improving the solubility of phosphate in controlled agroecosystems (Tan *et al.*, 2021).

8.3.1.5 *Raffinose*

Under benign circumstances, this oligosaccharide from the raffinose family is typically absent, but it builds up in the exudates of desert shrubs like *Haloxylonammodendron*, where it cooccurs with fatty acids. Plant resilience is strengthened when galactinol–raffinose biosynthesis is induced in roots by cold, osmotic stress, or drought. A portion of this biosynthesis is released to serve as a selective carbon source for Actinobacteria that can withstand stress and as an osmoprotectant. According to genetic research, raffinose synthase upregulation increases the exudation flux of raffinose in addition to raising internal raffinose pools.

8.3.2 Roles and mechanisms

- Rhizosphere microbes use hexoses and sucrose as an instantaneous source of carbon and energy. They can make up more than 60% of the total dissolved organic carbon surrounding roots, and copiotrophic bacteria oxidize them in a matter of minutes, promoting microbial growth and respiration that speeds up nutrient turnover.
- Aggregation of soil and enhancement of its structure. Using specific chemoreceptors, motile bacteria swim up the concentration field, adhere to the epidermis, and polymerize the sugars (or derived EPS precursors) into a protective biofilm that stabilizes early root colonization. This process is known as chemo-signaling.
- Role in microbial chemotaxis and biofilm formation. In order to increase porosity, water retention, and long-term carbon sequestration in the rhizosphere, microbes convert plant sugars into extracellular polysaccharides whose sticky fibrils link mineral particles and fungal hyphae into water-stable aggregates.

8.4 Amino Acids

8.4.1 Common amino acids

Glutamic acid (Glu) is of the most common nitrogen-rich solutes that seep from roots; glutamate is rapidly detected by soil bacteria. In *Bradyrhizobium japonicum*, Glu is among the top three chemoattractants that have been studied. Rhizobia and a number of *Pseudomonas* organisms employ Glu as a carbon skeleton and to trigger symbiotic genes by swimming up nanomolar gradients of the molecule. In addition to causing $Ca2^{+}$ transients through root glutamate receptor-like channels, which link root signaling and external

N availability, plants release more Glu when tissue N levels are high. When microbial deamination of Glu releases ammonium, which the plant can readily reassimilate, a rapid N-cycling loop is closed (Barbour *et al.*, 1991).

During drought or salinity, glycine (Gly), the smallest amino acid, often rises in exudates and diffuses easily into the rhizosphere. Gly also feeds the glycine–betaine pathway, which encourages the establishment of osmoprotectant producers near stressed roots and gives copiotrophic bacteria a first-hit C/N supply. Studies conducted on cucumbers and *Bacillus velezensis* showed that a Gly-enriched exudate cocktail promotes early root colonization and accelerates spore germination, both of which lead to increased shoot biomass. The steep gradients produced by Gly's fast metabolism direct chemotactic PGPR toward the root surface.

When roots undergo brief hypoxia or redox changes brought on by rare earth elements, alanine (Ala) is released into the rhizosphere. According to a new study on pak choi, foliar lanthanum caused Ala secretion, which in turn drew a cadmium-immobilizing *Pseudomonas* sp. This chemotactic loop reduced the vegetable's absorption of Cd and metal toxicity. Fast-growing bacteria have an advantage during early colonization because Ala is a rapidly absorbed dual C/N donor that can enter microbial pyruvate pools without the need for expensive deamination, even outside stress situations.

Serine (Ser) is a metabolically strategic substance that is often released at lower absolute concentrations. It provides precursors for the production of phospholipids, tryptophan, and folate in microorganisms associated with roots and supports one-carbon metabolism. In order to stabilize microaggregates and enhance water retention around root hairs, serine-rich exudates promote the synthesis of extracellular polymers and the development of multispecies biofilms on root surfaces. A reciprocal P-for-C exchange is created when plants increase Ser exudation under P-limitation. The amino acid also increases microbial phosphatase activity, which releases organic phosphate for plant uptake.

Lysine (Lys) is a cationic amino acid that tends to build up in exudates after pathogen attack or exposure to high light levels. It leaks through cationic amino acid transporters. It is rapidly decarboxylated to either pipecolic acid or cadaverine in the rhizosphere; the latter is a forerunner of systemically acquired resistance in leaves, while the former disrupts fungal cell walls. Thus, free Lys can serve as the carbon backbone of defense signals in addition to acting as an antibacterial. Additionally, its positive charge permits mild chelation of anionic clays, which somewhat enhances the stability of microaggregates (Wen *et al.*, 2022).

8.4.2 Functional roles

8.4.2.1 Supply of nitrogen for microorganisms

Three essential rhizosphere processes are supported by amino acids found in root exudates. A significant portion of the dissolved organic nitrogen surrounding roots may consist of glutamate, alanine, and other soluble

N-rich solutes. Copiotrophic bacteria quickly absorb these solutes and, by deamination, produce ammonium that plants can reabsorb, strengthening the microbially mediated N cycle.

8.4.2.2 *Signal molecules in interactions between microbes and plants*

Amino acids function as specific signaling molecules. In addition to activating plant GLR $Ca2^+$ channels, which relate external N status to systemic electrical and calcium waves, nanomolar gradients of glutamate, glycine, or alanine also lead motile rhizobacteria via specific chemoreceptors, speeding colonization (Yu *et al.*, 2023).

8.4.2.3 *Role in stress reactions and allelopathy*

A few amino acids that are released have a part in both stress reactions and allelopathy. In the rhizosphere, lysine is decarboxylated to either pipecolic acid or cadaverine, which respectively promote systemic acquired resistance and inhibit competing fungi, while alanine pulses under metal stress recruit CD-immobilizing. During drought or salinity, *Pseudomonas* and glycine promote the synthesis of osmoprotectants.

8.4.3 Rhizospheric effects

The rhizosphere biota is constantly being reshaped by the molecular "gatekeepers" of sugars, amino acids, organic acids, and phenolics released by roots. While specialized metabolites (like benzoxazinoids or coumarins) selectively enrich mutualists and repel pathogens, exudates preferentially recruit fast-growing copiotrophs like Proteobacteria and repress oligotrophs by providing labile C- and N-sources and generating precisely calibrated chemo-signals. The sensitivity of community assembly to exudate chemistry is demonstrated by comparative metabolomics across maize, sorghum, and rye cover crops, which have revealed that even minor changes in the ratio of sugars to phenolics can rearrange more than 30% of bacterial and fungal amplicon sequence variants in a matter of days.

The extracellular enzyme activity that controls the cycling of nutrients is likewise modulated by the same substances. Microbes are stimulated to synthesize hydrolases such β-glucosidase, phosphomonoesterase, and N-acetylglucosaminidase by the carbon in exudate; enzyme hotspots develop along root axes where the ratios of sugar to organic acid are highest. Exudate stoichiometry manipulation in microcosms showed that while increased citrate or malate levels upregulate phosphatases that release inorganic P, an elevated amino acid percentage increases protease and urease activity, speeding up N mineralization. Exudate-driven enzyme changes can increase soil organic matter turnover by 25% and strengthen disease suppression, according to recently published research on *Fusarium*-resistant cucurbits.

8.5 Organic Acids

8.5.1 Types of organic acids

8.5.1.1 Citric acid

When plants experience iron, phosphate, or aluminum stress, MATE-type transporters and FRD3/FRDL effluxers aggressively pump citrate, a tricarboxylate with a high metal-binding capacity, from the root apices. In acidic soils, the anion mobilizes sparingly soluble Fe- and Ca-phosphates, increasing local Pi availability, while also chelating Al^{3+} outside the root, reducing toxicity. Citrate also transports Fe through the xylem and feeds bacteria that produce siderophores, which improves micronutrient cycling because it is soluble over a broad pH range (Wang *et al.*, 2022).

8.5.1.2 Malic acid

Within minutes of Al^{3+} touching the root tip, malate is released, passing through ALMT channels that are opened by Al binding. Cereals and many legumes are resistant to aluminum because of the dicarboxylate, which sequesters the metal in nontoxic compounds near the rhizoplane. Beyond detoxification, nanomolar malate gradients serve as chemical beacons. For instance, the roots of tomatoes and watermelons employ malate (in combination with citrate) to draw in biofilm-forming strains of *Bacillus* and *Paenibacillus* that produce growth hormones and defend against infections.

8.5.1.3 Oxalic acid

In order to dissolve phosphate-bearing minerals or immobilize excess metals, oxalate easily forms stable complexes with Ca, Fe, and Al. Although it can increase the mobility of certain contaminants, its potent chelating ability also releases trace elements required by rhizobacteria. As a reminder that the same ligand-driven chemistry that mobilizes nutrients can also unintentionally mobilize pollutants, released oxalate in lettuce increased its absorption by decreasing soil sorption of perfluorooctanoic acid.

8.5.1.4 Fumaric acid

Fumarate is a powerful signaling chemical, although being emitted in tiny amounts. Along with malate and citrate, it is secreted by the roots of tomatoes, cucumbers, and bananas to direct beneficial *Bacillus* strains toward the rhizoplane, promoting swarming and the production of biofilms. On the other hand, it can attract *Ralstonia* pathogens in tobacco, demonstrating that the same chemoattractant can favor allies or adversaries based on the situation. Because of its quick turnover, it is a temporary yet important trigger in the recruitment of microbes.

8.5.1.5 Tartaric acid

The roots of grapes and sunflowers typically contain tartaric acid which is four-carbon dihydroxy dicarboxylate. It mobilizes micronutrients by binding Fe(III) and other multivalent cations, but it also has less evident effects on soil chemistry. For example, in contaminated soils, tartaric-rich exudates accelerated the desorption and microbial biotransformation of the brominated flame-retardant TBBPA, changing the structure of bacterial communities and enzyme activities. These results demonstrate that tartaric acid influences pollutant fate and nutrient fluxes (Xiang *et al.*, 2020).

8.5.2 Physiological roles

8.5.2.1 Solubilization of phosphates and micronutrients

When phosphorus or trace metals are scarce, roots increase the production of citrate, malate, and oxalate by the MATE, ALMT, and FRDL transporters. These anions form highly soluble metal–organic complexes that release PO_4^{3-}, Fe, and Zn for absorption, outcompeting oxides and carbonates for Ca^{2+}, Fe^{3+}, or Al^{3+}. Within 2 days, white lupin cluster roots can double the P inflow by releasing about 70 µmol of citrate per g^{-1} root d.w. A nutrient-rich microsite surrounding the root tip is strengthened by phosphate-solubilizing *Pseudomonas* and *Bacillus*, which thrive on the spilled carbon and contribute phosphatase activity.

8.5.2.2 Chelation of toxic metals (Al, Fe, Mn)

Soluble Al^{3+} and Mn^{2+} stop the growth of roots in acidic soils. ALMT1 channels and MATE pumps open minutes after Al^{3+} contact, releasing citrate or malate that binds Al^{3+} in 1:2 complexes at the rhizoplane and renders it innocuous. In 50µM Al, soybean and wheat lines that overexpress ALMT1 maintain 40% longer roots and release twice as much malate. Citrate chelates Fe^{3+} for safe xylem transport, preventing Fenton damage and feeding the shoot, while oxalate production detoxifies Mn^{2+} in buckwheat in the same manner.

8.5.2.3 pH alteration and nutrient mobilization

Proton equivalents are carried into the soil solution via organic acid efflux, which frequently lowers the pH of the rhizosphere by one unit within millimeters of the root. Acidification can reduce neutral pH infections, dissolve carbonate coatings, and desorb sorbed phosphate, Cu, Zn, and Fe. Malate increased Olsen-P by 35% and decreased pH from 7.8 to 6.9 in calcareous wheat soils, mimicking natural exudation. In addition to energizing microbial phosphatase and siderophore systems, the ensuing acidity speeds up organic-P mineralization and micronutrient mobilization, establishing a positive feedback loop between chemical and biological nutrient release.

8.5.2.4 *Stress responses*

Roots transition to a "chemical mining" mode when soil resources become scarce. A lack of phosphate or iron triggers PHR- and STOP1-dependent pathways that increase MATE and other exporters, increasing the release of citrate, malate, and coumarin by up to an order of magnitude. Together, the tricarboxylate complexes Ca- and Fe-phosphates and catecholic coumarins increase P and Fe absorption and attract bacteria that produce siderophores and phosphate solubilization by reducing Fe^{3+} to the bioavailable Fe^{2+} pool (Chutia *et al.*, 2019).

8.6 Phenolic Compounds

8.6.1 Major classes

8.6.1.1 *Flavonoids*

These C-glycosylated or O-methylated polyphenols (such as quercetin and naringenin) are released from border and epidermal cells by vesicle trafficking and ABC pumps. They function as "frequency channels" in the rhizosphere, promoting flagellar swimming, triggering nod genes in rhizobia, and increasing the production of biofilms by plant growth-promoting rhizobacteria (PGPR). Selective flavonoid release causes symbiotic nodulation signaling in legumes and attracts Oxalobacteraceae, which encourage root branching, in maize. Additionally, their antioxidant qualities shield roots from soil surface damage caused by UV or ROS (Ghitti *et al.*, 2024).

8.6.1.2 *Lignin-derived phenolics*

Monolignols and their methylated derivatives, including ferulic, p-coumaric, and methyl-ferulate, frequently show up in exudate profiles even when the lignin polymer itself is not exuded. These tiny aromatics can accumulate to micromolar concentrations and are produced via phenylpropanoid turnover. They have two purposes: they attract disease-suppressive *Bacillus* species and allelopathically inhibit competing plants and pathogens (methyl-ferulate inhibits *Phytophthora* respiration). Additionally, their aromatic rings adhere to the surfaces of minerals, reducing carbon loss and promoting organic matter linked to minerals.

8.6.1.3 *Tannins*

Under biotic stress, woody perennials and certain cereals emit hydrolyzable tannins (tannic acid) and condensed proanthocyanidins. Their multiidentate phenolic hydroxyls act as a first line of chemical defense by disrupting cell walls and precipitating microbial enzymes. Tolerant PGPR metabolize tannins and multiply, altering the microbiome to limit disease, while low nanomolar tannin concentrations restrict *Fusarium* hyphal development and reduce pathogen protease activity. Tannic acid can also chelate Fe^{3+} in calcareous soils, gradually increasing the host's solubility of micronutrients (Zhao *et al.*, 2018).

8.6.1.4 Coumarins

When roots detect low iron or high pH, the ABCG and PDR transporter families export scopoletin, fraxetin, and related benzopyrone lactones. Then, their catechol groups complex and reduce Fe^{3+} to Fe^{2+}, significantly increasing Fe absorption in calcareous environments. By suppressing soilborne pathogens and attracting siderophore-producing Actinobacteria, coumarins work in tandem as redox antimicrobials to form a guild that shares iron around the root.

8.6.1.5 Simple phenols

During regular phenylpropanoid cycling, low molecular aromatics including salicylic, caffeic, and p-hydroxybenzoic acids are either actively released or passively leaked. At submillimolar levels, they inhibit quorum sensing and membrane integrity in sensitive microbes but stimulate lignocellulolytic enzyme production in others, fine-tuning community structure. They connect defense chemistry with nutrient acquisition by mobilizing P, Cu, and Zn from oxide surfaces through their acidity and metal-binding ability.

8.6.2 Functions

8.6.2.1 Antioxidant and antimicrobial shield

Scopoletin and fraxetin are examples of root-secreted coumarins that include hydroxyl or catechol groups that scavenge reactive oxygen species and also function as selective antibiotics. Scopoletin, for instance, suppresses *Fusarium* and *Verticillium* but spares coumarin-tolerant *Pseudomonas* allies, thereby tilting the microbial balance in favor of beneficial taxa and buffering oxidative stress at the rhizoplane (Stassen *et al.*, 2021).

8.6.2.2 Frontline defense against soil pathogens

Phenolic exudates can both poison invaders and enlist microbial bodyguards. In addition to releasing enough of this ester to inhibit a variety of oomycete and fungal pathogens, tobacco roots that upregulate the phenylpropanoid methyl-ferulate also enrich rhizosphere *Bacillus* that reinforce disease suppression. This dual action reduced the incidence of black-shank in field tests by threefold. Similar pulses of lignin-derived acids or tannins from by pathogens produce a chemically hostile zone surrounding roots and encourage communities that can detoxify these phenolics instead of giving in to them.

8.6.2.3 Symbiotic signaling for nodulation

The rhizobial transcription factor NodD in legumes senses nanomolar combinations of flavonoids released from the roots (such as luteolin and daidzein) and activates nod genes that promote the synthesis of Nod-factor, curling of root hair, and nodule organogenesis. Small variations in the flavonoid profile can modify which rhizobial strains colonize and how many nodules form. The structural variety of the flavonoid cocktail functions as a compatibility code, guaranteeing partner specificity and efficient biological nitrogen fixation (Clúa *et al.*, 2018).

8.7 Secondary Metabolites

Secondary metabolites secreted by roots belong to a few chemically different types whose structures and rhizospheric activities are determined by their biosynthetic origins. Saponins mix terpenoid cores with sugars to obtain surfactant function, glucosinolates fuse glucose with sulfur-nitrogen moieties generated from amino acids, terpenoids are derived from isoprene units, and alkaloids are derived from amino acid precursors. Despite having different chemistries, these four classes work together to construct the plant's complex chemical handshake with its underground partners. They are selective antimicrobials, high-affinity signaling molecules, and modulators of nutrient dynamics.

8.7.1 Major classes

8.7.1.1 Terpenoids

Terpenoids range from bulky di- and triterpene phytoalexins to volatile mono- and sesquiterpenes. They are constructed from isopentenyl diphosphate via the mevalonate (MVA) or methyl-erythritol phosphate (MEP) pathways. For instance, the diterpenoids kauralexins and zealexins, which are released by young maize roots, restrict the growth of *Fusarium* hyphal cells while favoring Proteobacteria that are resistant to terpenoid, thereby combining microbiome filtering with direct pathogen inhibition. The carbon signature of desert halophytes is dominated by secreted triterpenes, which also correspond with bacterial consortia that have adapted to dryness, highlighting their dual function as community shapers and stress metabolites (Afridi *et al.*, 2024).

8.7.1.2 Alkaloids

Nitrogen-rich heterocycles that are produced constitutively or when roots are injured or diseased include nicotine (found in tobacco), camalexin (found in

Arabidopsis), and gramine (found in barley). These heterocycles are derived from amino acid backbones. Higher concentrations of nicotine become bacteriostatic to sensitive fungi and oomycetes, forming a dose-dependent microbial filter, while low micromolar nicotine attracts chemotactic *Pseudomonas* strains and promotes their rhizoplane colonization. Similarly, indole alkaloids selectively enrich alkaloid-resistant PGPR while inhibiting nitrifier action and discouraging herbivores (Lee *et al.*, 2022).

8.7.1.3 Saponins

A triterpenoid or steroid aglycone esterified with one or more sugar chains makes up these amphiphilic glycosides. *Panax notoginseng* secretes ginsenosides that are a prime example of their ecological polarity; while larger pulses, which are typical in fatigued monocultures, favor opportunistic fungi and impair natural disease suppression, modest effluxes permeabilize microbial membranes just enough to stimulate quorum sensing in beneficial *Bacillus* and *Streptomyces*. Drought or heat can exacerbate saponin exudation, leading to changes in the way carbon sources are used, which can repattern the soil microbiota.

8.7.1.4 Glucosinolates

Glucosinolates (GLSs), sulfur-rich thioglucosides that are retained in vacuoles and released either intact or as byproducts of enzymatic breakdown like isothiocyanates, are a characteristic of the Brassicaceae family. A structure–function relationship between GLS variety and microbiome assembly is demonstrated by *Arabidopsis* mutants with altered side-chain chemistry, which produce diverse GLS profiles and in turn attract different bacterial guilds with detoxification pumps or myrosinases. Hydrolyzed isothiocyanates function as strong biofumigants that reduce a variety of soil diseases and insects in addition to microbial filtering. They also release bound sulfur that microorganisms and plants can recycle (Unger *et al.*, 2024).

8.7.2 Ecological functions

The subsurface chemical arsenal of root exudates prevents or eliminates diseases and pests. Within hours following infection, phenolic esters such methyl-ferulate, coumarins, and glucosinolate hydrolysis products build up; they penetrate fungal membranes, prevent oomycete respiration, and simultaneously enhance disease-suppressive *Bacillus* and other antagonists in the rhizosphere. To illustrate how the same molecule delivers direct toxicity and attracts biocontrol allies, tobacco cultivars that oversecrete methyl-ferulate, for instance, reduce the severity of black-shank by more than 50% while boosting the abundance of beneficial microorganisms (Köllner *et al.*, 2008).

The exudate profile is likewise altered by herbivore attacks, whether they occur above or below ground. Feeding by foliar caterpillars or root-feeding beetle larvae increases the release of certain terpenes, flavonoids, and fatty acid derivatives that either attract the herbivore's natural enemies or repel it. When *Diabrotica* larvae injure maize roots, they release the sesquiterpene (E)-β-caryophyllene, which attracts entomopathogenic nematodes. In *Arabidopsis*, herbivory-altered exudates decrease arbuscular mycorrhizal signaling, which trades symbiosis for defense. Because of this adaptability, plants can integrate defense throughout the entire plant and adjust below-ground chemistry to herbivore pressure in real time.

Exudates serve as ecological gatekeepers that mold rhizosphere consortia by combining selective antimicrobials with labile carbohydrates and amino acids. Where hexose and glutamate leaks are strong, protobacteria and helpful *Pseudomonas* thrive, while phenolic-rich zones favor degraders that carry detoxifying pumps and suppress sensitive pathogens. Altering the sugar-to-phenolic ratio can reorder more than 30% of bacterial and fungal taxa in a matter of days, according to experimental swapping of exudate cocktails. Amino acid-enriched mixtures increase the activities of urease and protease, while citrate-rich exudates increase microbial phosphatases, which together speed up the mineralization of N and P. In this manner, plants maximize health and nutrient capture by feeding, editing, and mobilizing their microbial partners all at once.

8.8 Future Perspectives and Research Directions

Future perspectives and research directions in the study of plant root exudates highlight an exciting convergence of molecular insight, biotechnology, and ecological relevance. Expanding on recent advances, researchers are now prioritizing a deeper mechanistic understanding of how root exudates mediate plant–microbe communication and stress resilience, especially using high-throughput molecular techniques and nondestructive sampling approaches. Application-driven research is rapidly emerging, with targeted manipulation of exudate profiles through biotechnology aiming to boost nutrient acquisition, enhance soil health, and facilitate the recruitment of beneficial soil microorganisms. Particularly promising is the potential to leverage rhizosecretion for the bioproduction of pharmaceuticals and the use of exudate-guided phytoremediation to address environmental contamination. Yet notable knowledge gaps persist, such as limited databases that link diverse exudate compositions to crop species and agronomic outcomes, especially under varying conditions found in controlled environment agriculture.

Research is also pivoting toward real-world application, including the exogenous application of root exudate mimics to enhance soil restoration, stress tolerance, and yield in both field and controlled environments. As climate change continues to alter soil and atmospheric parameters, investigating how exudation patterns respond to elevated CO_2, drought, and other stresses is a frontier that will shape both basic science and practical innovation. Ultimately, advancing plant root exudate research will require a collaborative, interdisciplinary approach, linking fundamental biology to applied agricultural technology, to support resilient, sustainable food systems for the future.

8.9 Conclusion

Plant root exudates represent a highly dynamic interface between plants and their environment, mediating interactions that are central to both ecological balance and agricultural productivity. The diverse compound classes released – ranging from primary metabolites such as sugars, amino acids, and organic acids to more specialized secondary metabolites including flavonoids, phenolics, terpenes, and signaling peptides – play crucial roles in nutrient acquisition, soil microbial modulation, allelopathy, and stress adaptation. Together, these exudates shape the rhizosphere into a hotspot of biochemical communication and coevolution.

From an ecological perspective, root exudates sustain soil biodiversity, regulate energy flow, and contribute to the resilience of ecosystems. In agriculture, understanding and harnessing these compounds opens pathways for reducing chemical inputs, enhancing crop nutrition, improving resistance to pathogens, and optimizing beneficial plant–microbe interactions. Exudate-informed strategies can therefore become key drivers of sustainable intensification, adressing food security challenges while minimizing ecological trade-offs.

Looking forward, the transition from fundamental discoveries to practical applications will depend on integrating molecular biology, soil ecology, and agronomic practices. Advances in root phenotyping, metabolomics, and synthetic biology are already enabling more precise manipulation of exudate traits. Still, translating this knowledge into field-ready solutions requires interdisciplinary collaboration and consideration of context-specific variability in soils, climates, and microbial communities.

In essence, root exudate research sits at the crossroads of plant science, ecology, and agricultural innovation. By deepening our understanding of these chemical dialogues and strategically applying them in real-world systems, we can move closer to a future where crop productivity, environmental stewardship, and ecosystem health are mutually reinforcing rather than competing goals.

References

Afridi, M.S., Kumar, A., Javed, M.A., Dubey, A., de Medeiros, F.H.V. *et al.* (2024) Harnessing root exudates for plant microbiome engineering and stress resistance in plants. *Microbiological Research* 279, 127564. DOI: 10.1016/j.micres.2023.127564.

Akiyama, K., Matsuzaki, K. and Hayashi, H. (2005) Plant sesquiterpenes induce hyphal branching in arbuscular mycorrhizal fungi. *Nature* 435(7043), 824–827. DOI: 10.1038/nature03608.

Badri, D.V. and Vivanco, J.M. (2009) Regulation and function of root exudates. *Plant, Cell and Environment* 32(6), 666–681. DOI: 10.1111/j.1365-3040.2008.01926.x.

Bais, H.P., Weir, T.L., Perry, L.G., Gilroy, S. and Vivanco, J.M. (2006) The role of root exudates in rhizosphere interactions with plants and other organisms. *Annual Review of Plant Biology* 57, 233–266. DOI: 10.1146/annurev.arplant.57.032905.105159.

Barbour, W.M., Hattermann, D.R. and Stacey, G. (1991) Chemotaxis of *Bradyrhizobium japonicum* to soybean exudates. *Applied and Environmental Microbiology* 57(9), 2635–2639. DOI: 10.1128/aem.57.9.2635-2639.1991.

Canarini, A., Kaiser, C., Merchant, A., Richter, A. and Wanek, W. (2019) Root exudation of primary metabolites: Mechanisms and their roles in plant responses to environmental stimuli. *Frontiers in Plant Science* 10, 157. DOI: 10.3389/fpls.2019.00157.

Chen, L.-Q., Hou, B.-H., Lalonde, S., Takanaga, H., Hartung, M.L. *et al.* (2010) Sugar transporters for intercellular exchange and nutrition of pathogens. *Nature* 468(7323), 527–532. DOI: 10.1038/nature09606.

Chutia, R., Abel, S. and Ziegler, J. (2019) Iron and phosphate deficiency regulators concertedly control coumarin profiles in *Arabidopsis thaliana* roots during iron, phosphate, and combined deficiencies. *Frontiers in Plant Science* 10, 113. DOI: 10.3389/fpls.2019.00113.

Clúa, J., Roda, C., Zanetti, M.E. and Blanco, F.A. (2018) Compatibility between legumes and rhizobia for the establishment of a successful nitrogen-fixing symbiosis. *Genes* 9(3), 125. DOI: 10.3390/genes9030125.

Compant, S., Clément, C. and Sessitsch, A. (2010) Plant growth-promoting bacteria in the rhizo-and endosphere of plants: Their role, colonization, mechanisms involved and prospects for utilization. *Soil Biology and Biochemistry* 42(5), 669–678. DOI: 10.1016/j.soilbio.2009.11.024.

Cumming, G.S. (2016) Heterarchies: Reconciling networks and hierarchies. *Trends in Ecology and Evolution* 31(8), 622–632. DOI: 10.1016/j.tree.2016.04.009.

Dakora, F.D. and Phillips, D.A. (2002) Root exudates as mediators of mineral acquisition in low-nutrient environments. *Plant and Soil* 245(1), 35–47. DOI: 10.1023/A:1020809400075.

Eze, M.O. and Amuji, C.F. (2024) Elucidating the significant roles of root exudates in organic pollutant biotransformation within the rhizosphere. *Scientific Reports* 14(1), 2359. DOI: 10.1038/s41598-024-53027-x.

Fay, P.A. and Knapp, A.K. (1995) Stomatal and photosynthetic responses to shade in sorghum, soybean and eastern gamagrass. *Physiologia Plantarum* 94(4), 613–620. DOI: 10.1111/j.1399-3054.1995.tb00975.x.

Ghitti, E., Rolli, E., Vergani, L. and Borin, S. (2024) Flavonoids influence key rhizocompetence traits for early root colonization and PCB degradation potential of *Paraburkholderia xenovorans* LB400. *Frontiers in Plant Science* 15, 1325048. DOI: 10.3389/fpls.2024.1325048.

Godbold, D.L., Hoosbeek, M.R., Lukac, M., Cotrufo, M.F., Janssens, I.A. *et al.* (2006) Mycorrhizal hyphal turnover as a dominant process for carbon input into soil organic matter. *Plant and Soil* 281(1–2), 15–24. DOI: 10.1007/s11104-005-3701-6.

Hale, M.G. and Moore, L.D. (1980) Factors affecting root exudation Ii: 1970–1978. *Advances in Agronomy* 31, 93–124. DOI: 10.1016/S0065-2113(08)60137-6.

Hawes, M.C., Gunawardena, U., Miyasaka, S. and Zhao, X. (2000) The role of root border cells in plant defense. *Trends in Plant Science* 5(3), 128–133. DOI: 10.1016/s1360-1385(00)01556-9.
Ivarson, K.C. and Sowden, F.J. (1969) Free amino acid composition of the plant root environment under field conditions. *Canadian Journal of Soil Science* 49(1), 121–127. DOI: 10.4141/cjss69-014.
Jiao, H., Xu, W., Hu, Y., Tian, R. and Wang, Z. (2022) Citric acid in rice root exudates enhanced the colonization and plant growth-promoting ability of *Bacillus altitudinis* LZP02. *Microbiology Spectrum* 10(6), e0100222. DOI: 10.1128/spectrum.01002-22.
Jones, D.L., Hodge, A. and Kuzyakov, Y. (2004) Plant and mycorrhizal regulation of rhizodeposition. *New Phytologist* 163(3), 459–480. DOI: 10.1111/j.1469-8137.2004.01130.x.
Köllner, T.G., Held, M., Lenk, C., Hiltpold, I., Turlings, T.C.J. *et al.* (2008) A maize (e)-beta-caryophyllene synthase implicated in indirect defense responses against herbivores is not expressed in most American maize varieties. *Plant Cell* 20(2), 482–494. DOI: 10.1105/tpc.107.051672.
Lee, J.H., Anderson, A.J. and Kim, Y.C. (2022) Root-associated bacteria are biocontrol agents for multiple plant pests. *Microorganisms* 10(5), 1053. DOI: 10.3390/microorganisms10051053.
Li, X., Zhang, T., Wang, X., Hua, K., Zhao, L. *et al.* (2013) The composition of root exudates from two different resistant peanut cultivars and their effects on the growth of soil-borne pathogen. *International Journal of Biological Sciences* 9(2), 164–173. DOI: 10.7150/ijbs.5579.
Lipton, D.S., Blanchar, R.W. and Blevins, D.G. (1987) Citrate, malate, and succinate concentration in exudates from p-sufficient and p-stressed *Medicago sativa* L. seedlings. *Plant Physiology* 85(2), 315–317. DOI: 10.1104/pp.85.2.315.
Lopes, L.D., Wang, P., Futrell, S.L. and Schachtman, D.P. (2022) Sugars and jasmonic acid concentration in root exudates affect maize rhizosphere bacterial communities. *Applied and Environmental Microbiology* 88(18), e0097122. DOI: 10.1128/aem.00971-22.
Ma, W., Tang, S., Dengzeng, Z., Zhang, D., Zhang, T. *et al.* (2022) Root exudates contribute to belowground ecosystem hotspots: A review. *Frontiers in Microbiology* 13, 937940. DOI: 10.3389/fmicb.2022.937940.
Mandal, S.M., Chakraborty, D. and Dey, S. (2010) Phenolic acids act as signaling molecules in plant-microbe symbioses. *Plant Signaling and Behavior* 5(4), 359–368. DOI: 10.4161/psb.5.4.10871.
Neal, A.L., Ahmad, S., Gordon-Weeks, R. and Ton, J. (2012) Benzoxazinoids in root exudates of maize attract *Pseudomonas putida* to the rhizosphere. *PLOS ONE* 7(4), e35498. DOI: 10.1371/journal.pone.0035498.
Nguyen, C. (2009) Rhizodeposition of organic C by plant: Mechanisms and controls. *Sustainable Agriculture* 97–123. DOI: 10.1007/978-90-481-2666-8_9.
Rasmann, S. and Hiltpold, I. (2022) Root exudation of specialized molecules for plant-environment interaction. *Chimia* 76(11), 922–927. DOI: 10.2533/chimia.2022.922.
Ryan, P.R., Delhaize, E. and Jones, D.L. (2001) Function and mechanism of organic anion exudation from plant roots. *Annual Review of Plant Biology* 52, 527–560. DOI: 10.1146/annurev.arplant.52.1.527.
Stassen, M.J.J., Hsu, S.-H., Pieterse, C.M.J. and Stringlis, I.A. (2021) Coumarin communication along the microbiome-root-shoot axis. *Trends in Plant Science* 26(2), 169–183. DOI: 10.1016/j.tplants.2020.09.008.
Stringlis, I.A., Yu, K., Feussner, K., de Jonge, R., Van Bentum, S. *et al.* (2018) MYB72-dependent coumarin exudation shapes root microbiome assembly to promote plant health. *Proceedings of the National Academy of Sciences USA* 115(22), E5213–E5222. DOI: 10.1073/pnas.1722335115.

Tan, B., Li, Y., Liu, T., Tan, X., He, Y. *et al.* (2021) Response of plant rhizosphere microenvironment to water management in soil-and substrate-based controlled environment agriculture (CEA) systems: A review. *Frontiers in Plant Science* 12, 691651. DOI: 10.3389/fpls.2021.691651.

Taylor, L.P. and Grotewold, E. (2005) Flavonoids as developmental regulators. *Current Opinion in Plant Biology* 8(3), 317–323. DOI: 10.1016/j.pbi.2005.03.005.

Unger, K., Raza, S.A.K., Mayer, T., Reichelt, M., Stuttmann, J. *et al.* (2024) Glucosinolate structural diversity shapes recruitment of a metabolic network of leaf-associated bacteria. *Nature Communications* 15(1), 8496. DOI: 10.1038/s41467-024-52679-7.

Walker, T.S., Bais, H.P., Grotewold, E. and Vivanco, J.M. (2003) Root exudation and rhizosphere biology. *Plant Physiology* 132(1), 44–51. DOI: 10.1104/pp.102.019661.

Wang, Y.-L., Almvik, M., Clarke, N., Eich-Greatorex, S., Øgaard, A.F. *et al.* (2015) Contrasting responses of root morphology and root-exuded organic acids to low phosphorus availability in three important food crops with divergent root traits. *AoB Plants* 7, plv097. DOI: 10.1093/aobpla/plv097.

Wang, Z., Liu, Y., Cui, W., Gong, L., He, Y. *et al.* (2022) Characterization of GmMATE 13 in its contribution of citrate efflux and aluminum resistance in soybeans. *Frontiers in Plant Science* 13, 1027560. DOI: 10.3389/fpls.2022.1027560.

Watt, M. (2009) The rhizosphere: Biochemistry and organic substances at the soil–plant interface. *Annals of Botany* 104(4), ix–x. DOI: 10.1093/aob/mcp166.

Wen, T., Yu, G.-H., Hong, W.-D., Yuan, J., Niu, G.-Q. *et al.* (2022) Root exudate chemistry affects soil carbon mobilization via microbial community reassembly. *Fundamental Research* 2(5), 697–707. DOI: 10.1016/j.fmre.2021.12.016.

Xiang, L., Chen, X.-T., Yu, P.-F., Li, X.-H., Zhao, H.-M. *et al.* (2020) Oxalic acid in root exudates enhances accumulation of perfluorooctanoic acid in lettuce. *Environmental Science and Technology* 54(20), 13046–13055. DOI: 10.1021/acs.est.0c04124.

Yu, B., Sun, Y., Jin, X., Xie, Z., Li, X. *et al.* (2023) Rice glutamate receptor-like channel OsGLR3.4 modulates the root tropism growth towards amino acids via plasma membrane depolarization and ROS generation. *Environmental and Experimental Botany* 205, 105146. DOI: 10.1016/j.envexpbot.2022.105146.

Zhao, Y.-M., Cheng, Y.-X., Ma, Y.-N., Chen, C.-J., Xu, F.-R. *et al.* (2018) Role of phenolic acids from the rhizosphere soils of *Panax notoginseng* as a double-edge sword in the occurrence of root-rot disease. *Molecules* 23(4), 819. DOI: 10.3390/molecules23040819.

Zheng, L., Zhao, S., Zhou, Y., Yang, G., Chen, A. *et al.* (2024) The soybean sugar transporter GmSWEET6 participates in sucrose transport towards fungi during arbuscular mycorrhizal symbiosis. *Plant, Cell and Environment* 47(4), 1041–1052. DOI: 10.1111/pce.14772.

9 Comparative Analysis of the Microbial Diversity and Abundance in the Rhizosphere of Medicinal and Nonmedicinal Plants

PRADEEP KUMAR[1], KARISHMA KUMARI[2], ANKITA YADAV[3], RAJESH DANGAL[1], SAKSHI ANAND[4], ZAIN-UL ABIDEEN USMANI[1], SAURABH GANGOLA[5], SEYIT UGUZ[1,6], SHALINI TIWARI[1] AND XUFEI YANG[1,7]*

[1]*Department of Agricultural and Biosystems Engineering, South Dakota State University, Brookings, USA*
[2]*Department of Agronomy, Horticulture and Plant Science, South Dakota State University, Brookings, USA*
[3]*Dr B.R. Ambedkar Center for Biomedical Research, University of Delhi, Delhi, India*
[4]*Department of Biotechnology, Jaypee Institute of Information Technology, Noida, India*
[5]*Department of Microbiology, Graphic Era University, Dehradun, India*
[6]*Biosystems Engineering, Faculty of Agriculture, Uludag University, Bursa, Turkey*
[7]*Department of Biosystems and Agricultural Engineering and Animal Science, Michigan State University, East Lansing, USA*

9.1 Introduction

The rhizosphere is the narrow zone of soil under the direct influence of root exudates and rhizospheric microorganisms (Mendes *et al.*, 2013). Despite being only a few millimeters deep, it represents one of the most dynamic plant–microenvironmental interfaces. This biologically active region supports a high level of microbial activity and diversity, fostering a complex

*Corresponding author: yangxuf1@msu.edu

DOI: 10.1079/9781836990475.0009

web of interactions among bacteria, fungi, archaea, protozoa, and viruses. These interactions have profound implications for plant health, development, nutrient cycling, and resistance to both biotic and abiotic stresses. Therefore, rhizospheric microbial richness and diversity are essential not only for agricultural productivity and ecological sustainability but also for understanding the mechanisms behind plant adaptation, resilience, and secondary metabolism (Li *et al.*, 2024).

The term "rhizosphere" was first introduced by Lorenz Hiltner in 1904, referring to a zone of intense biological and chemical interactions around the plant root (Hiltner, 1904). In the light of modern molecular biology and high-throughput sequencing technologies, this concept has significantly expanded. The rhizosphere is now recognized as one of the most metabolically active zones of microbial colonization in terrestrial ecosystems. Root exudates, including sugars, amino acids, organic acids, fatty acids, and secondary metabolites, serve as nutrient sources and chemical signals that actively shape the structure and function of rhizospheric microbial communities (Canarini *et al.*, 2019). In return, microbes modulate plant physiology through hormone production, enhanced nutrient solubilization, and resistance to pathogens. This bidirectional interaction, often described as a *chemical dialogue*, is crucial for regulating plant–microbe relationships.

The microbial communities in the rhizosphere differ significantly from those in bulk soil, both compositionally and functionally. Their composition is highly responsive to various factors, including plant genotype, growth stage, soil type, climatic conditions, and anthropogenic inputs such as fertilizers and pesticides. The resulting rhizobiome is a complex, coevolved ecological system where symbiosis and competition coexist, contributing to both plant fitness and soil health.

Rhizospheric microbes play pivotal roles in ecosystem functions. They fix atmospheric nitrogen, solubilize phosphorus, mobilize potassium, and produce siderophores that scavenge iron, a nutrient often limited in soil environments. Furthermore, several plant growth-promoting rhizobacteria (PGPR) produce phytohormones such as indole-3-acetic acid (IAA), gibberellins, and cytokinins, which stimulate plant growth and development. Microbial antagonists in the rhizosphere also contribute to disease suppression by competing with pathogens, producing antimicrobial compounds, or inducing systemic resistance (ISR) in host plants. These beneficial associations are fundamental for achieving sustainable agriculture, as they reduce dependence on synthetic chemicals and improve plant resilience to environmental stressors (Thepbandit and Athinuwat, 2024).

Beyond agriculture, rhizospheric microorganisms are increasingly recognized for their roles in biogeochemical cycling, pollutant degradation, and carbon sequestration. The symbiotic relationship between mycorrhizal fungi and plant roots, for instance, enhances nutrient acquisition and improves soil structure and fertility.

In the case of medicinal plants, the rhizospheric microbiota may also play a critical role in the biosynthesis, accumulation, and regulation of bioactive phytochemicals. Recent studies indicate that microbial endophytes and rhizosphere-associated microbes can alter the metabolic pathways of medicinal plants, thereby influencing the yield and pharmacological efficacy of therapeutic compounds (Hnamte *et al.*, 2024). Medicinal plants, widely used in both traditional and modern medicine, produce a diverse range of secondary metabolites such as alkaloids, flavonoids, terpenoids, and phenolics. These compounds not only contribute to the plant's medicinal value but also influence the composition of rhizospheric microbial communities by exhibiting selective antimicrobial activity, altering nutrient exudation patterns, and participating in complex chemical communication.

In contrast, nonmedicinal plants, those cultivated primarily for food, fodder, fibre, or ornamental purposes, typically do not produce the same spectrum or concentration of secondary metabolites in their root exudates. This fundamental difference in chemical output can significantly affect the microbial community structure in the rhizosphere. For example, the presence of antimicrobial compounds in medicinal plant root exudates may suppress certain microbial taxa while favoring others that are tolerant or functionally beneficial (Castronovo *et al.*, 2021).

These distinctions raise important ecological and functional questions regarding the microbial dynamics in plant rhizospheres. For instance, do medicinal plants harbour more specialized or functionally diverse microbial assemblages compared to nonmedicinal counterparts? This leads to further inquiry into whether such assemblages are more efficient in promoting plant growth or enhancing the biosynthesis of secondary metabolites, which are often critical for the therapeutic properties of medicinal plants. Additionally, comparing the rhizosphere microbial communities of medicinal and non-medicinal plants in terms of richness, diversity, and functional capabilities may reveal unique interactions and adaptations shaped by plant chemistry and ecological roles. Understanding these differences is essential for leveraging microbiome functions in sustainable agriculture and natural product development.

Understanding these differences is crucial for enhancing plant growth, improving herbal drug quality, and uncovering novel microbial taxa with biotechnological potential. Such insights can guide the development of microbiome-based approaches for improving crop health and increasing the yield of pharmacologically valuable compounds.

This chapter aims to provide a comparative analysis of microbial diversity and abundance in the rhizospheres of medicinal and nonmedicinal plants (Table 9.1). The primary goal is to characterize the microbial communities, particularly bacteria, fungi, and other microbial taxa, present in the rhizospheres of selected plant species. This will be achieved using both high-throughput sequencing technologies and traditional culture-based methods to allow comprehensive profiling of microbial taxa. Diversity indices such as species richness, evenness, and Shannon–Wiener diversity will be evaluated to identify patterns and differences in microbial communities across plant types.

An essential part of this study is to understand how plant type influences microbial community structure, including the identification of dominant

Table 9.1. Comparative characteristics of rhizosphere microbial communities in medicinal and nonmedicinal plants.

Parameter	Medicinal plants	Nonmedicinal plants
Root exudate composition	High in secondary metabolites (alkaloids, flavonoids, terpenoids)	Rich in primary metabolites (sugars, amino acids)
Dominant microbial phyla	Actinobacteria, Proteobacteria, Firmicutes	Proteobacteria, Bacteroidetes, Firmicutes
Common bacterial genera	*Bacillus*, *Streptomyces*, *Pseudomonas*	*Rhizobium*, *Azotobacter*, *Flavobacterium*
Fungal associations	Arbuscular mycorrhizal fungi (AMF), *Trichoderma*, *Aspergillus*	AMF, *Fusarium*, *Penicillium*
Functional specialization	Antimicrobial production, IAA synthesis, xenobiotic degradation	Nitrogen fixation, phosphate solubilization, pathogen suppression
Microbial diversity (Shannon Index)	Moderate to high	High
Sensitivity to root exudates	High (selective enrichment based on secondary metabolites)	Moderate (broader tolerance)

IAA, indole-3-acetic acid.

taxa, keystone species, and functionally important microbial guilds. We will also assess the functional potential of rhizospheric microbiota through metagenomic, enzymatic, and biochemical analyses, focusing on traits related to plant growth promotion and secondary metabolite production, especially those relevant to medicinal plants.

Additionally, the chapter will explore the role of root exudates and phytochemicals in shaping rhizosphere microbial communities. By examining how these compounds influence microbial recruitment, survival, and activity, we aim to elucidate potential feedback mechanisms through which microbes in turn affect plant metabolism, health, and secondary metabolite biosynthesis.

Overall, this chapter seeks to highlight the ecological, agricultural, and biotechnological significance of rhizosphere microbiomes in medicinal and nonmedicinal plants. The knowledge gained from this comparative study can be applied to promote sustainable agriculture, optimize phytochemical yields, and harness microbial resources for future innovations in plant-based therapeutics and microbiome engineering.

9.2 Comparative Microbiology of the Rhizosphere of Medicinal and Nonmedicinal Plants

Medicinal plants are renowned for their ability to produce a wide array of bioactive secondary metabolites, including alkaloids, terpenoids, flavonoids, phenolics, and glycosides. These compounds play critical roles in plant defense and ecological communication. Many of these metabolites are exuded into the rhizosphere through root secretions, where they function as selective agents, either serving as substrates for specific microbial groups or exerting antimicrobial effects that suppress others. This biochemical selectivity fosters the enrichment of specialized microbial communities that are adapted to tolerate, transform, or degrade these complex compounds (Qu *et al.*, 2024).

In contrast, nonmedicinal plants such as cereals, legumes, and vegetables predominantly release primary metabolites, including simple sugars, amino acids, and organic acids, which serve as readily available energy sources for a broad spectrum of microbial taxa. As a result, while their rhizospheres may host a more taxonomically diverse microbial population, these communities are generally focused on functions related to nutrient acquisition, growth enhancement, and soil health improvement.

The rhizospheres of medicinal plants commonly support specialist microbial taxa capable of metabolizing secondary metabolites or surviving in chemically complex soil environments. Dominant bacterial genera include *Bacillus*, *Streptomyces*, and *Pseudomonas*, which not only withstand plant-derived antimicrobial compounds but also promote plant growth by producing bioactive substances such as IAA, siderophores, antibiotics, and ACC deaminase. For example, *Withania somnifera* (ashwagandha) is associated with *Streptomyces* and *Bacillus* species that enhance plant vigour and suppress root pathogens. Similarly, the rhizosphere of *Ocimum sanctum* (holy basil) supports beneficial fungi such as *Trichoderma* and arbuscular mycorrhizal fungi (AMF) that contribute to nutrient uptake and biotic stress tolerance (Pantigoso *et al.*, 2022).

Fungal communities also play vital roles in medicinal plant rhizospheres. AMF enhance phosphorus uptake, improve water absorption, and contribute to the biosynthesis of secondary metabolites. Other fungi such as *Aspergillus* and *Trichoderma* enrich functional diversity by aiding plant immunity, organic matter decomposition, and metabolite transformation. Overall, the microbial diversity in medicinal plant rhizospheres is moderate to high and influenced by factors such as plant species, exudate profile, soil type, and environmental conditions. These microbial assemblages not only contribute to plant growth and defense but also represent a reservoir of novel enzymes and bioactive compounds with pharmaceutical and industrial potential.

Although nonmedicinal plants may lack the chemical complexity of medicinal plants, they also maintain diverse and ecologically significant rhizosphere microbiomes. Leguminous crops such as soybean and chickpea form symbiotic relationships with nitrogen-fixing bacteria such as *Rhizobium* and *Bradyrhizobium*, which convert atmospheric nitrogen into bioavailable forms, reducing dependence on chemical fertilizers. Cereal crops including maize (*Zea mays*), rice (*Oryza sativa*), and wheat (*Triticum aestivum*) host phosphate-solubilizing bacteria such as *Pseudomonas* and *Bacillus*, and beneficial fungi that contribute to nutrient cycling and pathogen suppression (Luo *et al.*, 2025).

Vegetable crops such as tomato (*Solanum lycopersicum*) and cabbage (*Brassica oleracea*) also support generalist microbial taxa such as *Flavobacterium* and *Trichoderma*, which contribute to disease resistance, plant hormone modulation, and root development. While these communities are not necessarily specialized in secondary metabolite transformation, they are functionally diverse and play essential roles in maintaining soil health, enhancing crop yield, and buffering environmental stress. AMF are also frequently associated with nonmedicinal plants, particularly in nutrient-deficient or degraded soils, where they assist in the uptake of phosphorus, zinc, and copper.

Despite differences in the chemical nature of root exudates, both medicinal and nonmedicinal plants rely on their rhizosphere microbiota for optimal growth and development. However, the functional priorities of their microbial associations differ. Medicinal plants tend to recruit microbes capable of surviving chemically challenging environments and metabolizing bioactive compounds, often resulting in specialized consortia with unique enzymatic and biosynthetic potential. These consortia may also play roles in modulating secondary metabolite profiles, which have implications for pharmaceutical applications.

Conversely, nonmedicinal plants primarily associate microbes that enhance nutrient availability, improve soil structure, and defend against pathogens. These microbes often thrive on simpler carbon compounds and support high biomass production and yield, making them critical for agroecosystem sustainability.

A direct comparison reveals clear distinctions in rhizosphere microbial assembly between the two plant groups. Medicinal plant rhizospheres are enriched with Actinobacteria, *Bacillus*, and *Streptomyces*, known for their ability to detoxify complex secondary metabolites and produce biocontrol agents (Table 9.2). Fungal genera such as AMF, *Trichoderma*, and *Aspergillus* further support plant immunity and metabolite regulation. In contrast, nonmedicinal plant rhizospheres are often dominated by *Rhizobium*, *Azospirillum*, *Pseudomonas*, and *Flavobacterium*, which promote plant growth through nutrient solubilization, nitrogen fixation, and hormone production. These contrasting microbial profiles reflect the differing ecological niches and metabolic outputs of medicinal versus nonmedicinal plants. Understanding these microbial associations not only enriches our knowledge of plant–microbe interactions but also opens up new possibilities for sustainable agriculture, soil microbiome engineering, and natural product discovery.

9.3 Factors Influencing Microbial Diversity in the Rhizosphere

The microbial diversity of the rhizosphere is shaped by a complex interplay of plant-intrinsic factors and extrinsic environmental variables (Table 9.3). These factors collectively govern the structure, function, and dynamic equilibrium of microbial communities surrounding plant roots, ultimately influencing plant growth, productivity, and resistance to stress.

Table 9.2. Major microbes and their host medicinal plants.

Microorganism	Host plant
Actinosynnema pretiosum	*Putterlickia verrucosa*
Alternaria	*Podophyllum hexandrum*, *Ginkgo biloba*
Aspergillus fumigatus	*Juniperus communis*
Botryosphaeria rhodina	*Bidens pilosa*
Chaetomium sp.	*Salvia officinalis*
Colletotrichum gloeosporioides	*Artemisia mongolica*
Colletotrichum sp.	*Artemisia annua*
Cryptosporiopsis quercina	*Tripterigeum wilfordii*
Entrophospora infrequens	*Nothapodytes foetida*
Fusarium proliferatum	*Dysoxylum binectariferum*
Fusarium subglutinans	*Tripterigeum wilfordii*
Pestalotiopsis microspora	*Terminalia morobensis*
Phomopsis sp.	*Erythrina crista-galli*
Serratia marcescens	*Rhyncholacis penicillata*
Streptomyces sp.	*Kennedia nigriscans*, *Grevillea pteridifolia*, *Monstera* sp.
Taxomyces andreanae	*Taxus brevifolia*

9.3.1 Root exudates and plant genotype

One of the most crucial plant-intrinsic factors is the composition of root exudates, which includes a wide array of low and high molecular weight compounds such as sugars, amino acids, organic acids, fatty acids, phenolics, flavonoids, alkaloids, and secondary metabolites. These exudates serve as chemical signals and energy sources for rhizospheric microbes, significantly shaping microbial composition and activity. Medicinal plants, known for secreting bioactive secondary metabolites like terpenoids, alkaloids, and polyphenols, tend to selectively enrich microbial taxa with specialized metabolic capabilities (Rizwan and Gothalwal, 2021). For instance, the presence of antimicrobial exudates in medicinal plants like *Ocimum sanctum* or *Azadirachta indica* has been shown to recruit beneficial microbes such as *Pseudomonas*, *Bacillus*, and *Trichoderma* that tolerate or degrade these compounds while suppressing pathogenic organisms (Canarini *et al.*, 2019).

Additionally, plant genotype influences microbial diversity through root morphology, exudation patterns, and immune responses. Genotype-specific selection has been observed where genetically distinct cultivars of the same species recruit different microbial communities. This selective pressure contributes to the formation of a "core microbiome" adapted to a particular host genotype, affecting nutrient uptake efficiency, resistance to pathogens, and stress adaptation. For example, studies on *Arabidopsis thaliana* and rice cultivars demonstrate genotype-driven differentiation in rhizobacterial and fungal community structure. Therefore, understanding the role of genotype and exudation in microbial assembly is essential for leveraging host–microbe interactions in agriculture and biotechnology (Pantigoso *et al.*, 2022).

Table 9.3. Factors influencing the rhizosphere of microbial communities.

Overarching category	Specific factor	Key influences and mechanisms	Sources
Plant-specific factors	Root exudates and signaling	Plants emit sugars, amino acids, and secondary metabolites to nourish and selectively attract helpful microbes. This is a sophisticated way for plants to communicate chemically and actively manage their relationships with beneficial organisms	(Kovalchuk *et al.*, 2003; Bais *et al.*, 2006; Neu *et al.*, 2021)
	Secondary metabolites	Compounds found in medicinal plants often possess antimicrobial properties that can help shape the microbiome by suppressing harmful pathogens. Interestingly, microbes themselves can influence how these plants produce such bioactive compounds	(Pang *et al.*, 2021; Yuan *et al.*, 2022)
Soil and environment	Soil properties	pH is a key factor that helps microbial groups to thrive in a given environment. Moisture levels play a important role by affecting microbial activity and favoring species that can tolerate dry conditions. In addition, the presence of organic matter and nutrients provides the essential resources needed to support microbial growth	(Feng and Deng, 2025)
	Climate and season	Temperature plays an important role in the metabolic activity of both plants and microbes. Seasonal changes lead to noticeable shifts in microbial community structures throughout the year, reflecting the energetic nature of these ecosystems	(Waldrop and Firestone, 2006; Lahlali *et al.*, 2021)
Agronomic practices	Farming systems	Conventional: Synthetic fertilizers and pesticides can harm beneficial soil microbes and reduce microbial diversity Organic: Applying compost or manure supports a richer microbial community and promotes healthier soil overall	(Tuck *et al.*, 2014; Windisch *et al.*, 2021)
	Cropping systems	Monoculture: Planting the same crop repeatedly tends to create a less diverse microbiome that is more vulnerable to pests and diseases because the root exudates are all similar Polyculture or crop rotation: Mixing different crops or rotating them helps increase microbial diversity and builds a more resilient soil ecosystem	(Venter *et al.*, 2016)

9.3.2 Soil properties

Soil characteristics, including pH, texture, organic matter content, moisture, nutrient availability, and redox potential, play fundamental roles in determining rhizosphere microbial diversity. Soil pH is particularly influential; bacterial diversity typically peaks in neutral to slightly alkaline soils, while fungi exhibit greater resilience to pH fluctuations. Organic matter serves as a carbon reservoir that supports microbial biomass and activity. Clay-rich soils tend to harbour more bacteria due to enhanced nutrient retention and surface area, whereas sandy soils promote fast-growing, copiotrophic microbes due to lower water-holding capacity (Naylor *et al.*, 2022).

The nutrient profile of the soil also drives microbial community shifts. In nitrogen-deficient soils, diazotrophs like *Azospirillum*, *Rhizobium*, and *Bradyrhizobium* dominate the rhizosphere of legumes and nonlegumes alike. Similarly, phosphorus availability influences the abundance of phosphate-solubilizing bacteria such as *Pseudomonas* and *Bacillus*. The physical structure of the soil affects oxygen diffusion and microbial habitat availability, creating microniches that foster diverse microbial assemblages. The activities of soil enzymes, such as dehydrogenase, phosphatase, and urease, also indirectly reflect microbial functionality and diversity.

9.3.3 Climatic and environmental conditions

Climatic factors, including temperature, precipitation, humidity, and light intensity, exert a significant influence on microbial dynamics. Seasonal changes can shift microbial community composition by altering root growth rates, exudation profiles, and microbial metabolic activity. For example, during warmer seasons, elevated root exudation rates enhance microbial colonization, while water stress during dry periods may favor drought-tolerant taxa like Actinobacteria or spore-forming *Bacillus* spp. (Schnecker *et al.*, 2023).

Agricultural management practices such as crop rotation, tillage, organic amendments, irrigation, and pesticide application also impact rhizosphere microbiota. Organic farming, involving compost and manure application, promotes microbial richness and activity by increasing organic carbon availability. In contrast, excessive chemical fertilizer or pesticide use can suppress beneficial microbes, reduce functional diversity, and increase susceptibility to pathogens.

Geographical location and altitude introduce variability through their effects on temperature gradients, UV exposure, and atmospheric pressure. Studies have reported unique microbial signatures in high-altitude medicinal plant rhizospheres, often marked by psychrotrophic and UV-resistant bacteria. For example, plants like *Rheum australe* and *Picrorhiza kurroa*, growing at elevations above 3000 m, exhibit rhizospheres rich in cold-adapted genera such as *Pseudomonas*, *Micrococcus*, and *Paenibacillus* (Egamberdieva *et al.*, 2015).

9.3.4 Biotic interactions

The rhizosphere is a hotspot of biotic interactions, where microbial populations interact through competition, cooperation, and signaling. Quorum sensing, biofilm formation, allelopathy, and horizontal gene transfer are key mechanisms by which microbes influence each other's abundance and function. Beneficial microbes may outcompete pathogens through niche exclusion, production of antibiotics or siderophores, and induction of systemic resistance in the host plant.

In some cases, plants mediate these microbial interactions by modulating exudate composition in response to stress or microbial colonization. For instance, the presence of mycorrhizal fungi can induce the secretion of flavonoids or strigolactones, which attract specific bacterial partners that further enhance plant growth or immunity. Such plant-mediated feedback loops reinforce microbial community structure and functional resilience (Egamberdieva *et al.*, 2015).

9.3.5 Anthropogenic and land use factors

Human activities, such as urbanization, industrialization, and land use change, profoundly affect rhizosphere microbial diversity of the rhizosphere. Soil compaction, pollution, heavy metal accumulation, and habitat fragmentation reduce microbial richness and shift communities toward stress-tolerant species. Medicinal plants cultivated in contaminated or disturbed soils often show altered microbial composition, with reductions in plant growth-promoting taxa and increases in metal-resistant or opportunistic pathogens.

Conversely, ecological restoration and sustainable land management can improve microbial diversity. Agroforestry systems, intercropping, and rewilding of degraded lands enhance microbial functional redundancy and nutrient cycling. Therefore, human intervention can either compromise or enrich the rhizosphere biodiversity, depending on the intensity and nature of land use practices.

9.4 Techniques for Assessing Rhizosphere Microbial Diversity and Abundance

A variety of classic (culture-based) and molecular (culture-independent) methods are available for analyzing the diversity and abundance of microbial communities in the rhizosphere. Classic microbiological techniques typically involve cultivating microorganisms on nutrient media to isolate and identify specific taxa. However, it is estimated that less than 2% of soil microbes, including those in the rhizosphere, can be cultured using standard laboratory methods (Wade, 2002). This limitation has led to the increasing application of molecular biology approaches for comprehensive and culture-independent characterization of rhizosphere microbiota.

Depending on the target biomolecules, nucleic acids or proteins, either genomics- or proteomics-based tools are employed. Earlier genomics-based techniques, such as denaturing gradient gel electrophoresis (DGGE) and fluorescence *in situ* hybridization (FISH), were instrumental in providing insights into microbial community structure. On the proteomics front, matrix-assisted laser desorption/ionization time-of-flight mass spectrometry (MALDI-TOF MS) has been used for protein-based microbial identification.

Gene sequencing techniques are now widely utilized to explore the phylogenetic relationships and taxonomic composition of rhizosphere microbial communities. Among these, metagenomics, particularly next-generation sequencing (NGS), has become a cornerstone for studying uncultured microbial populations directly from root-associated soil. Amplicon-based NGS approaches targeting the 16S rRNA gene are commonly used to assess bacterial and archaeal diversity, whereas fungal communities are often profiled using 18S rRNA and internal transcribed spacer (ITS) regions.

For instance, studies using Illumina MiSeq® or HiSeq® platforms have revealed diverse microbial taxa associated with plant rhizospheres under different environmental and agronomic conditions. Beyond amplicon sequencing, shotgun metagenomic sequencing allows for simultaneous detection and functional annotation of bacteria, fungi, archaea, and viruses present in rhizospheric soil, offering a more holistic view of microbial ecology. However, its application remains relatively limited due to cost and data complexity.

In addition to genetic and proteomic analyses, the use of biochemical markers or biomarkers such as phospholipid fatty acids (PLFAs), extracellular enzymes, and microbial respiration indicators provides functional insights into rhizosphere microbial activity and biomass. Recent reviews, including Wang *et al.* (2022), have comprehensively summarized the molecular and biochemical techniques utilized to investigate the composition and functional potential of rhizosphere microbial communities.

9.4.1 Bioinformatics and statistical approaches

To unravel the taxonomic and functional complexity of rhizospheric microbes, high-throughput sequencing and bioinformatic analyses are indispensable. Advanced multi-omics approaches including metagenomics, metatranscriptomics, metaproteomics, and metabolomics enable comprehensive profiling of microbial diversity, gene expression patterns, protein activity, and metabolic interactions within the rhizosphere (Priya *et al.*, 2021). Each omics layer provides a unique perspective, from identifying microbial taxa to elucidating molecular mechanisms that mediate plant–microbe and microbe–microbe interactions. A variety of bioinformatics tools are employed to analyze these omics datasets. For instance, MEGAHIT is used for metagenomic assembly, DESeq2 for transcriptomic differential expression analysis, MaxQuant for proteomic data processing, and MetaboAnalyst

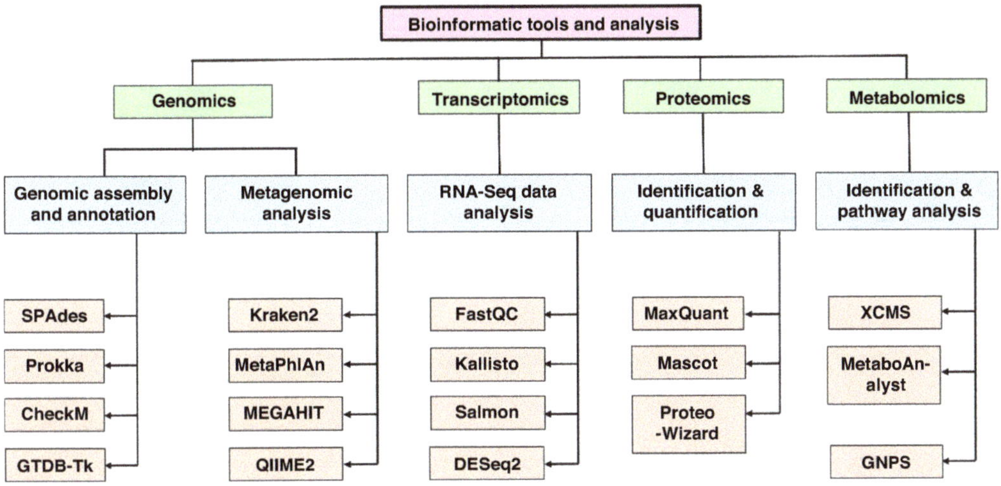

Fig. 9.1. Overview, approaches, and major bioinformatics tools for studying rhizospheric microbes.

for metabolomics data interpretation. Integrative platforms such as QIIME 2 facilitate microbial community profiling, while Cytoscape supports network-based visualization of functional interactions, aiding in the construction of a systems-level understanding of rhizosphere microbiomes. An overview of these bioinformatics tools and their applications in rhizosphere microbiome research is illustrated in Fig. 9.1.

9.5 Role of Artificial Intelligence, Machine Learning, and Modern Sensors in Studying the Rhizosphere Microbiome

Artificial intelligence (AI), machine learning (ML), and novel sensors have become vital for studying the rhizosphere microbiome of medicinal and nonmedicinal plants by accurately identifying, characterizing, and comparing microbial abundance and diversity. AI and machine learning can help with the study of large-scale metagenomic data, gene annotation, predictive modelling of plant traits (such as disease resistance and stress tolerance), and network analysis of plant–microbe interactions. Current sensors provide real-time monitoring of soil, plant health, and microbial activity, all of which contribute to precision agriculture through improved fertilization and irrigation, as well as spatial evaluation via remote sensing (Kumari *et al.*, 2025a, b). Collectively, these technologies contribute to the specification of how environmental conditions and plant species influence rhizosphere microbial communities, thus enhancing knowledge of sustainable agriculture and plant health, as stressed by the International Phytobiome Alliance.

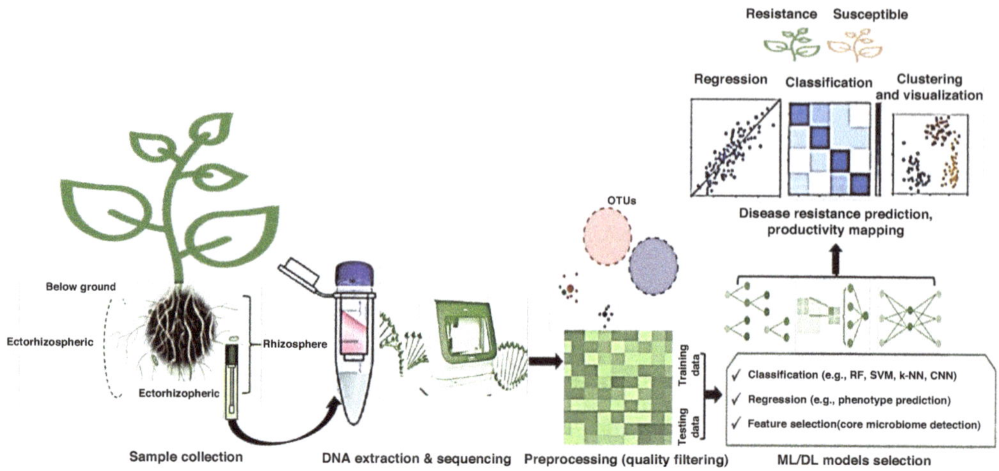

Fig. 9.2. Pipeline of plant microbiome analysis using ML and DL approaches.

9.5.1 Machine learning for microbiome analysis

Plant development, health, and stress tolerance are influenced by a wide variety of bacteria, archaea, fungi, and viruses known as the plant-associated microbiota, rhizosphere, phyllosphere, and endophytes (Durán *et al.*, 2018). However, the plants use genetic features and root exudates to control these populations (Xiong *et al.*, 2020). Through microbiome-based engineering, this interaction could be used to increase the plant's resistance and productivity (Toju *et al.*, 2018). NGS technologies enable functional annotation and microbial community characterization through techniques such as metagenomic shotgun sequencing (Sharpton, 2014) and 16S rRNA amplicon sequencing. Shotgun metagenomics can provide more functional information, although amplicon sequencing aids in taxonomic classification. Combined, they enhance functional interpretation and species-level resolution (Zhang *et al.*, 2016).

According to Wang and Jia (2016), ML methods including random forest (RF), support vector machines, and k-NN are utilized to handle the complexity of microbiome data for phenotype prediction, community comparison, and microbial function inference (Fig. 9.2). Machine learning-based techniques, including RF, have been used to investigate the influence of the microbiome on plant development in response to technical demand (Chang *et al.*, 2017). According to LaPierre *et al.* (2019), deep learning algorithms such as MetaPheno provide sophisticated pattern recognition and feature extraction. Individual microbial taxa are also linked to plant attributes using multiple hypothesis corrections and statistical testing in metagenome-wide association studies (MWAS) (Xu *et al.*, 2019).

Machine learning has enabled efficient analysis of microbiome data by classifying microbial taxa from 16S rRNA sequences, predicting plant–microbe interactions, and identifying rhizosphere biomarkers. These applications help uncover microbial roles in plant health, soil fertility, and environmental responses, offering valuable insights for sustainable agriculture and microbial ecology research.

9.5.2 Role of sensors and IoT in rhizosphere monitoring

The integration of modern sensors and Internet of Things (IoT) technology offers ground-breaking real-time solutions for rhizosphere observation of medicinal and nonmedicinal plants, which is particularly important given the escalating agricultural challenges imposed by climate change, inflation, and labor concerns. Important soil physicochemical characteristics, including moisture, pH, temperature, nutrients, and electrical conductivity, are monitored by precision agriculture using electrochemical, optical, and biosensors (Fig. 9.3). These parameters are essential for deciphering interactions between plants, microbes, and soil.

Smart sensors with microprocessors allow for data processing at the point of occurrence, autonomous decision making, and continuous communication by cellular technology, Bluetooth®, or wi-fi, enabling remote and broad agricultural farming (Kumari *et al.*, 2025b). IoT platforms collect and send high-resolution spatial and temporal data to cloud systems, which

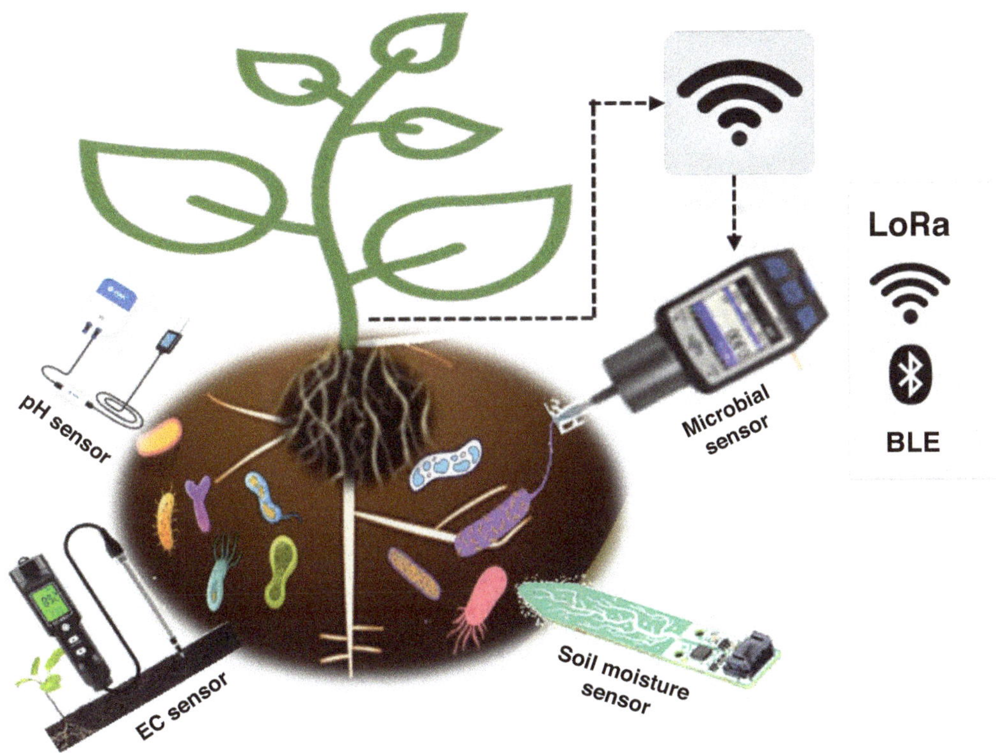

Fig. 9.3. Real-time rhizosphere monitoring with sensors via the IoT (LoRa, BLE).

improves operational effectiveness and decision making by enabling farmers to remotely monitor field conditions via mobile apps (Liang and Shah, 2023; Sharma and Shivandu, 2024). Robotic planters, GPS-guided agricultural equipment, and smart irrigation are examples of automated systems that adjust in real time to data inputs, maximizing resource utilization and reducing waste. Such databases are also used by AI and ML-based algorithms for predictive analytics, including assistance in forecasting insect outbreaks, estimating crop yields, and maintaining equipment (Kumari and Nafchi, 2024).

Strong connectivity made possible by emerging communication technologies like 5G, LoRaWAN®, and edge computing allows for the quick, reliable, and scalable implementation of IoT-based solutions in the agricultural sector (Liu *et al.*, 2024). Despite obstacles including high expenses and mismatches between systems, new standardization and cost-cutting initiatives are encouraging wider implementation. Subsequently, sensor integration with the IoT enables precise monitoring of rhizosphere conditions, advances our understanding of microbial diversity among plant species, and supports resilient, sustainable, and data-driven agricultural systems.

These sensors and IoT technologies enable real-time monitoring of soil parameters such as pH, electrical conductivity, moisture, temperature, and nutrient levels. They facilitate early detection of microbial community shifts linked to plant stress or growth stages and support automated data collection and transmission for ML-based analysis and decision making.

9.5.3 AI-driven multi-omics integration

Multi-omics is the large-scale, integrated study of many types of "omics" data, including genomes, transcriptomics, proteomics, and metabolomics, to create a full molecular profile of microbial communities (Goering *et al.*, 2016). When applied to rhizosphere microbiomes, this comprehensive approach allows for the reconstruction of microbial networks, the characterization of functional pathways, and the identification of specialized microbial metabolites or biosynthetic gene clusters (BGCs) that influence plant characteristics.

This systems biology concept is critical for mapping molecular interactions within the plant–microbe–soil triad. It offers mechanistic insights into metabolic exchanges, microbial signaling, and the biosynthetic potential of rhizospheric microorganisms, paving the way for precision agriculture and medical plant development. Integrating multi-omics data with AI models allows researchers to determine active microbial networks and their regulatory influence on plant physiology. For example, combining metagenomics with metatranscriptomics can assist in discovering active genes and microbial taxa involved in food cycling, hormone signaling, and pathogen suppression (Carthew, 2021). Metabolomics and proteomics help to refine these insights by identifying microbial products such as siderophores, phytohormones, and volatile organic molecules. AI techniques such as random forests, support vector machines, and neural networks (Fig. 9.4) can be trained on these

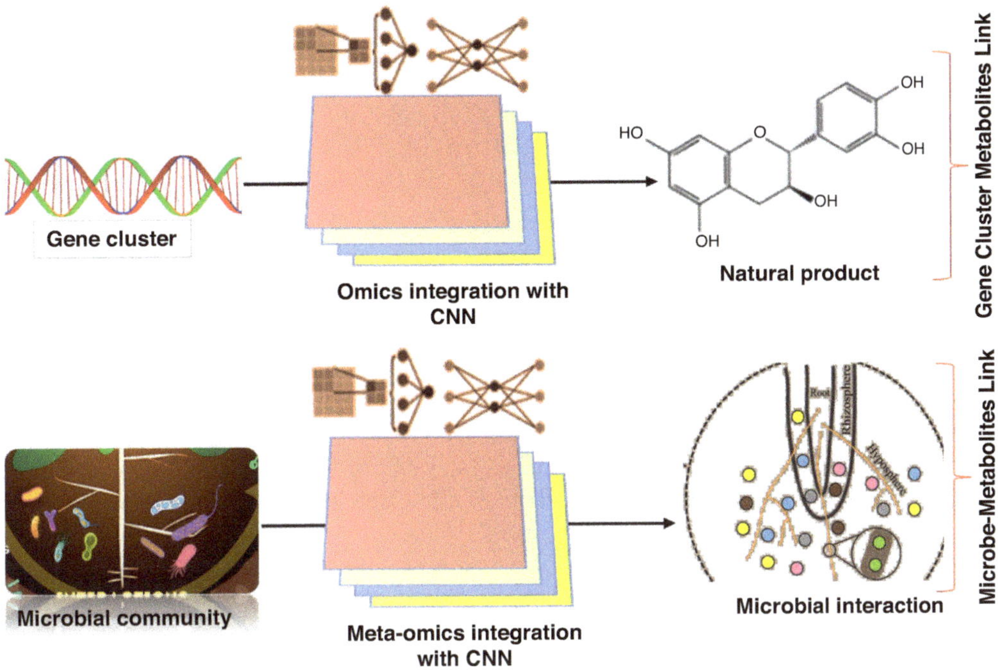

Fig. 9.4. Neural network-based deciphering of functional microbial networks and metabolite profiles.

integrated datasets to determine causation and forecast the microbial drivers of plant health outcomes (Tinikul *et al.*, 2020). These prediction models can divide microbial communities into functional modules and identify keystone species that act as bioindicators or bioeffectors in the rhizosphere ecosystem.

9.5.4 Discovery of plant-specific microbial metabolites and biosynthetic gene clusters

Advances in sequencing and metabolite detection have resulted in the identification of thousands of BGCs, many of which are associated with natural product biosynthesis. AI-based methods such as antiSMASH (Blin *et al.*, 2021), NP-Linker (Eldjárn *et al.*, 2021), and BiG-SCAPE (Navarro-Muñoz *et al.*, 2020) can predict the structure, function, and chemical products of clusters. Metabologenomics, or the integrated analysis of genomic and metabolomic data, allows for the discovery of plant-specific microbial metabolites by correlating mass spectrometry data to BGC sequences (Goering *et al.*, 2016). These metabolites frequently serve important roles in plant defense and adaptability. AI methods make it easier to dereplicate recognized molecules, identify novel metabolites, and associate specific metabolites with medicinal or nonmedicinal plant hosts.

Furthermore, curated databases such as MIBiG (Terlouw *et al.*, 2023), COCONUT (Sorokina *et al.*, 2021), and GNPS (Wang *et al.*, 2016) are combined with AI pipelines to improve metabolite annotation and contextualization. These efforts not only speed up drug development, but also give molecular fingerprints that differentiate medicinal plant rhizospheres from nonmedicinal equivalents.

9.5.5 Comparative meta-omics in medicinal versus nonmedicinal plants

Comparative meta-omics allows researchers to monitor changes in microbial composition and function across various plant species. Microbial consortia in medicinal plants are frequently enriched in taxa that may produce specialized secondary metabolites such as alkaloids, terpenoids, and phenolics. These consortia may also contain distinct BGCs that are not observed in nonmedicinal plant rhizospheres (Nützmann *et al.*, 2018). AI technologies assist in the deconvolution of these complicated datasets by finding microbial biomarkers and metabolic pathways that are preferentially expressed or suppressed in response to plant type. Unsupervised clustering and dimensionality reduction techniques, such as t-SNE and UMAP, enable the visualization of metagenomic differences, whereas supervised learning models help in classification and feature selection. By using AI to combine multi-meta-omics data, scientists can connect certain microorganisms and their metabolite products to the therapeutic qualities of host plants, opening up new possibilities for the creation of biofertilizers, tailored microbial inoculants, and natural product discovery.

9.5.6 Integrative platforms and future directions

Many platforms, such as PoDP (Schorn *et al.*, 2021), MetaMiner™ (Cao *et al.*, 2019), and GRAPE-GARLIC (Johnston *et al.*, 2016), now facilitate AI-driven multi-omics integration. These platforms provide the coanalysis of spectral, transcriptomic, metabolomic, and genomic data, frequently following the FAIR data principles to guarantee repeatability and interoperability. The creation of reliable, interpretable AI models will be essential going forward to fully utilize multi-omics in rhizosphere microbiome research. In particular, frameworks for explainable AI (XAI) can make it clear how choices are made and which characteristics—such as particular genes, taxa, or metabolites—have the most influence on phenotypes that are seen. The usefulness of these tools in field-based research may also be increased by federated learning and transfer learning strategies, which may enable models to generalize across datasets and contexts.

9.6 Functional and Ecological Implications of Rhizosphere Microbial Differences

The rhizosphere microbiome plays a crucial role in influencing plant productivity and health, with distinct differences observed between medicinal and nonmedicinal plants. Medicinal plants tend to host specialized microbial communities enriched with beneficial microbes that contribute to their unique bioactive compound profiles. These microorganisms can stimulate the synthesis of secondary metabolites, such as alkaloids, terpenoids, and phenolics, which are vital for the plants' therapeutic properties. The selective root exudates released by medicinal plants create a unique ecological niche that favors microbes capable of metabolizing these compounds or participating in their biosynthesis, thereby enhancing metabolite production (Pantigoso *et al.*, 2022).

In addition to boosting metabolite synthesis, rhizosphere microbes associated with medicinal plants also contribute to disease suppression through biocontrol mechanisms. Plant growth-promoting bacteria and fungi inhibit soilborne pathogens by antibiosis, competition, and ISR, leading to reduced disease incidence and improved plant health. In contrast, microbial communities in nonmedicinal plants are primarily oriented towards nutrient cycling and growth promotion, although they also play a role in pathogen resistance. Knowledge of how these microbial communities differ is important because it helps us use the natural soil microbes to keep plants healthy and boost production of useful plant compounds. This knowledge can support eco-friendly farming and help grow plants that produce medicines.

9.7 Applications in Agriculture, Medicine, and Biotechnology

Rhizosphere microbiomes play a vital role in agriculture, medicine, and biotechnology, especially when comparing medicinal and nonmedicinal plants. The microbial communities surrounding plant roots help promote growth by improving nutrient uptake, enhancing stress tolerance, and defending against pathogens. In medicinal plants, root exudates create a specialized environment that supports microbes capable of producing bioactive compounds or stimulating the plant's own secondary metabolite production, which boosts plant health and the yield of valuable medicinal compounds. In contrast, nonmedicinal plants tend to host microbial populations focused mainly on supporting growth and vigor under different environmental conditions.

Exploring the rhizosphere microbiome of medicinal plants reveals a rich source of novel microbes with potential pharmaceutical applications, including bacteria and fungi that produce antibiotics, anticancer agents, and other therapeutic compounds. These microbes are uniquely adapted to the chemical conditions created by medicinal plant root exudates, offering opportunities to discover compounds absent in nonmedicinal plant rhizospheres. Moreover, emerging technologies such as microbiome engineering and synthetic biology hold promise for manipulating these microbial communities to develop desirable traits. By understanding how microbial diversity and abundance differ between medicinal and nonmedicinal plants, scientists can develop targeted strategies to improve crop productivity, increase medicinal compound production, and create innovative biotechnological solutions for sustainable agriculture and new drug discovery.

9.8 Future Directions and Research Gaps

Growing interest in rhizosphere microbiomes has highlighted significant gaps in our understanding of the compositional and functional distinctions between microbial communities associated with medicinal and nonmedicinal plants. There is an urgent need for comprehensive comparative metagenomic studies to unravel the taxonomic structure and functional capabilities of these communities at a deeper resolution. Such analyses could reveal key microbial taxa and metabolic pathways uniquely linked to medicinal plants, particularly those involved in secondary metabolite biosynthesis and plant health promotion. Furthermore, the coevolutionary dynamics between plants and their associated microbiota remain largely unexplored.

Investigating how long-term plant–microbe interactions have shaped rhizospheric microbial assemblages, especially in medicinal species selected for their therapeutic traits, is a critical research frontier. Equally important is understanding the influence of climate change on these microbial communities. Alterations in temperature, precipitation patterns, and soil properties may significantly impact microbial diversity and stability, potentially disrupting essential plant–microbe symbiosis. To address these challenges, future studies should adopt multi-ecoregional and longitudinal designs, integrating multi-omics technologies with ecological and evolutionary frameworks. This integrative approach will not only deepen insights into rhizosphere microbiome composition and function but also facilitate the discovery of novel microbial resources for pharmaceutical development and sustainable plant production systems.

9.9 Conclusion

Comparative microbiome analysis of rhizospheres suggests the occurrence of specialized ecological and functional traits in medicinal and nonmedicinal plants. Medicinal plants are colonized by specialized microbiota that, apart from inducing nutrient cycling and growth promotion, also support biosynthesis of bioactive secondary metabolites. Such interactions are shaped by selective root exudation and coevolutionary pressures, inducing symbiosis with metabolite-producing microbes. Nonmedicinal plants are typically observed to be colonized by microbiota specialized in general growth promotion and stress tolerance.

Advances in AI and ML, sensor technologies, and multi-omics integration are revolutionizing our understanding of these complex plant–microbe systems. AI-driven tools enable high-throughput multi-omics data analysis, elucidating microbial functions, biosynthetic gene clusters, and host phenotypic outcomes. Meanwhile, real-time monitoring using IoT sensors is enhancing precision agriculture using data-driven decisions.

Future interdisciplinary research, incorporating comparative meta-omics, interpretable AI, and field-deployable sensors, will be squarely at the center of unlocking dynamic plant–microbiome interactions. Integrative approaches hold great promise for sustainable agriculture, environmental resilience, and the development of microbiome-based therapeutics in medicinal and nonmedicinal plant contexts.

References

Bais, H.P., Weir, T.L., Perry, L.G., Gilroy, S. and Vivanco, J.M. (2006) The role of root exudates in rhizosphere interactions with plants and other organisms. *Annual Review of Plant Biology* 57(1), 233–266. DOI: 10.1146/annurev.arplant.57.032905.105159.

Blin, K., Shaw, S., Kloosterman, A.M., Charlop-Powers, Z., Wezel, G.P. *et al.* (2021) antiSMASH 6.0: Improving cluster detection and comparison capabilities. *Nucleic Acids Research* 49(W1), W29–W35. DOI: 10.1093/nar/gkab335.

Canarini, A., Kaiser, C., Merchant, A., Richter, A. and Wanek, W. (2019) Root exudation of primary metabolites: Mechanisms and their roles in plant responses to environmental stimuli. *Frontiers in Plant Science* 10, 157. DOI: 10.3389/fpls.2019.00157.

Cao, L., Gurevich, A., Alexander, K.L., Naman, C.B., Leão, T. *et al.* (2019) MetaMiner: A scalable peptidogenomics approach for discovery of ribosomal peptide natural products with blind modifications from microbial communities. *Cell Systems* 9(6), 600–608. DOI: 10.1016/j.cels.2019.09.004.

Carthew, R.W. (2021) Gene regulation and cellular metabolism: An essential partnership. *Trends in Genetics* 37(4), 389–400. DOI: 10.1016/j.tig.2020.09.018.

Castronovo, L.M., Vassallo, A., Mengoni, A., Miceli, E., Bogani, P. *et al.* (2021) Medicinal plants and their bacterial microbiota: A review on antimicrobial compounds production for plant and human health. *Pathogens* 10(2), 106. DOI: 10.3390/pathogens10020106.

Chang, H.X., Haudenshield, J.S., Bowen, C.R. and Hartman, G.L. (2017) Metagenome-wide association study and machine learning prediction of bulk soil microbiome and crop productivity. *Frontiers in Microbiology* 8, 519. DOI: 10.3389/fmicb.2017.00519.

Durán, P., Thiergart, T., Garrido-Oter, R., Agler, M., Kemen, E. *et al.* (2018) Microbial interkingdom interactions in roots promote Arabidopsis survival. *Cell* 175(4), 973–983. DOI: 10.1016/j.cell.2018.10.020.

Egamberdieva, D., Shrivastava, S. and Varma, A. (eds) (2015) *Plant-Growth-Promoting Rhizobacteria (PGPR) and Medicinal Plants*, Vol. 42. Springer International Publishing, Cham, Switzerland. DOI: 10.1007/978-3-319-13401-7.

Eldjárn, G.H., Ramsay, A., Hooft, J.J.J., Duncan, K.R., Soldatou, S. *et al.* (2021) Ranking microbial metabolomic and genomic links in the NPLinker framework using complementary scoring functions. *PLOS Computational Biology* 17, 1008920. DOI: 10.1371/journal.pcbi.1008920.

Feng, K. and Deng, Y. (2025) Intertwined relationship between soil pH and microbes in biogeography. *Global Change Biology* 31(4), e70208. DOI: 10.1111/gcb.70208.

Goering, A.W., McClure, R.A., Doroghazi, J.R., Albright, J.C., Haverland, N.A. *et al.* (2016) Metabologenomics: Correlation of microbial gene clusters with metabolites drives discovery of a nonribosomal peptide with an unusual amino acid monomer. *ACS Central Science* 2(2), 99–108. DOI: 10.1021/acscentsci.5b00331.

Hiltner, L. (1904) Über neuere Erfahrungen und Probleme auf dem Gebiet der Bodenbakteriologie und unter besonderer Berücksichtigung der Gründüngung und Brache. *Arbeiten der Deutschen Landwirtschaftlichen Gesellschaft* 98, 59.

Hnamte, L., Kumar, A., Yadav, M.K. and Singh, P.K. (2024) An updated view of bacterial endophytes as antimicrobial agents against plant and human pathogens. *Current Research in Microbial Sciences* 7, 100241. DOI: 10.1016/j.crmicr.2024.100241.

Johnston, C.W., Skinnider, M.A., Dejong, C.A., Rees, P.N., Chen, G.M. *et al.* (2016) Assembly and clustering of natural antibiotics guides target identification. *Nature Chemical Biology* 12(4), 233–239. DOI: 10.1038/nchembio.2018.

Kovalchuk, I., Kovalchuk, O., Kalck, V., Boyko, V., Filkowski, J. *et al.* (2003) Pathogen-induced systemic plant signal triggers DNA rearrangements. *Nature* 423(6941), 760–762. DOI: 10.1038/nature01683.

Kumari, K. and Nafchi, A.M. (2024) Sustainable agriculture with AI, machine learning, deep learning, and IoT for future farming. In: *American Society of Agricultural and Biological Engineers Conference*, Anaheim, California, July 28–31, p. 1. DOI: 10.13031/aim.202400906.

Kumari, K., Nafchi, A.M., Mirzaee, S. and Abdalla, A. (2025a) AI-driven future farming: Achieving climate-smart and sustainable agriculture. *AgriEngineering* 7(3), 89. DOI: 10.3390/agriengineering7030089.

Kumari, K., Parray, R., Basavaraj, Y.B., Godara, S., Mani, I. *et al.* (2025b) Spectral sensor-based device for real-time detection and severity estimation of groundnut bud necrosis virus in tomato. *Journal of Field Robotics* 42(1), 5–19. DOI: 10.1002/rob.22391.

Lahlali, R., Ibrahim, D.S.S., Belabess, Z., Kadir Roni, M.Z., Radouane, N. *et al.* (2021) High-throughput molecular technologies for unraveling the mystery of soil microbial community: Challenges and future prospects. *Heliyon* 7(10), e08142. DOI: 10.1016/j.heliyon.2021.e08142.

LaPierre, N., Ju, C.J.T., Zhou, G. and Wang, W. (2019) MetaPheno: A critical evaluation of deep learning and machine learning in metagenome-based disease prediction. *Methods* 166, 74–82. DOI: 10.1016/j.ymeth.2019.03.003.

Li, H., Chang, L., Liu, H. and Li, Y. (2024) Diverse factors influence the amounts of carbon input to soils via rhizodeposition in plants: A review. *Science of the Total Environment* 948, 174858. DOI: 10.1016/j.scitotenv.2024.174858.

Liu, Y., Yan, Y., Chen, J. and Lu, X. (2024) Is the response of plant root systems to precipitation primarily driven by the rhizosphere soil? *Ecological Processes* 13(1), 84. DOI: 10.1186/s13717-024-00554-0.

Liang, C. and Shah, T. (2023) IoT in agriculture: The future of precision monitoring and data-driven farming. *Eigenpub Review of Science and Technology* 7(1), 85–104.

Luo, C., He, Y. and Chen, Y. (2025) Rhizosphere microbiome regulation: Unlocking the potential for plant growth. *Current Research in Microbial Sciences* 8, 100322. DOI: 10.1016/j.crmicr.2024.100322.

Mendes, R., Garbeva, P. and Raaijmakers, J.M. (2013) The rhizosphere microbiome: Significance of plant beneficial, plant pathogenic, and human pathogenic microorganisms. *FEMS Microbiology Reviews* 37(5), 634–663. DOI: 10.1111/1574-6976.12028.

Navarro-Muñoz, J.C., Selem-Mojica, N., Mullowney, M.W., Kautsar, S.A., Tryon, J.H. *et al.* (2020) A computational framework to explore large-scale biosynthetic diversity. *Nature Chemical Biology* 16(1), 60–68. DOI: 10.1038/s41589-019-0400-9.

Naylor, D., McClure, R. and Jansson, J. (2022) Trends in microbial community composition and function by soil depth. *Microorganisms* 10(3), 540. DOI: 10.3390/microorganisms10030540.

Neu, A.T., Allen, E.E. and Roy, K. (2021) Defining and quantifying the core microbiome: Challenges and prospects. *Proceedings of the National Academy of Sciences USA* 118(51), e2104429118. DOI: 10.1073/pnas.2104429118.

Nützmann, H.W., Scazzocchio, C. and Osbourn, A. (2018) Metabolic gene clusters in eukaryotes. *Annual Review of Genetics* 52(1), 159–183. DOI: 10.1146/annurev-genet-120417-031237.

Pang, Z., Chen, J., Wang, T., Gao, C., Li, Z. *et al.* (2021) Linking plant secondary metabolites and plant microbiomes: A review. *Frontiers in Plant Science* 12, 621276. DOI: 10.3389/fpls.2021.621276.

Pantigoso, H.A., Newberger, D. and Vivanco, J.M. (2022) The rhizosphere microbiome: Plant-microbial interactions for resource acquisition. *Journal of Applied Microbiology* 133(5), 2864–2876. DOI: 10.1111/jam.15686.

Priya, P., Aneesh, B. and Harikrishnan, K. (2021) Genomics as a potential tool to unravel the rhizosphere microbiome interactions on plant health. *Journal of Microbiological Methods* 185, 106215. DOI: 10.1016/j.mimet.2021.106215.

Qu, P., Wang, B., Qi, M., Lin, R., Chen, H. *et al.* (2024) Medicinal plant root exudate metabolites shape the rhizosphere microbiota. *International Journal of Molecular Sciences* 25(14), 7786. DOI: 10.3390/ijms25147786.

Rizwan, R. and Gothalwal, R. (2021) Microbiomes of medicinal plants and their potential avenues as bioresources – A review. *International Journal of Current Research in Biosciences and Plant Biology* 8(11), 29–40. DOI: 10.20546/ijcrbp.2021.811.004.

Schnecker, J., Baldaszti, L., Gündler, P., Pleitner, M., Sandén, T. *et al.* (2023) Seasonal dynamics of soil microbial growth, respiration, biomass, and carbon use efficiency in temperate soils. *Geoderma* 440, 116693. DOI: 10.1016/j.geoderma.2023.116693.

Schorn, M.A., Verhoeven, S., Ridder, L., Huber, F., Acharya, D.D. *et al.* (2021) A community resource for paired genomic and metabolomic data mining. *Nature Chemical Biology* 17(4), 363–368. DOI: 10.1038/s41589-020-00724-z.

Sharma, K. and Shivandu, S.K. (2024) Integrating artificial intelligence and Internet of Things (IoT) for enhanced crop monitoring and management in precision agriculture. *Sensors International* 5, 100292. DOI: 10.1016/j.sintl.2024.100292.

Sharpton, T.J. (2014) An introduction to the analysis of shotgun metagenomic data. *Frontiers in Plant Science* 5, 209. DOI: 10.3389/fpls.2014.00209.

Sorokina, M., Merseburger, P., Rajan, K., Yirik, M.A. and Steinbeck, C. (2021) COCONUT online: Collection of open natural products database. *Journal of Cheminformatics* 13(1), 2. DOI: 10.1186/s13321-020-00478-9.

Terlouw, B.R., Blin, K., Navarro-Muñoz, J.C., Avalon, N.E., Chevrette, M.G. *et al.* (2023) MIBiG 3.0: A community-driven effort to annotate experimentally validated biosynthetic gene clusters. *Nucleic Acids Research* 51(D1), D603–D610. DOI: 10.1093/nar/gkac1049.

Thepbandit, W. and Athinuwat, D. (2024) Rhizosphere microorganisms supply availability of soil nutrients and induce plant defense. *Microorganisms* 12(3), 558. DOI: 10.3390/microorganisms12030558.

Tinikul, R., Chunthaboon, P., Phonbuppha, J. and Paladkong, T. (2020) Bacterial luciferase: Molecular mechanisms and applications. *Enzymes* 47, 427–455. DOI: 10.1016/bs.enz.2020.06.001.

Toju, H., Peay, K.G., Yamamichi, M., Narisawa, K., Hiruma, K. *et al.* (2018) Core microbiomes for sustainable agroecosystems. *Nature Plants* 4(5), 247–257. DOI: 10.1038/s41477-018-0139-4.

Tuck, S.L., Winqvist, C., Mota, F., Ahnström, J., Turnbull, L.A. *et al.* (2014) Land-use intensity and the effects of organic farming on biodiversity: A hierarchical meta-analysis. *Journal of Applied Ecology* 51(3), 746–755. DOI: 10.1111/1365-2664.12219.

Venter, Z.S., Jacobs, K. and Hawkins, H.-J. (2016) The impact of crop rotation on soil microbial diversity: A meta-analysis. *Pedobiologia* 59(4), 215–223. DOI: 10.1016/j.pedobi.2016.04.001.

Wade, W. (2002) Unculturable bacteria – the uncharacterized organisms that cause oral infections. *Journal of the Royal Society of Medicine* 95(2), 81–83. DOI: 10.1177/014107680209500207.

Waldrop, M.P. and Firestone, M.K. (2006) Response of microbial community composition and function to soil climate change. *Microbial Ecology* 52(4), 716–724. DOI: 10.1007/s00248-006-9103-3.

Wang, J. and Jia, H. (2016) Metagenome-wide association studies: Fine-mining the microbiome. *Nature Reviews Microbiology* 14(8), 508–522. DOI: 10.1038/nrmicro.2016.83.

Wang, M., Carver, J.J., Phelan, V.V., Sanchez, L.M., Garg, N. *et al.* (2016) Sharing and community curation of mass spectrometry data with global natural products social molecular networking. *Nature Biotechnology* 34(8), 828–837. DOI: 10.1038/nbt.3597.

Wang, T., Yang, K., Ma, Q., Jiang, X., Zhou, Y. *et al.* (2022) Rhizosphere microbial community diversity and function analysis of cut chrysanthemum during continuous monocropping. *Frontiers in Microbiology* 13, 801546. DOI: 10.3389/fmicb.2022.801546.

Windisch, S., Sommermann, L., Babin, D., Chowdhury, S.P., Grosch, R. *et al.* (2021) Impact of long-term organic and mineral fertilization on rhizosphere metabolites, root-microbial interactions and plant health of lettuce. *Frontiers in Microbiology* 11, 597745. DOI: 10.3389/fmicb.2020.597745.

Xiong, W., Song, Y., Yang, K., Gu, Y., Wei, Z. *et al.* (2020) Rhizosphere protists are key determinants of plant health. *Microbiome* 8(1), 27. DOI: 10.1186/s40168-020-00799-9.

Xu, Y., Xing, L., Su, J., Zhang, X. and Qiu, W. (2019) Model-based clustering for identifying disease-associated SNPs in case-control genome-wide association studies. *Scientific Reports* 9(1), 13686. DOI: 10.1038/s41598-019-50229-6.

Yuan, Y., Zuo, J., Zhang, H., Zu, M. and Liu, S. (2022) The Chinese medicinal plants rhizosphere: Metabolites, microorganisms, and interaction. *Rhizosphere* 22, 100540. DOI: 10.1016/j.rhisph.2022.100540.

Zhang, Y., Ji, P., Wang, J. and Zhao, F. (2016) RiboFR-seq: A novel approach to linking 16S rRNA amplicon profiles to metagenomes. *Nucleic Acids Research* 44(10), e99. DOI: 10.1093/nar/gkw165.

10 Exploration of the Unique Microbiomes of Medicinal Plants, Driven by Their Bioactive Secondary Metabolites

Varun Kumar Sharma[1*], Pragati Saxena[1], Shilpy Singh[1], Abhishek Awasthi[2], Lomas Kumar Tomar[3], Ved Vrat Verma[4] and Dharmsheel Shrivastav[1*]

[1]*Department of Biotechnology and Microbiology, School of Sciences, Noida International University, Gautam Budh Nagar, India*
[2]*School of Studies in Biotechnology, Shaheed Mahendra Karma Vishwavidyalaya, Jagdalpur, Bastar, India*
[3]*School of Biomedical Engineering and Health Sciences, Shobhit University, Meerut, India*
[4]*Department of Biotechnology, School of Engineering and Technology, Noida International University, Gautam Budh Nagar, India*

10.1 Introduction

Medicinal plants have been utilized for centuries in traditional medicine and represent a rich source of bioactive compounds with significant therapeutic potential. These plants produce a wide variety of phytochemicals, such as alkaloids, flavonoids, terpenoids, polyphenols, and glycosides, which demonstrate a broad spectrum of pharmacological activities, including antimicrobial, antiinflammatory, antioxidant, anticancer, and immunomodulatory effects (Roy *et al.*, 2022; Mehra *et al.*, 2023). The increasing interest in plant-derived bioactive compounds has led to extensive research into their pharmacological applications, particularly in drug discovery and development (Mehra *et al.*, 2023; Bulakarima *et al.*, 2025a).

*Corresponding authors: varungenetics@gmail.com and Dharmsheel.shrivastav1@gmail.com

(eds M.K. Mishra and S. Tiwari)
DOI: 10.1079/9781836990475.0010

Plants produce secondary metabolites as a defense mechanism against pathogens, herbivores, and environmental stresses. These metabolites not only protect the plants but also provide medicinal benefits to humans (Divekar *et al.*, 2022; Mehra *et al.*, 2024). Advances in biotechnology, genomics, and metabolomics have enhanced our ability to identify and isolate novel bioactive compounds from medicinal plants. Many pharmaceutical drugs, such as aspirin (from willow bark), morphine (from opium poppy), and quinine (from cinchona bark), are derived from plant sources, highlighting their significance in modern medicine (Mehra *et al.*, 2023, 2024; Bulakarima *et al.*, 2025a).

Given the rapid loss of biodiversity due to deforestation and habitat destruction, there is an urgent need to conserve medicinal plants and explore sustainable cultivation practices. Integrating traditional knowledge with modern scientific approaches can further enhance the discovery of novel bioactive compounds, ensuring a continuous supply of plant-based therapeutics for future generations (Mehra *et al.*, 2023; Chaachouay and Zidane, 2024; Mishra, 2024).

10.1.1 Importance of microbiomes in plant health and secondary metabolite synthesis

Plant microbiomes, consisting of diverse communities of bacteria, fungi, and other microorganisms, play a crucial role in plant health, growth, and metabolism (Olanrewaju *et al.*, 2024; Jagtap *et al.*, 2025). These microbial communities colonize different plant compartments, including roots (rhizosphere), leaves (phyllosphere), and internal tissues (endosphere), contributing to nutrient uptake, stress tolerance, and disease resistance (Olanrewaju *et al.*, 2024; Enea *et al.*, 2025) (Fig. 10.1).

Microbiomes are particularly significant in the regulation of plant secondary metabolites and control their biosynthesis. Beneficial microbes enhance the production of bioactive compounds by influencing metabolic pathways through various mechanisms (Pang *et al.*, 2021; Koza *et al.*, 2022; Nimbulkar *et al.*, 2025).

- *Inducing systemic resistance:* Certain microbes activate plant defense mechanisms, resulting in enhanced production of antimicrobial and antioxidant compounds.
- *Enhancing nutrient availability:* Microbial symbionts improve nutrient uptake, which indirectly supports secondary metabolite synthesis.
- *Modulating hormonal pathways:* Plant-associated microbes regulate hormone levels, affecting growth and metabolic responses.
- *Producing bioactive molecules:* Some microbes produce bioactive compounds that synergize with plant metabolites, enhancing their pharmacological properties.

Recent studies have highlighted the role of endophytic fungi and bacteria in boosting secondary metabolite production, often improving the medicinal

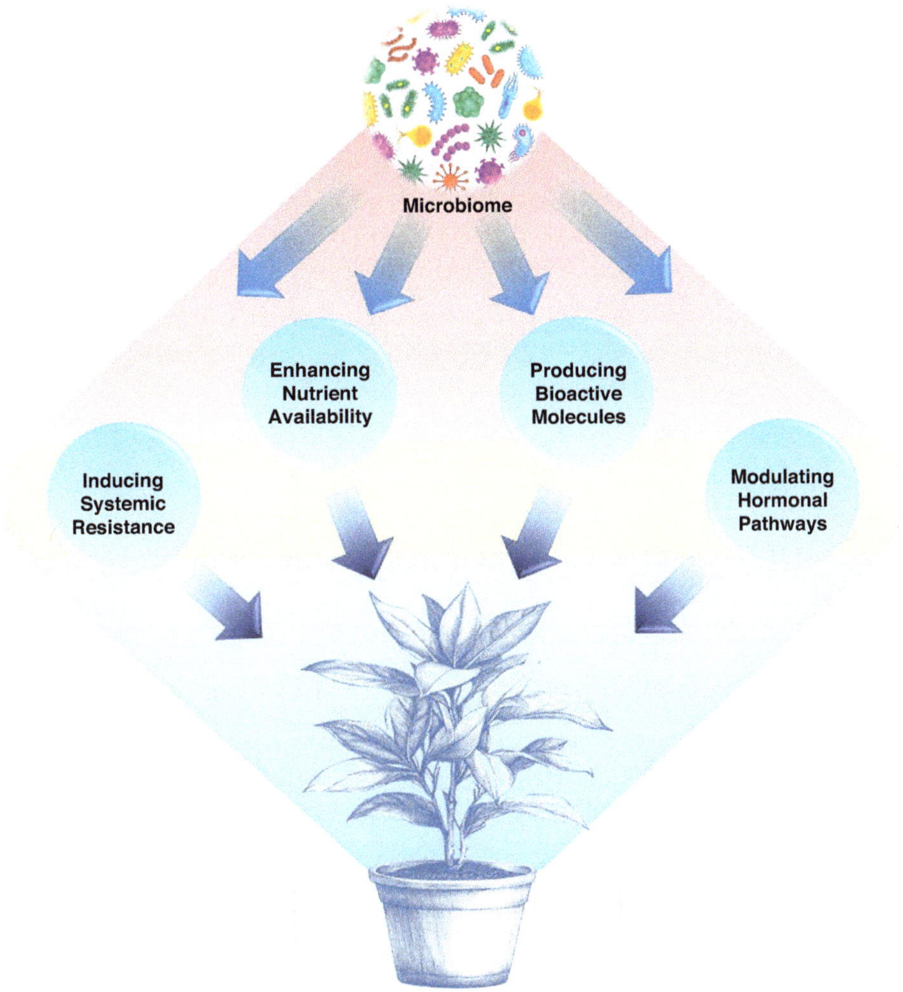

Fig. 10.1. Positive impact of microbiome on plant health and growth.

value of plants. For example, the interaction between *Artemisia annua* and endophytic fungi has been shown to enhance production of artemisinin, a key antimalarial compound. Similarly, rhizobacteria associated with medicinal plants contribute to increased alkaloid and flavonoid biosynthesis (Pang *et al.*, 2021; Koza *et al.*, 2022; Tripathi *et al.*, 2025).

Harnessing plant microbiomes for sustainable agriculture and biopharmaceutical applications is a promising approach. Understanding these complex interactions can lead to innovative strategies for improving crop resilience, enhancing bioactive compound yields, and developing microbiome-based therapeutics (Mehra *et al.*, 2023; Mishra, 2024).

10.1.2 Exploring plant-associated microbiomes for pharmaceutical and agricultural advancements

Plant-associated microbiomes, composed of diverse microorganisms such as bacteria, fungi, and archaea, play a critical role in plant health and productivity (Koza *et al.*, 2022; Nimbulkar *et al.*, 2025). The exploration of these microbiomes holds immense potential for both pharmaceutical and agricultural advancements, offering sustainable solutions to pressing global challenges, including food security, environmental sustainability, and novel drug discovery (Nadarajah and Abdul Rahman, 2023; Bulakarima *et al.*, 2025b; Misu *et al.*, 2025).

In agriculture, beneficial microbes promote plant growth by facilitating nutrient uptake, synthesizing essential phytohormones, and improving soil structure. Certain microorganisms act as natural biocontrol agents against pathogens, reducing the need for chemical pesticides (Jadhav *et al.*, 2025; Misu *et al.*, 2025). Additionally, plant-associated microbiomes enhance plant tolerance to abiotic stresses such as drought, salinity, and extreme temperatures, thereby playing a vital role in promoting climate-resilient agriculture. These microbial communities also enhance soil fertility by decomposing organic matter, fixing nitrogen, and recycling essential nutrients, contributing to long-term agricultural sustainability (Nadarajah and Abdul Rahman, 2023; Misu *et al.*, 2025).

In pharmaceuticals, many plant-associated microbes produce bioactive compounds with antibacterial, antifungal, and antiviral properties, which can be harnessed for developing new pharmaceuticals (Davidova *et al.*, 2024). Endophytic microorganisms inhabiting plant tissues synthesize secondary metabolites with promising therapeutic potential, including anticancer, antiinflammatory, and immunomodulatory properties (Wijesekara and Xu, 2023). Genetic engineering and synthetic biology approaches can leverage plant microbiomes for the biosynthesis of valuable compounds, including antibiotics, enzymes, and biofuels. Some plant-associated microbes also have probiotic potential, contributing to gut health and immune system regulation (Nazir *et al.*, 2024).

The integration of advanced technologies such as metagenomics, metabolomics, and synthetic biology has accelerated the exploration of plant microbiomes. Future research should focus on optimizing microbial applications for large-scale agricultural and pharmaceutical use (Nam *et al.*, 2023). Sustainable strategies involving plant–microbe interactions could revolutionize crop production, disease management, and drug discovery, providing a basis for a healthier and more stable and viable future. Exploring plant-associated microbiomes is a promising frontier in both agriculture and medicine. By harnessing the potential of these microbial communities, we can develop innovative solutions for food security, environmental sustainability, and human health, ultimately transforming global industries for the better (Yusuf *et al.*, 2025).

10.2 Diversity of Microbiomes in Medicinal Plants

Microbes are found everywhere in the environment, existing not only in the human body but also in water, soil, air, and food. Airborne microbes travel with the wind and establish new colonies. The plant microbiome is essential for maintaining and promoting plant health. Since plants are stationary, they face significant challenges in obtaining adequate nutrients, protecting themselves against herbivores and pathogens, and coping with abiotic stresses, including drought, salinity, and pollutants (Afridi *et al.*, 2022). The plant-associated microbiome plays a pivotal role in boosting plant health and productivity by reducing the occurrence of plant diseases, minimizing reliance on chemical fertilizers and pesticides, and mitigating greenhouse gas emissions (Sessitsch *et al.*, 2023).

Microbes contribute significantly to ecosystem processes by decomposing organic matter, breaking down toxic substances, recycling nutrients, and facilitating carbon sequestration, which helps maintain ecological balance (Iqbal *et al.*, 2025). The plant microbiome exhibits distinct diversity depending on specific microhabitats, which are generally classified into the phyllosphere, caulosphere, carposphere, anthosphere, endosphere, and rhizosphere. Microbes in these regions influence the composition, structure, diversity, and overall productivity of plant communities (Sivakumar *et al.*, 2020; Iqbal *et al.*, 2025). Plants and microbes form different types of relationships, including parasitic, mutualistic, and commensalistic interactions. Parasitic interactions harm the plant, mutualistic interactions promote plant growth, and commensalistic interactions are neutral, having no significant impact on plant health (Nadarajah and Abdul Rahman, 2023).

10.2.1 Types of plant microbiome (phytobiome)

The term "phytobiome" combines "phyto" (plants) and "biome," which refers to distinct ecological regions. The phytobiome plays a critical role in various plant tissues, including the phyllosphere, rhizosphere, and endosphere (Fig. 10.2). It consists of a broad array of microorganisms such as fungi, bacteria, viruses, protists, and nematodes, all of which interact with plant structures (Sivakumar *et al.*, 2020; Enea *et al.*, 2025). The microbiome, composed of these microbes and their genetic material, forms a complex and dynamic relationship with plants. These interactions are influenced by environmental conditions and contribute to the plant's capability to withstand environmental stress (Meel and Saharan, 2024; Enea *et al.*, 2025).

10.2.1.1 Phyllosphere

The phyllosphere, which includes aerial surfaces such as stems, leaves, flowers, and fruits, represents the largest biological interface on Earth. It plays a critical role in essential processes like carbon dioxide fixation, oxygen production, and biomass generation (Enea *et al.*, 2025). The microbial community

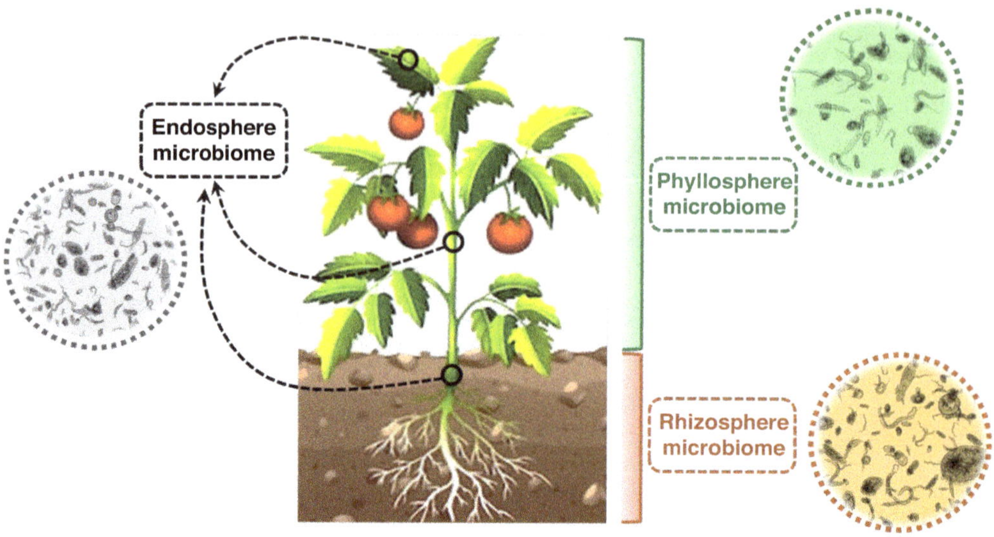

Fig. 10.2. Demonstration of different plant microbiome and their localization on the plant.

in this zone includes bacteria, filamentous fungi, yeasts, algae, protozoa, and nematodes. These microbes contribute to nitrogen fixation, disease resistance, metabolite transformation, and the production of hormones involved in plant growth. Their composition is influenced by factors such as humidity, radiation, temperature, and environmental fluctuations, which impact plant processes like photosynthesis and respiration. Additionally, wind and rain contribute to functional variations in the phyllosphere microbiome over time (Sivakumar *et al.*, 2020; Enea *et al.*, 2025).

10.2.1.2 Rhizosphere

The rhizosphere refers to the underground soil zone directly surrounding plant roots, where microbial activity is particularly intense. This region includes the soil volume affected by roots, as well as root tissues that alter the surrounding physical, chemical, and biological properties (Thepbandit and Athinuwat, 2024; Wankhade *et al.*, 2025). Plants actively influence the composition and functionality of their rhizosphere microbiome by altering soil structure, pH and oxygen availability, while releasing carbon-rich exudates that serve as energy sources for microbial communities. Microbial diversity is highest near the root surface (rhizoplane) and gradually decreases with distance (Saeed *et al.*, 2021).

10.2.1.3 Endosphere

The endosphere is composed of the innermost plant tissues colonized by microorganisms. These microbes colonize internal structures such as roots, stems, and leaves, often forming beneficial relationships with the host plant (Babalola and Adedayo, 2023). Endophytic microorganisms contribute to plant growth, enhance resistance against pathogens, and produce essential metabolites that support plant health and development. The term "endophyte" refers to organisms that live within plant tissues without causing harm. These microbes can be mutualistic, providing benefits to the plant, or commensal, existing without significant impact (Babalola and Adedayo, 2023).

Endophytic bacteria primarily belong to groups such as Proteobacteria, Firmicutes, and Actinobacteria, while fungal endophytes are commonly found in the Ascomycota, Basidiomycota, and Zygomycota groups (Muthu Narayanan *et al.*, 2022; Kumari *et al.*, 2023). The diversity and abundance of endophytes vary based on plant tissue type, developmental stage, soil composition, and environmental conditions. Generally, microbial diversity is highest in roots, followed by stems and leaves, with variations observed across different plant species. Factors such as plant age, organic matter content in soil, and environmental stress can influence the distribution of endophytic microbes (Xia *et al.*, 2022).

10.2.2 Factors influencing microbiome composition in medicinal plants

The microbiome of medicinal plants plays a vital role in supporting their growth, health, and the production of secondary metabolites. The structure and function of these plant-associated microbial communities are influenced by a variety of biotic and abiotic factors that collectively shape microbial diversity, abundance, and activity (Fig. 10.3). Understanding these factors is essential for optimizing plant–microbe interactions to enhance medicinal properties and sustainable cultivation (Köberl *et al.*, 2013).

10.2.2.1 Plant species

The host plant species plays a pivotal role in determining the composition of its associated microbiome. Even when growing in the same environment, different plant species can harbor distinct microbial communities (Negre Rodríguez *et al.*, 2025). Studies comparing root microbiomes of various crops, such as wheat, sorghum, and maize, have demonstrated differences in microbial community composition. Similarly, research on grapevines and nearby weed species has shown that microbiome compositions in roots and rhizospheres vary, with more pronounced differences observed in root-associated communities (Negre Rodríguez *et al.*, 2025).

Distantly related plant species tend to have greater variations in their microbiome compositions, suggesting that plant phylogeny plays a role in structuring microbial communities (Liu *et al.*, 2017; Negre Rodríguez *et al.*, 2025). This influence extends to endophytic microbes, as different plant species tend to host unique bacterial and fungal endophytes. For example, bacterial communities found in potato roots may be absent or rare in eucalyptus roots, indicating that plants selectively recruit their endophytic microbes (Liu *et al.*, 2017).

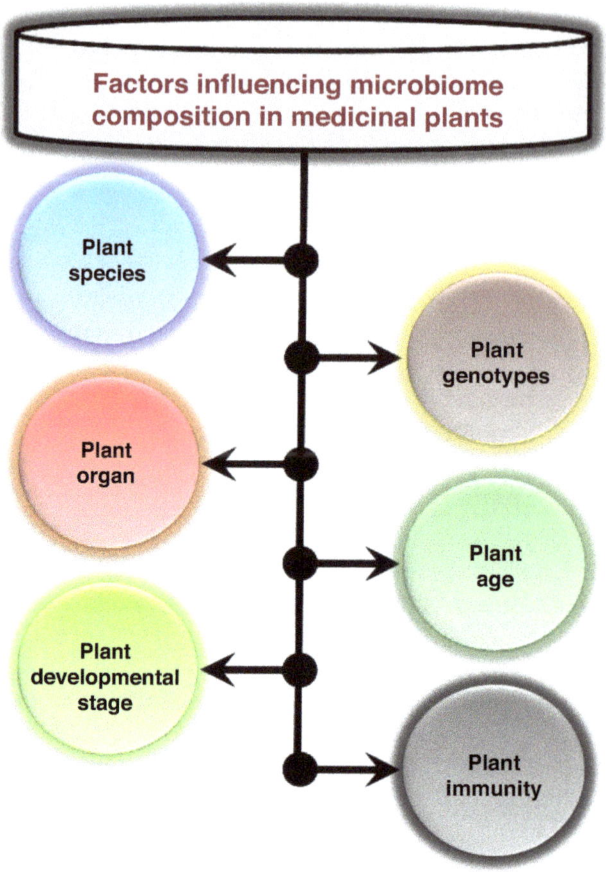

Fig. 10.3. Association of factors influencing microbiome composition in plants.

In addition to endophytic and rhizospheric microbiomes, the phyllosphere microbiome (microorganisms on leaf surfaces) is also shaped by plant identity (Enea *et al.*, 2025). Research suggests that leaf microbiome composition correlates with plant evolutionary relationships, mirroring patterns observed in endospheric microbial communities (De Mandal and Jeon, 2023; Enea *et al.*, 2025).

The ability of plants to recruit specific microbes from their environment suggests an evolutionary adaptation to shape root microbiome assemblages. Microbial community composition in the rhizosphere and endosphere has been linked to plant taxonomy (Sivakumar *et al.*, 2020; Enea *et al.*, 2025). Some studies suggest that while soil type is a main factor shaping rhizosphere and root microbiomes, root and nodule endophyte compositions are more predisposed by plant species. Differences in microbiome assemblage across plant species are likely due to variations in resource consumption. Additionally, factors such as leaf permeability, surface properties, root exudates, antimicrobial compound production, and plant immune responses contribute to microbiome diversity (Fitzpatrick *et al.*, 2018).

10.2.2.2 Plant genotypes

Genetic variations within a plant species can also influence microbiome composition. Different plant genotypes, including various cultivars of a species, have been shown to host distinct microbial communities. For instance, research on different potato varieties found that certain microbial species were cultivar specific. Similar genotype-dependent effects have been observed in the rhizosphere communities of potato, maize, wheat, pea, oat, and sweet potato (Olanrewaju *et al.*, 2024; Negre Rodríguez *et al.*, 2025).

Studies on microbial diversity in maize have demonstrated that plant genotype influences both microbial richness and diversity. It has been estimated that host genotype contributes to a measurable proportion of microbiome variation in the rhizosphere. Additionally, genetic differences between transgenic plant and wild-type plant lines are linked to distinctive rhizosphere microbial communities (Saeed *et al.*, 2021; Negre Rodríguez *et al.*, 2025). Within-species genetic variation can also impact the microbiome composition of plant leaves. Research on bacterial communities in leaves and roots of certain plants suggests that while leaf microbiomes are strongly influenced by host genotype, root microbiomes tend to vary depending on environmental conditions (Peiffer *et al.*, 2013; Saeed *et al.*, 2021; Negre Rodríguez *et al.*, 2025).

Some researchers propose that plant genotype affects key microbial species, which in turn influence the broader microbiome through microbe–microbe interactions. These effects may impact plant fitness and growth. While host specificity in microbiomes may be partly driven by nutrient preferences of plants, further research is needed to determine whether these associations are heritable, how strong they are, and whether they can be utilized in plant breeding strategies (Brachi *et al.*, 2022).

10.2.2.3 Plant organ

Distinct microbial communities inhabit different plant tissues, with variations observed between surface and internal plant structures. Each plant tissue type plays a greater role in determining microbiome composition than factors such as host species or soil microbiome (Brown *et al.*, 2020; Sessitsch *et al.*, 2023). This is likely due to the unique adaptation strategies required by microbes to colonize different plant tissues. For example, surface tissues such as leaves are subjected to continuous environmental fluctuations and offer a comparatively limited nutrient supply relative to roots or internal tissues. As a result, microbes that colonize leaf surfaces must be adapted to these harsher conditions (Sessitsch *et al.*, 2023).

While some studies suggest that plant organ type has little influence on microbial community composition, others highlight significant differences. Research on certain plant species has shown that bacterial communities are strongly affected by specific plant compartments, with the rhizosphere, endosphere, and phyllosphere hosting distinct microbial populations that differ from those found in surrounding soil (Enea *et al.*, 2025). In contrast, fungal communities appear to be more influenced by the geographic location of the host plant, which may be due to dispersal limitations in fungi, as they share similarities with plants and animals in this regard, whereas bacteria exhibit different dispersal patterns (Sivakumar *et al.*, 2020; Peng *et al.*, 2022; Enea *et al.*, 2025).

10.2.2.4 Plant age and developmental stage

Plant age and developmental stage significantly influence microbiome composition. In plant–pathogen interactions, resistance linked to plant age is commonly observed, with older plants often exhibiting increased resistance. Studies on symbiosis also indicate that plant development plays a key role in shaping microbial communities (Turner *et al.*, 2013). Research on rhizosphere bacterial communities has shown that seedlings recruit distinct microbiomes at different developmental stages.

Plants secrete a diverse array of compounds and phytochemicals through their root exudates, some of which are specific to particular developmental stages and play a key role in shaping the root-associated microbiome. Analysis of microbial communities in different plant species has demonstrated that plant age strongly affects bacterial populations in the rhizosphere (Pantigoso *et al.*, 2022; Yang *et al.*, 2025). Factors such as root growth, morphology, architecture, physiology, and exudate composition are all linked to age-related microbiome shifts. Furthermore, research is required to identify the specific root exudates involved in microbiome assembly at different developmental stages. Understanding how plants communicate with their microbial communities in the rhizosphere could lead to advancements in agricultural practices by optimizing beneficial microbial applications for crop improvement (Ma *et al.*, 2022; Yang *et al.*, 2025).

10.2.2.5 Plant immunity and signaling

Plant health plays a pivotal role in shaping the composition of the plant's microbiome. To defend against pathogenic threats, plants have developed two major immune strategies: pattern-triggered immunity (PTI) and effector-triggered immunity (ETI). PTI is initiated when pattern recognition receptors (PRRs) on the plant cell surface detect conserved microbial or pathogen-associated molecular patterns (MAMPs/PAMPs) (Masood *et al.*, 2024). However, the mechanisms by which plants recognize and respond to beneficial or nonpathogenic microbes remain less well understood, particularly in terms of how immune responses are modulated to distinguish between beneficial and harmful organisms (Nguyen *et al.*, 2021).

In response to herbivory or pathogen attack, plants emit specific hormones and volatile organic compounds that influence the composition of root exudates. These chemical changes can in turn alter the microbial community structure in the rhizosphere (Sharma *et al.*, 2023). For example, infestations by aphids or microbial pathogens have been associated with an increased

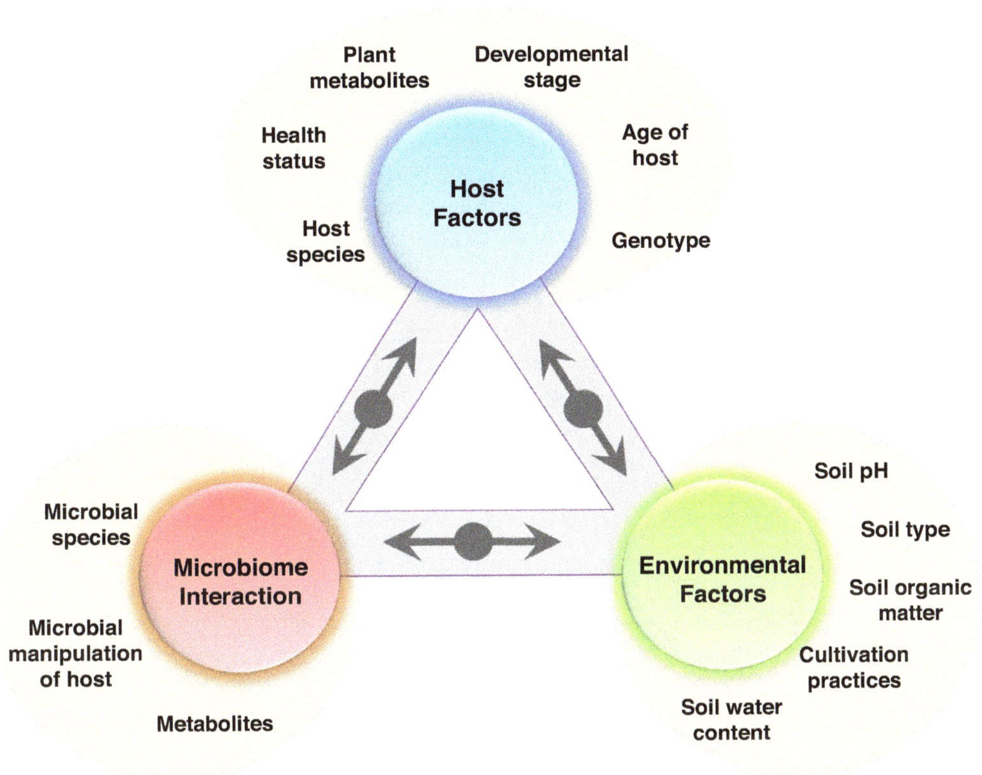

Fig. 10.4. The dynamic and interconnected relationships among three major components: host factors, environmental factors, and microbiome interaction.

presence of beneficial rhizobacteria in plant roots. Under such stress conditions, plants may adjust the secretion of organic acids, such as malic acid and citric acid, to attract protective microbes capable of suppressing disease. Additionally, certain exudates stimulate biofilm formation in beneficial bacteria, enhancing their colonization efficiency and their capacity to defend the plant (Sharma *et al.*, 2023).

Most research on plant–microbe interactions has focused on one-to-one relationships, yet in natural environments, plants are simultaneously exposed to multiple pathogens and herbivores. Understanding how these multiple interactions collectively shape the plant microbiome remains an important area of study (Köberl *et al.*, 2013; Kumari *et al.*, 2023). Recent findings suggest that specific compounds in root exudates, such as sugars and sugar alcohols, can influence microbiome composition, further demonstrating the complex relationship between plants and associated microbial communities (Köberl *et al.*, 2013; Kumari *et al.*, 2023; Yusuf *et al.*, 2025) (Fig. 10.4).

Plants utilize various strategies to respond to pathogen infections and insect attacks. One such strategy is activating defense responses in plant roots that can affect microbial recruitment in the rhizosphere region. Induced systemic resistance (ISR) and systemic acquired resistance (SAR) are two key signaling pathways that regulate plant defense. For instance, insect feeding on leaves can trigger these systemic responses, leading to the attraction of beneficial bacteria to the root zone (Jung *et al.*, 2024). Similarly, increased jasmonic acid signaling due to either injury or external application has been linked to higher colonization of beneficial mycorrhizal fungi. Altering salicylic acid signaling also results in distinct microbial community compositions in plant tissues. Some studies suggest that plants deficient in jasmonate-mediated defense pathways tend to harbor a more diverse epiphytic microbiome (Jung *et al.*, 2024).

Plant-released chemical signals, such as flavonoids, can trigger specific responses in rhizosphere microbial communities. Additionally, certain compounds, like strigolactones, influence mycorrhizal fungal growth and even promote seed germination in parasitic plants (Wankhade *et al.*, 2025). These findings highlight the intricate ways in which plant defense systems regulate microbiome composition, although the outcomes can vary depending on environmental and biological factors (Fig. 10.4).

10.3 Bioactive Secondary Metabolites and Microbiome Interaction

10.3.1 Types of secondary metabolites in medicinal plants

Metabolism encompasses all biochemical reactions within an organism, producing metabolites that are generally small molecules. These metabolites are broadly categorized into two groups: primary metabolites, which are essential for core cellular processes such as growth, respiration, and reproduction; and secondary metabolites, which are not directly involved in these fundamental functions but serve important ecological roles (Fernie *et al.*, 2024). The term "secondary metabolite" was first introduced in 1891, signifying compounds that are not universally present in all living cells but provide survival advantages, particularly in response to environmental stresses. Although secondary metabolites originate from primary metabolism, they do not contribute directly to the structural or functional integrity of an organism. However, their presence significantly enhances an organism's ability to survive, particularly under ecological pressures (Mehra *et al.*, 2022; Fernie *et al.*, 2024).

The distinction between primary and secondary metabolism is not always clear-cut, as several biochemical pathways overlap. For instance, amino acids considered primary metabolites also serve as precursors for numerous secondary metabolites. Likewise, sterols, which are structural components of cellular membranes, are often classified as secondary metabolites (Fig. 10.5). This overlap suggests a shared biochemical network between primary and secondary metabolism. Moreover, secondary metabolites function as a biochemical buffer system, allowing excess carbon and nitrogen from primary metabolism to be stored and later converted back into primary metabolites

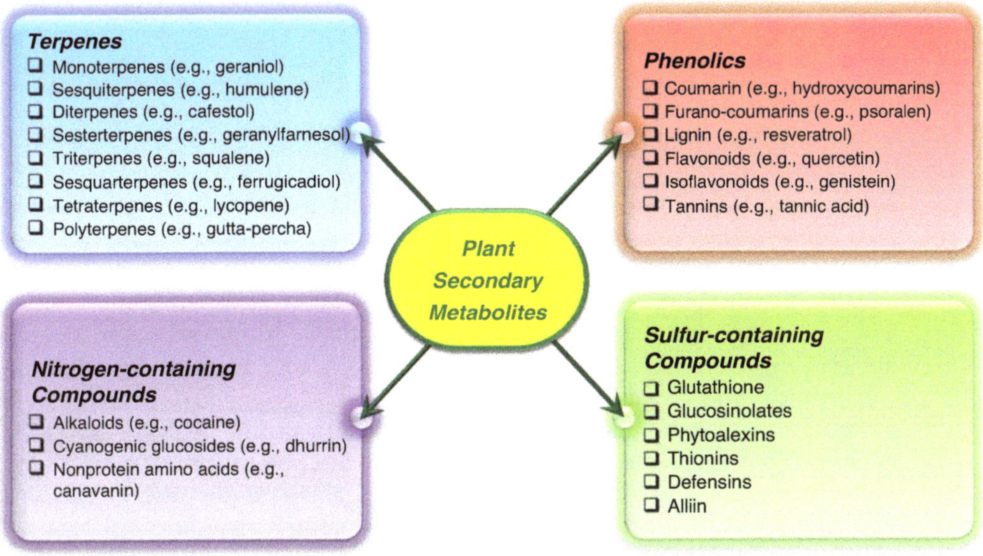

Fig. 10.5. Classes of secondary metabolites and their specific compounds.

when needed. Their production is influenced by factors such as growth, tissue differentiation, and external environmental pressures (Pott *et al.*, 2019).

10.3.2 Functions and applications of secondary metabolites

Although secondary metabolites are not directly required for plant growth and development, they play crucial roles in plant defense, communication, and ecological interactions. These compounds protect plants against herbivores and pathogens through antimicrobial and insecticidal molecules such as alkaloids, phenolics, and terpenoids. They also facilitate symbiotic relationships, as demonstrated by flavonoids that enable nitrogen-fixing bacteria to interact with leguminous plants (Divekar *et al.*, 2022; Praveen *et al.*, 2023). In addition, certain secondary metabolites contribute to metal transport and stress adaptation; for example, phytochelatins help plants tolerate and detoxify heavy metals. Furthermore, these compounds influence plant–plant interactions through allelopathy, affecting the growth and development of neighboring plants (Divekar *et al.*, 2022; Praveen *et al.*, 2023).

Humans have long utilized these bioactive compounds across diverse fields due to their wide range of beneficial properties. In medicine, many secondary metabolites act as antibiotics, enzyme inhibitors, immunomodulators, and anticancer agents, playing crucial roles in disease prevention and treatment (Alvarez-Leite, 2025). In agriculture, they function as biopesticides, insecticides, and plant growth regulators, supporting sustainable crop management and protection. Within the food and nutrition sector, these compounds contribute as natural pigments, nutraceuticals, and flavoring agents, enhancing both the nutritional value and sensory quality of food products (Alvarez-Leite, 2025).

10.3.3 Classification of secondary metabolites

Secondary metabolites are broadly categorized into three groups based on their biosynthetic origins: Phenolic compounds are mainly derived from simple sugars and characterized by benzene rings. They include flavonoids, tannins, lignins, and coumarins, which play roles in plant pigmentation, UV protection, and defense (Al-Khayri *et al.*, 2023). Terpenoids and steroids are mainly composed of carbon and hydrogen, and include essential oils, carotenoids, and sterols, which contribute to plant fragrance, coloration, and membrane structure. Nitrogen-containing compounds include alkaloids, cyanogenic glycosides, and glucosinolates, which often function as defense molecules against herbivores and pathogens. Understanding the biosynthesis and ecological roles of secondary metabolites enhances their application in medicine, agriculture, and industry, emphasizing their significance beyond plant survival (Zagoskina *et al.*, 2023).

The major classes of bioactive secondary metabolites and their functional applications in medicinal plants are discussed below.

10.3.3.1 Flavonoids

Flavonoids represent one of the largest and most diverse groups of plant secondary metabolites, with over 4000 distinct compounds identified to date. They are widely distributed across green plants and are predominantly present in glycosylated forms within leaves, flowers, stems, and roots (Dias *et al.*, 2021). Structurally, flavonoids are characterized by two benzene rings connected by a three-carbon bridge, and they are classified into several major subclasses, including chalcones, flavones, flavanols, flavanones, anthocyanins, and isoflavones. Many flavonoids are vividly pigmented, contributing to the coloration of various plant tissues (Panche *et al.*, 2016).

10.3.3.2 Phenylpropanoids

Phenylpropanoids are specialized metabolites produced exclusively by plants and certain microorganisms, derived from the shikimate pathway (Ninkuu *et al.*, 2025). This metabolic route is responsible for the biosynthesis of essential aromatic amino acids, including phenylalanine and tyrosine, which are vital for numerous biological processes in both plants and animals. As humans are unable to synthesize these amino acids endogenously, they must acquire them through dietary intake, primarily from plant-based sources (Ferrer *et al.*, 2008: Ninkuu *et al.*, 2025).

10.3.3.3 Terpenes and terpenoids

Terpenes represent a highly diverse class of metabolites, with more than 35,000 different compounds characterized to date. Structurally, they are composed of isoprene units, which can be modified by cyclization reactions. Based on the number of isoprene units in their carbon skeletons, terpenes are classified into various groups (Khanam *et al.*, 2025). Terpenoids, a modified form of terpenes, exhibit a range of biological activities, including antitubercular effects, anticancer properties, anxiolytic activity and mutagenic effects. Some well-known terpenoid-derived anticancer agents include paclitaxel (Taxol®) from the Pacific yew tree (*Taxus brevifolia*), which inhibits cell proliferation and induces apoptosis. Betulinic acid, found in *Betula* spp., has shown anticancer effects by targeting multiple cell signaling pathways (Khanam *et al.*, 2025).

10.3.3.4 Nitrogen-containing compounds (alkaloids)

Alkaloids are nitrogen-containing cyclic organic compounds that, although relatively uncommon in nature, exhibit potent biological activity. This diverse group includes several well-known compounds such as caffeine, nicotine, cocaine, and morphine, many of which have pronounced effects on the central nervous system, including anxiolytic, analgesic, and hallucinogenic properties (Heinrich *et al.*, 2021). Despite representing a relatively small subset of plant secondary metabolites, alkaloids contribute to nearly 50% of all plant-derived pharmaceuticals, underscoring their immense pharmacological significance. Anticancer alkaloids include vincristine and vinblastine (from *Catharanthus roseus*, Madagascar periwinkle), which are used in chemotherapy for leukemia and lymphoma, and camptothecin (from *Camptotheca acuminata*), which inhibits topoisomerase I, an enzyme crucial for DNA replication and repair (Heinrich *et al.*, 2021).

10.3.3.5 Polyphenols

Polyphenols are bioactive compounds found in many medicinal plants, with strong anticancer properties. Some well-researched polyphenols include resveratrol (found in grapes and berries), which suppresses angiogenesis, metastasis, and cell proliferation while promoting apoptosis, and epigallocatechin gallate (EGCG) (from green tea), which influences multiple cell survival, proliferation, and angiogenesis pathways (Roszkowski, 2023).

10.3.3.6 Sulfur-containing compounds

Sulfur-containing compounds in medicinal plants have demonstrated promising anticancer effects. Notable examples include sulforaphane (found in cruciferous vegetables like broccoli and cauliflower), which activates phase II detoxifying enzymes, suppresses tumor growth, and promotes apoptosis, and allyl isothiocyanate (from *Allium* species such as garlic), which inhibits tumor growth, induces apoptosis, and reduces metastasis (Miękus *et al.*, 2020).

In addition, plant-derived secondary metabolites, including flavonoids, phenylpropanoids, terpenoids, alkaloids, polyphenols, and sulfur-containing compounds, play critical roles in plant defense and human health. Their broad-spectrum anticancer, antimicrobial, and therapeutic effects make them valuable for medicine, nutrition, and agriculture. Further research into their mechanisms and interactions with the human body could lead to new pharmaceutical and therapeutic applications (Al-Khayri *et al.*, 2023).

10.3.4 Role of secondary metabolites in shaping plant-associated microbiomes

10.3.4.1 Plant secondary metabolites

Secondary metabolism involves metabolic pathways that produce small molecular compounds nonessential for plant growth and reproduction. In plants, these secondary metabolites (PSMs) derive from primary metabolites or intermediate products of biosynthetic pathways. Based on their biosynthetic origins, PSMs can be classified into phenolics, terpenes, steroids, alkaloids, and flavonoids (Dias *et al.*, 2021; Pang *et al.*, 2021; Al-Khayri *et al.*, 2023). These compounds serve multiple functions, including plant growth, development, immunity, defense signaling, and responses to environmental stresses (Fig. 10.6). Additionally, they contribute to pest and pathogen resistance, mediate plant–microbe interactions, and influence microbial community composition (Pang *et al.*, 2021). Many PSMs also hold significance in human health and agriculture, impacting economic productivity. Despite significant advances in omics technologies that have deepened our understanding of plant metabolomes, the functions of many PSMs remain largely unknown.

10.3.4.2 Plant microbiome

The plant microbiome is composed of microbial communities in the rhizosphere, phyllosphere, and endosphere. These microbial consortia aid host plants by enhancing immunity, suppressing diseases, supplying nutrients, and protecting against environmental stressors (Sivakumar *et al.*, 2020). Recent advances in sequencing-based technologies have expedited the classification of plant microbiomes, including epiphytic, seed, core microbiomes, synthetic communities (SynComs), and DefenseBiomes (van Leeuwen *et al.*, 2023). These microbiomes are dynamic, influenced by biotic and abiotic factors, with plant-secreted PSMs playing a key role in modulating microbial interactions. For example, maize secondary metabolites like benzoxazinoids influence bacterial composition, aiding plant adaptation to environmental conditions (Pantigoso *et al.*, 2022) (Fig. 10.6).

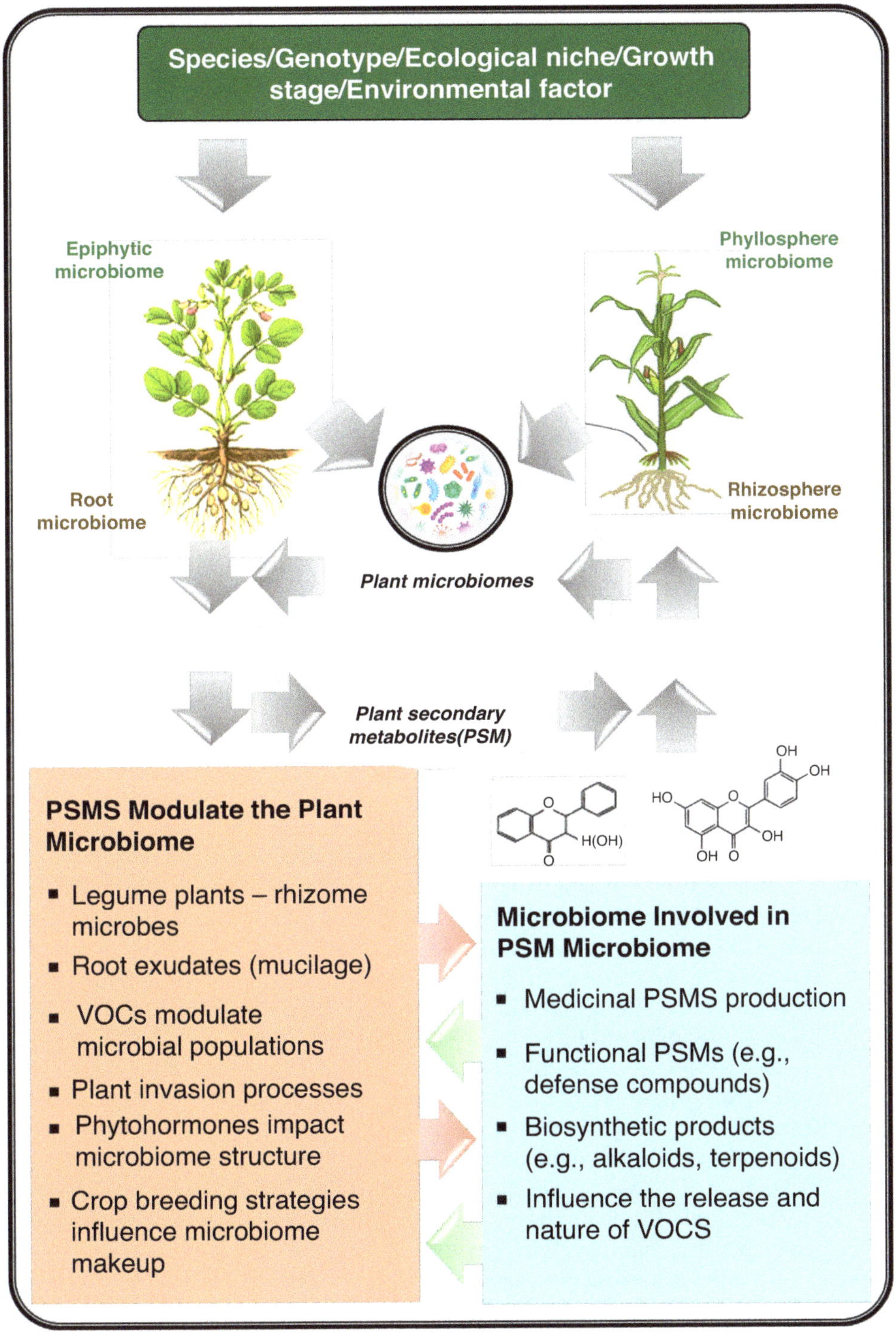

Fig. 10.6. Dynamic interaction between plant microbiomes and plant secondary metabolites (PSMs).

10.3.4.3 Synthetic communities and microbiome analysis

SynComs simplify complex microbial assemblies, allowing researchers to analyze plant genotype effects on microbial associations. These communities range from small taxonomic assemblies to multi-kingdom communities exceeding 200 strains (Rodriguez *et al.*, 2019; van Leeuwen *et al.*, 2023). Although larger SynComs better reflect microbiome diversity, their stability remains a concern. Experiments with SynComs have demonstrated that plant metabolites influence microbial community composition, impacting functions such as root microbiome assembly, nutrient availability, and pathogen resistance (van Leeuwen *et al.*, 2023).

10.3.4.4 Exometabolomics and microbial interactions

Exometabolomics, also called metabolic footprinting, identifies extracellular metabolites using techniques like LC-MS, GS-MS, or NMR. This approach provides insights into microbial communication and metabolism, aiding microbiome studies (Kao *et al.*, 2021). For instance, research using *Arabidopsis* root exudates has revealed variations in metabolite utilization among microbial strains, affecting microbial community structure. These findings demonstrate the potential of exometabolomics in understanding metabolic dependencies in the interactions plant and microorganisms (Kosina *et al.*, 2018).

10.3.4.5 Genome-wide association studies (GWAS)

Genetic variation among plant species and accessions influences microbiome composition. GWAS enables the identification of genes linked to microbiome assembly without prior candidate selection, allowing unbiased discovery of plant–microbe interaction mechanisms (Deng *et al.*, 2021; Singh *et al.*, 2024). Studies have shown that root microbiomes are shaped by genetic factors related to cell wall integrity, immunity, and secondary metabolism. In particular, *Arabidopsis* studies have linked plant genotypes to variations in microbial activity, emphasizing the functional significance of host genetics in microbiome interactions (Deng *et al.*, 2021).

Genome-wide association studies have also identified plant genes affecting microbiome-related functions, such as plant growth promotion (PGP). Research has demonstrated that different *Arabidopsis* accessions vary in their ability to host beneficial microbes and resist pathogens. These findings underscore the role of genetic diversity in shaping microbial communities and highlight the potential of GWAS in uncovering novel plant–microbe interaction mechanisms (Horton *et al.*, 2014).

Moreover, secondary metabolites play a crucial role in shaping plant-associated microbiomes by influencing microbial composition and interactions. The use of synthetic communities, exometabolomics, and GWAS has significantly advanced our understanding of these complex interactions (Jacoby *et al.*, 2021).

10.3.5 Microbial modulation of secondary metabolite biosynthesis

Microbiomes, or microbial communities, play a crucial role in ecosystem health, influencing processes from global nutrient cycling to host disease dynamics. These communities are highly complex due to their diverse species and dynamic interactions. Microorganisms produce a wide array of metabolites that mediate interspecies interactions, shaping ecosystem functions through competition, cooperation, and communication (Boon *et al.*, 2014).

Secondary metabolites, produced by bacteria, exhibit significant biological activities, serving as antivirals, antimicrobials, anticancer agents, immune-suppressants, and pesticides. In natural environments, these compounds mediate microbial interactions, ranging from competitive to cooperative, influencing both narrow- and broad-spectrum molecular targets (Pardo-Esté *et al.*, 2024). However, characterizing the *in situ* activities of these metabolites remains challenging, as they can trigger coordinated community responses and regulate secondary metabolism (Pardo-Esté *et al.*, 2024).

The genes responsible for secondary metabolite biosynthesis, transport, and regulation are often organized into biosynthetic gene clusters (BGCs) within bacterial genomes. These gene clusters are linked to bacterial developmental processes, including sporulation, as well as responses to antibiotics and various environmental stressors (Covington *et al.*, 2021). Many BGCs remain silent under traditional laboratory conditions, complicating their functional characterization. Genome mining suggests that the majority of BGCs remain unexplored, and current screening methods for activating silent BGCs are often untargeted (Covington *et al.*, 2021). Despite incomplete knowledge of BGC roles in pure cultures, understanding their functions within microbial communities is even more limited. Microbial communities exhibit immense complexity, with each member potentially harboring numerous BGCs, which can influence metabolite expression across species. Investigating BGC regulation and community interactions requires reductionist approaches to unravel the intricate networks governing microbial secondary metabolism (Chevrette *et al.*, 2022).

10.3.5.1 In vitro *approaches for the sustainable production of secondary metabolites*

In vitro metabolic engineering represents a powerful and innovative tool for the sustained production of commercially valuable secondary metabolites and bioengineered compounds. These approaches enable the large-scale synthesis of targeted bioactive molecules and support the discovery and development of novel natural products with pharmaceutical and nutritional relevance (Alamgir, 2018; Mipeshwaree Devi *et al.*, 2023). By strategically altering the metabolic pathways of plants, the concentration and diversity of secondary metabolites (SMs) can be significantly enhanced. Bioactive compounds derived from medicinal, ornamental, and aromatic plants through undifferentiated organogenic and callus culture techniques contribute substantially to the pharmaceutical and food industries. Moreover, the application of abiotic stressors and elicitors has been shown to play a pivotal role in stimulating metabolite accumulation in *in vitro* plant cultures across various species (Najmi *et al.*, 2022).

10.3.5.2 Tissue culture approaches

Tissue culture techniques have significantly advanced the production of PSMs by enabling precise control over growth conditions, nutrient composition, and the induction of genetic and epigenetic modifications (Chevrette *et al.*, 2022; Pardo-Esté *et al.*, 2024). Through the strategic manipulation of environmental and physiological parameters, such as nutrient availability, hormone concentrations, and stress factors, these *in vitro* methods effectively stimulate and enhance the biosynthesis of secondary metabolites under controlled conditions (Mipeshwaree Devi *et al.*, 2023). Genetic and epigenetic alterations activate silent biosynthetic pathways, enhancing metabolite production. Additionally, subjecting plant tissues to abiotic or biotic stressors triggers defense responses, further boosting SM synthesis. The establishment of cell suspension or organ cultures in bioreactor systems enables continuous, scalable, and controlled production of secondary metabolites (Ozyigit *et al.*, 2023).

Tissue culture techniques provide flexible and efficient strategies for augmenting the production of high-value plant-derived compounds, offering both cost-effective and environmentally sustainable alternatives to conventional methods. *In vitro* cultures are maintained under strictly controlled, aseptic conditions, thereby eliminating the influence of external variables such as seasonal changes, geographical constraints, and environmental fluctuations (Mipeshwaree Devi *et al.*, 2023).

A variety of culture systems, including callus, hairy root, shoot, and suspension cultures, are employed to optimize the biosynthesis of targeted secondary metabolites in commercially important plant species (Wijerathna-Yapa and Hiti-Bandaralage, 2023). Among these approaches, callus cultures are particularly effective in producing bioactive compounds from rare and endangered medicinal plants. Enhanced phytochemical profiles have been reported in several species, including *Coryphantha macromeris*, *Echinacea purpurea*, and *Ziziphora persica*, through the application of *in vitro* culture techniques (Sagharyan *et al.*, 2020). Notably, hairy root cultures often demonstrate superior metabolite productivity compared to other culture types. For example, *Plumbago zeylanica* exhibited a threefold increase in plumbagin production in liquid root suspension cultures, and a 6.5-fold increase in semisolid cultures. Similarly, in *Silybum marianum*, the application of γ-ray elicitation in cell suspension cultures significantly boosted the production of silymarin (Kavi Kishor *et al.*, 2025).

CULTURE MEDIA MODIFICATION AND ELICITOR APPLICATIONS. A robust and well-defined technological framework for the production of plant-based natural products is established through the regulated *in vitro* cultivation of plant cells and tissues (Alamgir, 2018; Mipeshwaree Devi *et al.*, 2023). Critical factors such as the pH of the culture medium, the concentration of additives, the type and availability of carbon sources, and the composition of basal media play pivotal roles in influencing cellular growth and metabolite biosynthesis. For instance, in *Cnidium officinale* callus cultures, varying concentrations of benzyladenine (BA) and 2,4-dichlorophenoxyacetic acid (2,4-D) were shown to significantly impact the accumulation of phenolic compounds and flavonoids. Similarly, in *Dysosma pleiantha*, modifications to B5 medium components, including peptone and 2,4-D, resulted in marked increases in secondary metabolite yields, with kaempferol, podophyllotoxin, and quercetin levels rising by 16.3-fold, 12.39-fold, and 5.03-fold, respectively (Espinosa-Leal *et al.*, 2018).

The application of stress-inducing elicitors is a widely adopted strategy to further enhance the production of secondary metabolites. These elicitors include biotic and abiotic agents such as chitosan, plant growth regulators (PGRs), hormones, microbial derivatives, and physical stressors (Petrova *et al.*, 2024). Chemical elicitors like polyethylene glycol (PEG) and sodium chloride (NaCl) are known to activate metabolic pathways that promote secondary metabolite synthesis. For example, in callus cultures of *Rhodiola imbricata*, a rare adaptogenic herb, red light exposure significantly improved biomass accumulation, whereas blue light conditions led to the highest salidroside production after 21 days of culture (Petrova *et al.*, 2024).

10.3.5.3 Molecular basis of plant metabolite synthesis

Abiotic and biotic stressors are key environmental factors that significantly influence plant growth, development, and overall agricultural productivity. Interestingly, plants continuously encounter stressors that trigger complex response interactions, ultimately reducing agricultural output. Given changing climate conditions, a deeper understanding of plant stress responses is necessary to improve crop resilience (Pandey *et al.*, 2017; Afoma *et al.*, 2025a).

Plant responses to stress involve complex cellular, molecular, physiological, and biochemical adaptations, largely governed by key transcription factors (TFs) that regulate stress-responsive genes and signaling pathways. Acting as master regulators, TFs play a central role in modulating plant responses to a variety of environmental stimuli, hormones, and developmental cues. Extensive research has identified several major TF families, such as NAC, bZIP, WRKY, MYB, and ERF, that are critical for enhancing stress tolerance in both model systems and agriculturally important crops (Joshi *et al.*, 2016). Under stress conditions, TFs bind to specific cis-regulatory elements within target gene promoters, leading to altered gene expression patterns. Their ability to interact with DNA and form transcriptional complexes enables the coordinated regulation of a wide range of downstream genes involved in stress adaptation. According to the Plant Transcription Factor Database, WRKY, MYB, bZIP, ERF, and NAC genes have been implicated in stress-responsive pathways. Advances in genetic engineering targeting these transcription factors have demonstrated significant potential for enhancing plant resilience and developing stress-tolerant crop varieties (Sheshadri *et al.*, 2016).

10.3.5.4 Stress signal transduction and metabolite biosynthesis

Upon sensing environmental stress, plant cells activate sophisticated signal transduction pathways initiated by receptors or sensors located on the cell wall or plasma membrane. These pathways involve a cascade of intracellular events mediated by cytoplasmic kinases, reactive oxygen species (ROS), calcium ions (Ca^{2+}), and key phytohormones such as abscisic acid, ethylene, salicylic acid, strigolactones, and jasmonic acid. These signaling molecules coordinate and amplify stress responses through networks involving protein kinases and phosphatases (Tuteja and Sopory, 2008; Shrivastav *et al.*, 2026). Central to these responses are calcium-dependent protein kinase and mitogen-activated protein kinase signaling pathways, which regulate the activity of transcription factors responsible for modulating stress-responsive genes. This regulation leads to the enhanced biosynthesis of secondary metabolites, which play critical roles in plant defense and adaptation. A deeper understanding of these molecular mechanisms holds substantial promise for improving the production of bioactive compounds and engineering plant varieties with enhanced stress tolerance (Dekomah *et al.*, 2022).

10.3.6 Bioreactors in the production of plant secondary metabolites

Bioreactors have become indispensable tools for the scalable and controlled production of PSMs. These specialized vessels support the growth of plant cells, tissues, or microorganisms in liquid nutrient media under sterile and precisely regulated environmental conditions. Unlike conventional reactors, bioreactors are engineered to optimize specific biological processes and metabolic pathways, ensuring minimal contamination, consistent productivity, and reduced variability (Verdú-Navarro *et al.*, 2023).

Bioreactors can function under several operational modes, each designed to optimize production according to specific bioprocess requirements. In batch culture, all necessary nutrients are introduced at the start of the process, with no further additions until the harvest phase (Palladino *et al.*, 2024). Fed-batch culture involves the intermittent addition of nutrients without removing the culture medium, thereby prolonging the production phase and preventing nutrient depletion. In a chemostat culture, fresh medium is continuously supplied while an equal volume of spent medium and cells is simultaneously removed, maintaining a steady-state environment ideal for controlled and consistent production. Perfusion culture operates similarly, with a continuous inflow of fresh medium and outflow of waste at equal rates, enabling long-term cell viability and productivity until nutrient resources are exhausted (Xu *et al.*, 2023; Palladino *et al.*, 2024).

Among these strategies, continuous culturing has emerged as a particularly effective approach for enhancing bioprocess efficiency and maximizing yield. For example, continuous cell cultures of *Taxus cuspidata* and semicontinuous cultures of *Taxus canadensis* have demonstrated substantial improvements in paclitaxel production (Yin *et al.*, 2024). Remarkably, systems employing 100% cell recycling achieved sustained metabolite output over extended periods, underscoring the potential of continuous bioprocessing for industrial-scale secondary metabolite production (Ozyigit *et al.*, 2023).

10.3.6.1 Efficiency of bioreactor systems

The efficiency of bioreactor systems depends on two fundamental components.

- *Macrokinetic control*: Refers to the regulation of microbial or plant cell growth and the synthesis of secondary metabolites.
- *Operational parameters*: Include optimization of pH, temperature, carbon sources, oxygen supply, water content, vitamin availability, gas exchange, and removal of byproducts (Lv *et al.*, 2024).

Modern bioreactors function as biological factories, offering numerous benefits such as year-round, independent plant material supply; consistent and reproducible production; enhanced nutrient uptake; and streamlined harvesting. These advantages contribute to greater efficiency, reduced environmental impact, and lower production costs (Mitra and Murthy, 2022). Technological innovations and genetic engineering have further enhanced SM yields. *Agrobacterium rhizogenes*-mediated transformation has gained prominence for its ability to induce genetically stable and fast-growing hairy root cultures (Liu *et al.*, 2025a). These root cultures are now widely used in commercial SM production due to their high biosynthetic capacity. Both gas-phase and liquid-phase bioreactor systems are being optimized to increase metabolite yields by improving agitation, aeration, and nutrient delivery, while minimizing shear stress and ensuring homogeneous culture conditions (Liu *et al.*, 2025b).

Recent studies have demonstrated that bioreactor-based cultivation significantly enhances the production of high-value compounds, including alkaloids, flavonoids, terpenoids, and polyphenols (Dias *et al.*, 2021; Pang *et al.*, 2021; Al-Khayri *et al.*, 2023). These advancements have far-reaching implications across pharmaceuticals, agriculture, and cosmetics sectors that increasingly rely on bioactive plant compounds. As demand grows, bioreactors offer a sustainable, efficient, and controlled alternative to traditional field cultivation, paving the way for next-generation plant-based biomanufacturing (Kumar *et al.*, 2023b).

10.4 Case Studies: Medicinal Plants and Their Microbiomes

10.4.1 *Withania somnifera* (ashwagandha): Strategies for enhanced withanolide production

Withania somnifera, commonly known as ashwagandha or Indian ginseng, is a medicinally significant plant widely utilized in traditional systems such as Ayurveda, Siddha, Unani, and Traditional Chinese Medicine. Plant parts including roots, stems, bark, leaves, flowers, and seeds exhibit a broad spectrum of therapeutic properties, including antihyperglycemic, neuropharmacological, immunomodulatory, cardioprotective, hepatoprotective, chemoprotective, antiaging, and aphrodisiac effects (Dar *et al.*, 2015).

The plant's medicinal efficacy is primarily attributed to steroidal lactones known as withanolides, particularly withanolide-A and withaferin-A. Withaferin-A demonstrates potent antileukemic, antimetastatic, apoptotic, antiinflammatory, radiosensitizing, and antidiabetic activities, while withanolide-A exhibits neuroprotective, immunomodulatory, and anti-stress properties (Dar *et al.*, 2015; Kumar *et al.*, 2023a). These bioactive compounds are differentially distributed, with withanolide-A concentrated in the roots and withaferin-A more abundant in the leaves, and are biosynthesized via the mevalonate (MVA) and methylerythritol phosphate (MEP) pathways (Kumar *et al.*, 2023a).

Biotechnological interventions have been employed to enhance withanolide production, including the overexpression of key biosynthetic genes such as squalene synthase (SQS) and cycloartenol synthase (CAS), leading to 4–4.5-fold and 1.06–1.66-fold increases in withanolide levels, respectively. *In vitro* systems, such as cell suspension and hairy root cultures, have also been effective, particularly when treated with elicitors like salicylic acid and methyl jasmonate (Grover *et al.*, 2013; Ahmad *et al.*, 2023). Additionally, the application of seaweed extracts has been shown to upregulate key enzymes involved in withanolide biosynthesis, including squalene synthase (SQS), squalene epoxidase (SQE), 3-hydroxy-3-methylglutaryl coenzyme A reductase (HMGR), and farnesyl pyrophosphate synthase (FPPS).

Despite these advances, practical limitations related to scalability, cost, and regulatory concerns necessitate the exploration of more sustainable strategies. One promising alternative is the use of endophytes, which can enhance withanolide production through symbiotic interactions, offering a natural and eco-friendly approach to improving the yield of these valuable bioactive compounds (Grover *et al.*, 2013; Ahmad *et al.*, 2023).

Endophytes, microbial organisms residing within plant tissues, promote plant growth, enhance stress resistance, and contribute to secondary metabolite biosynthesis. In *W. somnifera*, endophytes have been identified as key contributors to withanolide production (Pandey *et al.*, 2018). Endophytic microorganisms were isolated from the roots, leaves, and seeds of *W. somnifera* to explore their diversity and physiological influence on the host plant. Using 16S rRNA gene sequencing, a total of 40 endophytic isolates were identified, comprising 11 fungal and 29 bacterial species. Specifically, the roots harbored 18 bacterial and seven fungal endophytes, the leaves contained eight bacterial and four fungal isolates, while the seeds yielded three bacterial endophytes (Pandey *et al.*, 2018).

These endophytes demonstrated a significant impact on the photosynthetic performance and overall physiology of *W. somnifera*. Notably, fungal endophytes were found to enhance photosynthetic efficiency by improving several key physiological parameters. Inoculated plants exhibited a 17–29% increase in chlorophyll content, a 29–46% rise in carotenoid concentration, a 17–43% improvement in CO_2 assimilation, an 18–31% enhancement in transpiration rate, and a 12–40% increase in stomatal conductance. These findings underscore the vital symbiotic role of endophytes in promoting plant growth, optimizing photosynthetic processes, and strengthening the physiological resilience of *W. somnifera* under varying environmental conditions (Pandey *et al.*, 2018).

Bacterial endophytes showed minimal impact on these physiological parameters. Moreover, bioreactors have revolutionized production of secondary metabolites in plants by offering a scalable and effectual platform for enhancing growth conditions and genetic manipulation. Advances in endophyte-assisted enhancement of bioactive compounds, particularly in medicinal plants like *W. somnifera*, present a sustainable and innovative strategy for increasing therapeutic metabolite yield. Further research in bioreactor technology and plant–microbe interactions is crucial for achieving cost-effective and large-scale production of valuable plant-derived compounds with applications in agriculture, nutraceuticals, and pharmaceuticals (Kumari *et al.*, 2023).

10.4.1.1 Growth-promoting potential of endophytes in W. somnifera

Several fungal endophytes isolated from *W. somnifera* demonstrated a notable ability to enhance both shoot and root biomass. In contrast, inoculation with bacterial endophytes did not lead to significant improvements in plant biomass. Specifically, plants inoculated with isolates WPLF4, WPRF5, WPRF6, WPRF7, WPRF8, WPRF9, WPRF10, and WPRF11 exhibited a 32–189% increase in shoot biomass compared to noninoculated controls (Pandey *et al.*, 2018). Furthermore, WPRF7, WPRF8, and WPRF9 significantly enhanced root biomass by 62–130%, with these three isolates showing the greatest potential in promoting both shoot and root growth.

To further investigate the plant growth-promoting potential of selected endophytes, traits such as phosphate solubilization and indole-3-acetic acid (IAA) production were assessed (Pandey *et al.*, 2018). Several fungal isolates, WPLF1 to WPRF10, exhibited phosphate solubilization activity, while WPLF3, WPRF6, WPRF7, WPRF9, WPRF10, and WPRF11 were positive for IAA production. Among the bacterial isolates, WPR12, WPR16, WPR17, WPS23, and WPR32 showed IAA production and nitrate reduction capabilities but lacked phosphate solubilization activity. These findings highlight the superior growth-promoting potential of specific fungal endophytes, particularly WPRF7, WPRF8, and WPRF9, in enhancing both biomass and functional traits in *W. somnifera* (Pandey *et al.*, 2018; Jiang *et al.*, 2025).

10.4.2 Curcumin bioavailability and formulations

Curcumin, a polyphenol compound, contains at least one aromatic ring structure with a hydroxyl group. Polyphenols, derived from the shikimic acid pathway, are present in various vegetables and secondary metabolites. Several factors influence the oral bioavailability of curcumin, including food processing, macronutrient composition, and the origin of the food. The polyphenol content in food is affected by processes such as heating, drying, grinding, climate conditions, and plant stress (Jabczyk *et al.*, 2021). Due to its poor bioavailability, various formulations have been developed to enhance curcumin absorption, including nanoparticles, lecithin-phosphatidylcholine carriers, and solid lipid nanoparticles.

Dietary lipids play a crucial role in curcumin solubility and absorption. Curcumin's lipophilic nature and hydroxyl groups contribute to its rapid metabolism in the liver and kidneys. Because of its instability in bodily fluids, consuming curcumin with oil or milk can improve its absorption. Additionally, lecithin-rich foods, such as vegetable oils or eggs, have been suggested to enhance curcumin absorption, while lecithin-based formulations have been found to improve the efficiency of curcumin nanoparticles (Tabanelli *et al.*, 2021; Mishra *et al.*, 2025).

Piperine, a natural alkaloid found in black pepper, increases curcumin bioavailability by inhibiting its biotransformation, particularly glucuronidation, in the small intestine and liver. The combination of piperine with curcumin has been shown to significantly enhance absorption. Differences in curcumin absorption have also been observed between men and women, possibly due to variations in drug efflux enzyme activity, body weight, and fat distribution (Tripathi *et al.*, 2022).

10.4.2.1 Curcumin and the microbiota

The human intestinal microbiota, composed of over 1000 bacterial species, plays a crucial role in physiology and metabolism, influencing conditions such as diabetes, obesity, inflammatory bowel disease, liver disorders, depression, psoriasis, and neurodegenerative diseases. Despite its poor systemic bioavailability, curcumin preferentially accumulates in the intestine following oral or intraperitoneal administration (Sasidharan Pillai *et al.*, 2024; Malik *et al.*, 2025; Shen *et al.*, 2025). Several studies have explored curcumin's regulatory effects on the microbiota. The human intestinal microbiota metabolizes curcumin through pathways involving active metabolites, reduction of the heptadienone backbone, and demethylation by certain bacterial species (Jabczyk *et al.*, 2021). *Escherichia coli* plays a role in curcumin metabolism by converting it into dihydrocurcumin (DHC) and tetrahydrocurcumin (THC). Additionally, other bacterial strains, including *Bifidobacteria*, *Enterococcus faecalis*, *Lactobacillus acidophilus*, and *L. casei*, contribute to curcumin metabolism, reducing its parent compound significantly (Jabczyk *et al.*, 2021).

Analysis of intestinal microbiota in curcumin-administered animals has revealed significant alterations in gut bacterial composition. A diet supplemented with curcumin has been associated with an increased abundance of certain bacterial families, while reducing others. Although no significant biodiversity changes were observed, curcumin administration appears to influence gut microbiota composition. Further investigation is warranted to evaluate its potential therapeutic applications in promoting gut health (Zam, 2018; Zhu and He, 2024).

Curcumin has also been studied in animals affected by infections, where its administration was linked to a reduction in proinflammatory bacteria and an increase in valuable bacteria, such as lactobacilli and bifidobacteria (Zhu and He, 2024). Despite the unclear mechanisms of its influence on the gut microbiota, evidence suggests that curcumin undergoes metabolic transformations, including demethylation, hydroxylation, and demethoxylation, resulting in metabolites that may be more bioactive than the parent compound. Additionally, curcumin appears to regulate gut microbiota composition, offering potential indirect health benefits (Zhu and He, 2024).

10.5 Techniques for Studying Medicinal Plant Microbiomes

10.5.1 Advanced high-throughput sequencing techniques and analytical approaches in microbiome research

First-generation DNA sequencing technology was based on the selective incorporation of chain-terminating dideoxynucleotides, with the introduction of the first automated sequencer in the late 1980s marking a significant advancement in genetic research. This technology was instrumental in sequencing the first bacterial genome and played a critical role in the Human Genome Project, laying the foundation for future innovations (Heather and Chain, 2016). The emergence of high-throughput sequencing platforms in the mid-2000s led to the development of second-generation sequencing techniques, which enabled massively parallel analysis, increased throughput, and significantly reduced costs, transforming genome sequencing from a process that took months or years to one that could be completed in days or hours (Satam *et al.*, 2023). While second-generation methods primarily rely on sequencing-by-synthesis, third-generation sequencing technologies have further advanced the field by allowing the direct analysis of single DNA molecules without the need for prior amplification (Satam *et al.*, 2023).

When considering high-throughput sequencing for microbiome studies, researchers must evaluate whether such methods are necessary for their specific experiments. Although high-throughput sequencing is fast, reliable, and increasingly cost-effective, alternative techniques such as microarrays may still be suitable for some applications (Tiwari *et al.*, 2009). Selecting the appropriate sequencing platform depends on experimental goals, as different platforms vary in throughput, speed, read length, and accuracy. In some cases, combining multiple sequencing techniques can enhance genome coverage (Tiwari *et al.*, 2011; Di Bella *et al.*, 2013). To optimize efficiency in time and cost, microbial sequencing often employs sample barcoding and multiplexing. Understanding sample complexity is essential in determining whether multiplexing is required and identifying the number of samples that can be analyzed in a single sequencing process. When using multiplexing, maintaining similar sample concentrations ensures equal data output for each sample (Smith *et al.*, 2010).

10.5.2 Metagenomics in microbiome research

Metagenomics is an advanced technique that enables the comprehensive analysis of entire microbial genomes directly extracted from environmental samples, offering detailed insights into species composition, genetic diversity, interspecies interactions, and the evolutionary dynamics of microbial communities (Nam *et al.*, 2023). Over recent decades, high-throughput sequencing technologies such as pyrosequencing and Illumina® sequencing have been extensively utilized in metagenomic and metatranscriptomic studies (Oulas *et al.*, 2015; Aguiar-Pulido *et al.*, 2016). For instance, soil samples can be processed through pyrosequencing, a sequencing-by-synthesis method that detects pyrophosphate release via luciferase activity, generating real-time sequence data. Both pyrosequencing and Illumina platforms follow essential steps of template preparation, library preparation, and sequencing, and their ability to perform massively parallel processing allows for the generation of vast sequencing data in a single run (Oulas *et al.*, 2015).

The sensitivity of next-generation sequencing enables direct, unbiased profiling of microbial DNA from complex samples, with data quality closely linked to the efficiency of DNA extraction methods. While pyrosequencing provides longer read lengths, Illumina sequencing offers higher throughput, making it particularly suitable for filling sequence gaps and improving genome assemblies (Tan *et al.*, 2015). These advancements have significantly accelerated microbiome research, particularly in exploring interactions between medicinal plants and their associated microbial communities. Complementing this, metabolomics focuses on the comprehensive analysis of small molecule metabolites within a biological system, encompassing both primary and secondary metabolites (Tan *et al.*, 2015). In plants, secondary metabolites often involved in defense mechanisms play a crucial role in deterring herbivores and pathogens. Metabolomic profiling enables the identification of metabolic fingerprints and potential biomarkers. Integrating metabolomic data with microbiome studies provides a holistic view of plant–microbe interactions, enhancing our understanding of the functional dynamics within plant-associated microbial ecosystems (Divekar *et al.*, 2022).

10.5.3 Analytical platforms and data processing in metabolomics

Metabolomics relies on advanced analytical technologies, with nuclear magnetic resonance (NMR), gas chromatography-mass spectrometry (GC-MS), and liquid chromatography-mass spectrometry (LC-MS) being the most employed (Liu *et al.*, 2021; Wishart *et al.*, 2022). Among these, mass spectrometry (MS)-based methods offer higher sensitivity for metabolite detection, though they require extensive sample preparation and are limited to compounds that ionize within the detectable mass range. In contrast, NMR is advantageous for analyzing metabolites that are difficult to ionize or dissolve, and it does not require derivatization (Liu *et al.*, 2021). Traditionally, metabolomics is categorized into targeted and untargeted approaches, though this distinction may become less defined as methodologies evolve.

Data generated through these platforms undergo several stages, including preprocessing, annotation, postprocessing, and statistical analysis, each tailored to the specific analytical method used. Preprocessing corrects for variations in peak shape due to noise, sample inconsistencies, or instrumental factors, though no universally accepted standard exists (Al-Sulaiti *et al.*, 2023). Accurate metabolite identification generally requires comparison to at least two orthogonal properties of a known reference compound analyzed under identical conditions. Because many metabolites lack well-defined chemical standards, MS-based annotation tools assign confidence levels across multiple tiers. NMR-based identification relies on spectral database comparisons, although its utility is limited by the breadth and quality of available databases (Xiao *et al.*, 2012).

10.6 Challenges in Identifying Functional Microbiome Roles

10.6.1 Towards a multi-omics approach

Traditional microbiome studies typically focus on a single aspect, such as microbial profiling through 16S rRNA amplicon sequencing or whole-genome shotgun sequencing. However, the microbiome's role in disease pathogenesis involves complex structural and functional interactions with the host, mediated by small molecules such as metabolites and signaling compounds.

A multi-omics approach integrates metagenomics with metatranscriptomics, metaproteomics, and metabolomics to provide a comprehensive view of these interactions (Aguiar-Pulido *et al.*, 2016; Kumar *et al.*, 2024). Metatranscriptomics involves sequencing microbial mRNA, usually after filtering abundant ribosomal RNA, to quantify gene expression activity. Metaproteomics measures all proteins synthesized by both microbiota and the host, offering insights into cellular processes. Metabolomics assesses the small molecules present in the microenvironment, shedding light on metabolic and biochemical interactions. An example of multi-omics application is the study of dopamine metabolism in marine yeast, where whole-genome sequencing identified key biosynthesis genes, while transcriptomics and proteomics assessed pathway activation (Xiong *et al.*, 2015; Singh *et al.*, 2024).

Despite its advantages, multi-omics faces challenges such as variability in results due to different analytical pipelines and reliance on limited microbiome databases. These constraints restrict findings to well-characterized organisms, transcripts, proteins, and metabolites. Nonetheless, multi-omics provides a holistic view of microbiota–host interactions, allowing for deeper insights into cellular processes and biomolecular relationships (Morgan *et al.*, 2022).

10.6.2 Challenges for amplicon sequencing analysis

One major challenge in gene marker-based microbiome studies is distinguishing sequencing errors from true nucleotide variations. Two major categories of tools exist for this purpose: operational taxonomic unit (OTU)-based and amplicon sequence variant (ASV)-based methods (Bharti and Grimm, 2021). OTU-based approaches group reads into clusters based on a predefined identity threshold, typically 97%, while ASV-based methods use denoising algorithms to identify true biological sequences before sequencing errors occur. Comparisons between these approaches indicate that OTUs offer lower taxonomic resolution compared to ASVs, which can significantly impact diversity estimations. Comprehensive 16S rRNA sequencing workflows utilize both OTU- and ASV-based methodologies to ensure robust and precise characterization of microbial communities (Bharti and Grimm, 2021; Mehra *et al.*, 2022; Afoma *et al.*, 2025a).

10.6.3 Challenges in metagenomic sequencing analysis

The rapid expansion of computational tools and algorithms for metagenomic analysis presents a significant challenge in selecting the most appropriate and effective methods. Critical steps in a typical metagenomic workflow include quality control, sequence assembly, binning of metagenome-assembled genomes, taxonomic profiling, and functional annotation. Each of these stages requires careful consideration to ensure accuracy, reproducibility, and biological relevance (Oulas *et al.*, 2015; Verma *et al.*, 2025).

10.6.3.1 Quality control

Quality control is a fundamental step in metagenomic sequencing workflows, essential for filtering low-quality sequences and eliminating contaminants from raw reads. Trimming procedures remove adapter sequences and low-quality bases, while contamination removal focuses on discarding host-derived or nontarget sequences (Pérez-Cobas *et al.*, 2020). These processes significantly improve the accuracy and efficiency of downstream analyses, particularly sequence assembly. A variety of specialized tools are available for quality trimming and contamination filtering, ensuring that only high-quality data are retained for subsequent computational and biological interpretation (Pérez-Cobas *et al.*, 2020; Verma *et al.*, 2025).

10.6.4 Challenges in short-read metagenomics

Short-read sequencing offers high throughput and cost-effectiveness, making it a widely used technology. The Illumina platform, a prominent short-read algorithm-based technology, generates billions of reads in parallel in a single process (Bleidorn *et al.*, 2025). The sequencing process involves adapter-ligated DNA libraries, bridge amplification, and fluorescently labeled reversible terminators (Satam *et al.*, 2023; Bleidorn *et al.*, 2025). The stepwise nature of this protocol minimizes sequencing errors and enhances data accuracy. A structured workflow for metagenomics processing includes quality control, assembly, binning, and taxonomic and functional profiling. Selecting optimal tools and algorithms for each step enhances data reliability and ensures accurate interpretation of microbiome dynamics (Satam *et al.*, 2023).

10.7 Applications and Biotechnological Implications

10.7.1 Harnessing microbiomes for sustainable agriculture and plant growth promotion

10.7.1.1 Optimizing plant selection to enhance beneficial microbial associations

Integrating plant microbiomes into crop breeding strategies represents a promising frontier in improving stress resilience, nutrient efficiency, and overall productivity. While different plant genotypes are known to cultivate distinct microbial communities, the underlying genetic mechanisms governing these interactions are not yet fully understood (Hanif *et al.*, 2024). Domestication has narrowed both plant genetic diversity and their associated microbiomes, reducing modern crops capacity to engage with beneficial microbes. However, recent studies particularly in sugarcane demonstrate substantial variation in microbial diversity and composition among cultivars (Hanif *et al.*, 2024; Yusuf *et al.*, 2025). These findings underscore the potential for selecting and breeding crop varieties that foster advantageous microbial relationships. Incorporating microbiome considerations into breeding programs is essential for promoting sustainable agriculture and maximizing crop performance (Yusuf *et al.*, 2025).

10.7.1.2 Impact of agricultural practices on plant–microbe interactions

Farming practices such as intercropping, organic cultivation, and reduced tillage play a pivotal role in shaping plant–microbe dynamics (Hanif *et al.*, 2024). Organic farming has been associated with increased microbial diversity and abundance, contributing to improved ecosystem stability and plant health. For instance, sugarcane–legume intercropping enhances soil fertility and microbial richness without negatively impacting yields (Sun *et al.*, 2022; Xiao *et al.*, 2023). Furthermore, the application of bacterial biological control agents has shown promising results in ratoon sugarcane systems, improving soil nutrient profiles, enriching beneficial microbial populations, and boosting crop productivity. These examples highlight the importance of aligning agricultural management practices with microbiome-friendly strategies to support long-term soil health and crop sustainability (Sun *et al.*, 2022).

10.7.2 Development of bio-inoculants and microbial formulations from medicinal plant-associated microbiomes

10.7.2.1 Role of bio-inoculants in advancing green agriculture

Bio-inoculant microbial formulations designed to promote plant health are key tools in the development of sustainable, environmentally friendly agricultural systems. Microorganisms within the phytomicrobiome form dynamic holobionts with their plant hosts, supporting growth and development through a range of synergistic interactions (Maitra *et al.*, 2021). Rhizospheric microbiomes contribute significantly to nutrient assimilation, enhancement of soil physicochemical properties, and the production of secondary metabolites, antibiotics, and phytohormones. Additionally, these microbes secrete biostimulants that regulate plant physiological and metabolic processes. The integration of beneficial microorganisms into agricultural practices is central to green agriculture, offering improved nutrient availability, enhanced stress tolerance, and strengthened plant immunity without the environmental costs of synthetic inputs (Maitra *et al.*, 2021).

10.7.2.2 Microbial contributions to nutrient assimilation and biofortification

Plants rely on a balanced supply of essential inorganic minerals to support their nutritional, developmental, and physiological needs. Key macronutrients include nitrogen, phosphorus, potassium, calcium, magnesium, and sulfur, while critical micronutrients comprise iron, zinc, copper, manganese, boron, molybdenum, and chlorine (White and Brown, 2010). Optimizing plant–microbe interactions can significantly improve the uptake of these nutrients and support biofortification efforts, thereby enhancing crop quality and yield. Microbial partners not only facilitate nutrient solubilization and mobilization but also contribute to the long-term sustainability of agricultural systems by reducing dependency on chemical fertilizers and improving soil health (White and Brown, 2010; Jadhav *et al.*, 2025).

10.8 Future Perspectives and Challenges

10.8.1 Integration of omics approaches for comprehensive microbiome exploration

Recent advances in molecular biology and high-resolution analytical technologies have significantly expanded our ability to investigate plant–microbe interactions through a suite of omics approaches, including genomics, metagenomics, transcriptomics, proteomics, and metabolomics (Jain *et al.*, 2024). These techniques provide multidimensional insights into microbial community structure, gene function, protein interactions, and metabolic processes, forming the foundation for innovative, microbiome-informed strategies in sustainable agriculture (Diwan *et al.*, 2022; Jain *et al.*, 2024). Web-based tools and automated pipelines have further transformed the landscape of omics data analysis. Platforms supporting metagenomic, proteomic, metabolomic, and metatranscriptomic studies offer integrated functionalities for phylogenetic profiling, functional annotation, gene expression analysis, and pathway mapping. These tools streamline complex data interpretation, equipping researchers with robust resources to decode intricate plant–microbe interactions and accelerate the application of microbiome science in agriculture and environmental management (Aguiar-Pulido *et al.*, 2016; Afoma *et al.*, 2025b).

10.8.2 Applications of omics in agriculture and beyond

The integration of genomic, transcriptomic, proteomic, and metabolomic technologies has reshaped multiple domains, including agriculture, medicine, food science, and environmental research. In agriculture, omics-based insights into plant–microbe symbioses have enabled the development of targeted interventions such as biofertilizers, biocontrol agents, and microbial biostimulants contributing to enhanced soil fertility, reduced reliance on agrochemicals, and improved crop resilience and productivity (Table 10.1) (Aguiar-Pulido *et al.*, 2016; Sahoo *et al.*, 2025). Precision agriculture increasingly harnesses omics data to tailor microbial consortia for specific environmental conditions, optimizing crop performance and resource use efficiency (Yusuf *et al.*, 2025). Combined with advanced computational modeling, these data facilitate the design of predictive frameworks for microbial community engineering and informed agricultural management.

Beyond the field, the plant microbiome has emerging implications for human health. Microbial communities associated with dietary plants may influence the human gut microbiome, immune responses, metabolism, and disease risk. Omics approaches hold potential to uncover plant–microbe–human interactions, opening avenues for personalized nutrition and novel therapeutic strategies (Amin *et al.*, 2025).

Table 10.1. Omics-based research on microbiome and their applications.

Omics discipline	Microbiomes	Primary role	Major insights and applications	Microbial groups studied	Benefits	Reference
Genomic analysis	*Azotobacter vinelandii*, *Klebsiella pneumoniae*, *Rhodobacter capsulatus*, *Rhodospirillum rubrum*	Investigating gene structures and functions	Enables comprehensive genome mapping, transcriptional profiling, and metabolic modelling	Nitrogen-fixing bacteria in symbiosis	Cost-effective sequencing, full-genome accessibility	(Jain *et al.*, 2024)
Metagenomic profiling	*Alternaria alternata*, apple host response, *Rhizobium*–legume symbiosis	Identifying genetic material from entire microbial populations	Overcomes culture-based limitations, discovers novel genes and microbial functions	Symbiotic nitrogen fixers	Captures both organismal and ecological interactions	(Medha *et al.*, 2025)
Epigenomic research	*Neobacillus* (associated with Naxos tubers), plant–microbe interactions involving *Pseudomonas syringae*, epigenomic correlations in fungi	Examining heritable modifications in genetic expression	Investigates environmental adaptation mechanisms and transgenerational inheritance	Mycorrhizal fungi, endophytes, PGPR, biocontrol microbes	Explores DNA methylation, histone modification, and noncoding RNAs	(Samantara *et al.*, 2021)
Transcriptome analysis	*Pseudomonas fluorescens* in modulating *Arabidopsis* immunity, the impact of *Penicillium expansum* on postharvest fruit decay, and the association of *Delftia acidovorans* with crop root systems	Measuring RNA expression levels under specific conditions	Enables differential gene expression profiling, mutation tracking, and alternative splicing studies	PGPR, biocontrol microbes	Deciphers regulatory pathways via whole-transcriptome sequencing	(Liu *et al.*, 2025b)
Metatranscriptomic approaches	*Xylella fastidiosa* in grapevine Pierce's disease	Assessing microbial gene expression in environmental contexts	Highlights active metabolic pathways and dynamic microbial functions	Biocontrol agents, pathogenic response microbes	Captures real-time RNA activity within microbial communities	(Kyrkou *et al.*, 2018)

Proteomic studies	*Sinorhizobium meliloti* in symbiosis, *Pseudomonas syringae* as a plant pathogen	Investigating protein synthesis and interactions	Identifies proteins linked to stress response, pathogenicity, and symbiosis	Symbiotic microbes	Facilitates protein function characterization and network analysis	(Chen *et al.*, 2003)
Metaproteomic investigations	*Trichoderma harzianum* in plant disease control	analysing functional protein roles in microbial ecology	Connects proteins to microbial carbon metabolism and phyllosphere interactions	PGPR, biocontrol microbes	Enables identification of proteins involved in disease suppression and host defense	(Chen *et al.*, 2025)
Metabolomic profiling	*Glomus intraradices* assisting nutrient uptake in tomatoes	Studying microbial metabolites and plant–microbiome interactions	Provides insights into metabolic pathways and biochemical exchanges between plants and microbes	Mycorrhizal fungi, endophytes	Supports development of metabolite-based plant growth strategies	(Gupta *et al.*, 2022)

PGPR, plant growth-promoting rhizobia.

In the realm of food science, omics technologies enhance food safety and traceability, refine fermentation processes, and support the development of functional foods with improved nutritional profiles. In conservation biology, these technologies are being applied to assess the impact of habitat degradation, climate change, and pollution on plant–microbe dynamics, informing targeted ecological restoration strategies that support biodiversity and ecosystem resilience (Rocha and Silva, 2023).

Overall, as omics technologies continue to advance, they are poised to revolutionize our understanding of plant–microbe systems and their applications in diverse sectors. The integration of multi-omics data, supported by computational innovation, will drive breakthroughs in sustainable agriculture, environmental restoration, human health, and biotechnology. However, key challenges remain, including data standardization, crossdomain data integration, cost-effective implementation, and ethical considerations. Addressing these challenges will be critical to fully realizing the transformative potential of omics-driven research in addressing complex biological and ecological problems.

10.9 Conclusion

Medicinal plants and their microbiomes hold immense promise for sustainable agriculture, pharmaceutical advancements, and ecological resilience. The intricate interactions among the plants and microbial communities play a pivotal role in plant health, secondary metabolite biosynthesis, and environmental adaptation. By harnessing these natural associations, scientists can develop innovative strategies to improve stress resilience, enhance crop productivity, and optimize bioactive compound production for therapeutic applications.

Advancements in metagenomics, synthetic biology, and multi-omics tactics have significantly expanded our understanding and knowledge of plant microbiomes, offering new insights into microbial diversity, functional dynamics, and biosynthetic potential. However, challenges such as biodiversity loss, environmental fluctuations, and analytical limitations necessitate further research and conservation efforts. The synergy between traditional knowledge and contemporary scientific innovations will be pivotal in unlocking the full potential of medicinal plants and their microbiome-associated benefits

Future research should focus on optimizing microbial applications, refining cultivation techniques, and identifying novel plant–microbe interactions to enhance the production of therapeutically valuable compounds. The microbiome-driven enhancement of bioactive compound biosynthesis presents a promising avenue for sustainable metabolite production, addressing challenges related to seasonal variability and resource limitations. Furthermore, emerging technologies such as metabolic engineering, bioreactor systems, and precision agriculture can revolutionize large-scale cultivation and pharmaceutical applications.

Harnessing plant microbiomes for agricultural and medical advancements requires an interdisciplinary approach, integrating biotechnology, microbiology, and systems biology. Collaborative efforts among researchers, conservationists, and policymakers will be essential in ensuring the sustainable application of medicinal plants and associated microbiomes. By embracing cutting-edge research and conservation strategies, we can unlock the vast biochemical prospective of these natural resources, contributing to global health, environmental sustainability, and the future of biotechnological innovations.

References

Afoma, U.M., Malik, J., Singh, S., Mishra, A.K., Tomar, L.K. *et al.* (2025a) Characterization of azo dye-degrading *Tistrella mobilis* isolated from industrial effluents: A molecular perspective. *Biologia* 80(8), 2103–2115. DOI: 10.1007/s11756-025-01957-w.

Afoma, U.M., Singh, S., Mishra, A.K., Sharma, C.K., Gupta, K. *et al.* (2025b) Integrating artificial intelligence in environmental monitoring: A paradigm shift in data-driven sustainability. *EcoHealth* 23, 38–57. DOI: 10.1007/s10393-025-01752-8.

Afridi, M.S., Javed, M.A., Ali, S., De Medeiros, F.H.V., Ali, B. *et al.* (2022) New opportunities in plant microbiome engineering for increasing agricultural sustainability under stressful conditions. *Frontiers in Plant Science* 13, 899464. DOI: 10.3389/fpls.2022.899464.

Aguiar-Pulido, V., Huang, W., Suarez-Ulloa, V., Cickovski, T., Mathee, K. *et al.* (2016) Metagenomics, metatranscriptomics, and metabolomics approaches for microbiome analysis. *Evolutionary Bioinformatics Online* 12(Suppl 1), 5–16. DOI: 10.4137/EBO.S36436.

Ahmad, Z., Shaheen, A., Wasi, A., Rehman, S.U., Tahseen, S. *et al.* (2023) Biotechnological intervention and withanolide production in *Withania coagulans*. *Agronomy* 13(8), 1997. DOI: 10.3390/agronomy13081997.

Alamgir, A.N.M. (2018) Biotechnology, *in vitro* production of natural bioactive compounds, herbal preparation, and disease management (treatment and prevention). *Therapeutic Use of Medicinal Plants and Their Extracts* 2, 585–664. DOI: 10.1007/978-3-319-92387-1_7.

Al-Khayri, J.M., Rashmi, R., Toppo, V., Chole, P.B., Banadka, A. *et al.* (2023) Plant secondary metabolites: The weapons for biotic stress management. *Metabolites* 13(6), 716. DOI: 10.3390/metabo13060716.

Al-Sulaiti, H., Almaliti, J., Naman, C.B., Al Thani, A.A. and Yassine, H.M. (2023) Metabolomics approaches for the diagnosis, treatment, and better disease management of viral infections. *Metabolites* 13(8), 948. DOI: 10.3390/metabo13080948.

Alvarez-Leite, J.I. (2025) The role of bioactive compounds in human health and disease. *Nutrients* 17(7), 1170. DOI: 10.3390/nu17071170.

Amin, A., Zaman, W. and Park, S. (2025) Harnessing multi-omics and predictive modeling for climate-resilient crop breeding: From genomes to fields. *Genes* 16(7), 809. DOI: 10.3390/genes16070809.

Babalola, O.O. and Adedayo, A.A. (2023) Endosphere microbial communities and plant nutrient acquisition toward sustainable agriculture. *Emerging Topics in Life Sciences* 7(2), 207–217. DOI: 10.1042/ETLS20230069.

Bharti, R. and Grimm, D.G. (2021) Current challenges and best-practice protocols for microbiome analysis. *Briefings in Bioinformatics* 22(1), 178–193. DOI: 10.1093/bib/bbz155.

Bleidorn, C., Sandberg, F., Martin, S., Vogler, A.P. and Podsiadlowski, L. (2025) The untapped potential of short-read sequencing in biodiversity research. *Trends in Genetics* 42, 137–149. DOI: 10.1016/j.tig.2025.09.001.

Boon, E., Meehan, C.J., Whidden, C., Wong, D.H.-J., Langille, M.G.I. *et al.* (2014) Interactions in the microbiome: Communities of organisms and communities of genes. *FEMS Microbiology Reviews* 38(1), 90–118. DOI: 10.1111/1574-6976.12035.

Brachi, B., Filiault, D., Whitehurst, H., Darme, P., Le Gars, P. *et al.* (2022) Plant genetic effects on microbial hubs impact host fitness in repeated field trials. *Proceedings of the National Academy of Sciences USA* 119(30), e2201285119. DOI: 10.1073/pnas.2201285119.

Brown, S.P., Grillo, M.A., Podowski, J.C. and Heath, K.D. (2020) Soil origin and plant genotype structure distinct microbiome compartments in the model legume *Medicago truncatula*. *Microbiome* 8(1), 139. DOI: 10.1186/s40168-020-00915-9.

Bulakarima, A.U., Sharma, V.K., Isa, M.A. and Dudha, N. (2025a) Phytochemical analysis and antioxidant evaluation of a crude leaves extract of *Nerium indicum* (Mill). *Journal of Information Systems Engineering and Management* 10(17s), 554–563. DOI: 10.52783/jisem.v10i17s.2759.

Bulakarima, A.U., Sharma, V.K., Singh, S., Isa, M.A. and Dudha, N. (2025b) Investigating the antibacterial and antioxidant properties of *Zingiber officinale* (Ginger) extracts. *Cuestiones de Fisioterapia* 54(3), 4449–4471. DOI: 10.48047/mk3jsc21.

Chaachouay, N. and Zidane, L. (2024) Plant-derived natural products: A source for drug discovery and development. *Drugs and Drug Candidates* 3(1), 184–207. DOI: 10.3390/ddc3010011.

Chen, H., Teplitski, M., Robinson, J.B., Rolfe, B.G. and Bauer, W.D. (2003) Proteomic analysis of wild-type *Sinorhizobium meliloti* responses to N-acyl homoserine lactone quorum-sensing signals and the transition to stationary phase. *Journal of Bacteriology* 185(17), 5029–5036. DOI: 10.1128/JB.185.17.5029-5036.2003.

Chen, X., Lu, Y., Liu, X., Gu, Y. and Li, F. (2025) Trichoderma: Dual roles in biocontrol and plant growth promotion. *Microorganisms* 13(8), 1840. DOI: 10.3390/microorganisms13081840.

Chevrette, M.G., Thomas, C.S., Hurley, A., Rosario-Meléndez, N., Sankaran, K. *et al.* (2022) Microbiome composition modulates secondary metabolism in a multispecies bacterial community. *Proceedings of the National Academy of Sciences USA* (42), e2212930119. DOI: 10.1073/pnas.2212930119.

Covington, B.C., Xu, F. and Seyedsayamdost, M.R. (2021) A natural product chemist's guide to unlocking silent biosynthetic gene clusters. *Annual Review of Biochemistry* 90, 763–788. DOI: 10.1146/annurev-biochem-081420-102432.

Dar, N.J., Hamid, A. and Ahmad, M. (2015) Pharmacologic overview of *Withania somnifera*, the Indian ginseng. *Cellular and Molecular Life Sciences* 72(23), 4445–4460. DOI: 10.1007/s00018-015-2012-1.

Davidova, S., Galabov, A.S. and Satchanska, G. (2024) Antibacterial, antifungal, antiviral activity, and mechanisms of action of plant polyphenols. *Microorganisms* 12(12), 2502. DOI: 10.3390/microorganisms12122502.

De Mandal, S. and Jeon, J. (2023) Phyllosphere microbiome in plant health and disease. *Plants* 12(19), 3481. DOI: 10.3390/plants12193481.

Dekomah, S.D., Bi, Z., Dormatey, R., Wang, Y., Haider, F.U. *et al.* (2022) The role of CDPKs in plant development, nutrient and stress signaling. *Frontiers in Genetics* 13, 996203. DOI: 10.3389/fgene.2022.996203.

Deng, S., Caddell, D.F., Xu, G., Dahlen, L., Washington, L. *et al.* (2021) Genome wide association study reveals plant loci controlling heritability of the rhizosphere microbiome. *ISME Journal* 15(11), 3181–3194. DOI: 10.1038/s41396-021-00993-z.

Di Bella, J.M., Bao, Y., Gloor, G.B., Burton, J.P. and Reid, G. (2013) High throughput sequencing methods and analysis for microbiome research. *Journal of Microbiological Methods* 95(3), 401–414. DOI: 10.1016/j.mimet.2013.08.011.

Dias, M.C., Pinto, D.C.G.A. and Silva, A.M.S. (2021) Plant flavonoids: Chemical characteristics and biological activity. *Molecules* 26(17), 5377. DOI: 10.3390/molecules26175377.

Divekar, P.A., Narayana, S., Divekar, B.A., Kumar, R., Gadratagi, B.G. *et al.* (2022) Plant secondary metabolites as defense tools against herbivores for sustainable crop protection. *International Journal of Molecular Sciences* 23(5), 2690. DOI: 10.3390/ijms23052690.

Diwan, D., Rashid, M.M. and Vaishnav, A. (2022) Current understanding of plant-microbe interaction through the lenses of multi-omics approaches and their benefits in sustainable agriculture. *Microbiological Research* 265, 127180. DOI: 10.1016/j.micres.2022.127180.

Enea, M., Beauregard, J., De Bellis, T., Faticov, M. and Laforest-Lapointe, I. (2025) The temperate forest phyllosphere and rhizosphere microbiome: A case study of sugar maple. *Frontiers in Microbiology* 15, 1504444. DOI: 10.3389/fmicb.2024.1504444.

Espinosa-Leal, C.A., Puente-Garza, C.A. and García-Lara, S. (2018) *In vitro* plant tissue culture: Means for production of biological active compounds. *Planta* 248(1), 1–18. DOI: 10.1007/s00425-018-2910-1.

Fernie, A.R., de Vries, S. and de Vries, J. (2024) Evolution of plant metabolism: The state-of-the-art. *Philosophical Transactions of the Royal Society of London B: Biological Sciences* 379(1914), 20230347. DOI: 10.1098/rstb.2023.0347.

Ferrer, J.-L., Austin, M.B., Stewart Jr, C. and Noel, J.P. (2008) Structure and function of enzymes involved in the biosynthesis of phenylpropanoids. *Plant Physiology and Biochemistry* 46(3), 356–370. DOI: 10.1016/j.plaphy.2007.12.009.

Fitzpatrick, C.R., Copeland, J., Wang, P.W., Guttman, D.S., Kotanen, P.M. *et al.* (2018) Assembly and ecological function of the root microbiome across angiosperm plant species. *Proceedings of the National Academy of Sciences USA* 115(6), E1157–E1165. DOI: 10.1073/pnas.1717617115.

Grover, A., Samuel, G., Bisaria, V.S. and Sundar, D. (2013) Enhanced withanolide production by overexpression of squalene synthase in *Withania somnifera*. *Journal of Bioscience and Bioengineering* 115(6), 680–685. DOI: 10.1016/j.jbiosc.2012.12.011.

Gupta, S., Schillaci, M. and Roessner, U. (2022) Metabolomics as an emerging tool to study plant-microbe interactions. *Emerging Topics in Life Sciences* 6(2), 175–183. DOI: 10.1042/ETLS20210262.

Hanif, M.S., Tayyab, M., Baillo, E.H., Islam, M.M., Islam, W. *et al.* (2024) Plant microbiome technology for sustainable agriculture. *Frontiers in Microbiology* 15, 1500260. DOI: 10.3389/fmicb.2024.1500260.

Heather, J.M. and Chain, B. (2016) The sequence of sequencers: The history of sequencing DNA. *Genomics* 107(1), 1–8. DOI: 10.1016/j.ygeno.2015.11.003.

Heinrich, M., Mah, J. and Amirkia, V. (2021) Alkaloids used as medicines: Structural phytochemistry meets biodiversity-an update and forward look. *Molecules* 26(7), 1836. DOI: 10.3390/molecules26071836.

Horton, M.W., Bodenhausen, N., Beilsmith, K., Meng, D., Muegge, B.D. *et al.* (2014) Genome-wide association study of *Arabidopsis thaliana* leaf microbial community. *Nature Communications* 5, 5320. DOI: 10.1038/ncomms6320.

Iqbal, S., Begum, F., Nguchu, B.A., Claver, U.P. and Shaw, P. (2025) The invisible architects: Microbial communities and their transformative role in soil health and global climate changes. *Environmental Microbiome* 20(1), 36. DOI: 10.1186/s40793-025-00694-6.

Jabczyk, M., Nowak, J., Hudzik, B. and Zubelewicz-Szkodzińska, B. (2021) Curcumin and its potential impact on microbiota. *Nutrients* 13(6), 2004. DOI: 10.3390/nu13062004.

Jacoby, R.P., Koprivova, A. and Kopriva, S. (2021) Pinpointing secondary metabolites that shape the composition and function of the plant microbiome. *Journal of Experimental Botany* 72(1), 57–69. DOI: 10.1093/jxb/eraa424.

Jadhav, A.V., Pati, P., Patel, K.G., Sharma, V.K., Shewale, M. *et al.* (2025) Biotechnological approaches for enhancing biological control agents against crop pests. *Journal of Entomological Research* 49(3), 788–795. DOI: 10.5958/0974-4576.2025.00139.7.

Jagtap, A.D., Khandual, A., Krishnakumar, S., Veer, M. and Sharma, V.K. (2025) Advances in fungal and viral pathogens for biological control of insect pests. *Journal of Entomological Research* 49(3), 837–844. DOI: 10.5958/0974-4576.2025.00145.X.

Jain, A., Sarsaiya, S., Singh, R., Gong, Q., Wu, Q. *et al.* (2024) Omics approaches in understanding the benefits of plant-microbe interactions. *Frontiers in Microbiology* 15, 1391059. DOI: 10.3389/fmicb.2024.1391059.

Jiang, Z., Peng, F., Yu, J. and Li, Q. (2025) Plant growth-promoting effects and possible mechanisms of a plant endophytic fungus *Aureobasidium* sp. JRF1. *Plant Physiology and Biochemistry* 222, 109724. DOI: 10.1016/j.plaphy.2025.109724.

Joshi, R., Wani, S.H., Singh, B., Bohra, A., Dar, Z.A. *et al.* (2016) Transcription factors and plants response to drought stress: Current understanding and future directions. *Frontiers in Plant Science* 7, 1029. DOI: 10.3389/fpls.2016.01029.

Jung, J., Ahn, S., Kim, D.H. and Riu, M. (2024) Triple interactions for induced systemic resistance in plants. *Frontiers in Plant Science* 15, 1464710. DOI: 10.3389/fpls.2024.1464710.

Kao, C.Y., Kuo, P.Y. and Liao, H.W. (2021) Untargeted microbial exometabolomics and metabolomics analysis of *Helicobacter pylori* J99 and *jhp0106* mutant. *Metabolites* 11(12), 808. DOI: 10.3390/metabo11120808.

Kavi Kishor, P.B., Thaddi, B.N., Guddimalli, R., Nikam, T.D., Sambasiva Rao, K.R.S. *et al.* (2025) The occurrence, uses, biosynthetic pathway, and biotechnological production of plumbagin, a potent antitumor naphthoquinone. *Molecules* 30(7), 1618. DOI: 10.3390/molecules30071618.

Khanam, S., Mishra, P., Faruqui, T., Alam, P., Albalawi, T. *et al.* (2025) Plant-based secondary metabolites as natural remedies: A comprehensive review on terpenes and their therapeutic applications. *Frontiers in Pharmacology* 16, 1587215. DOI: 10.3389/fphar.2025.1587215.

Köberl, M., Schmidt, R., Ramadan, E.M., Bauer, R. and Berg, G. (2013) The microbiome of medicinal plants: Diversity and importance for plant growth, quality and health. *Frontiers in Microbiology* 4, 400. DOI: 10.3389/fmicb.2013.00400.

Kosina, S.M., Greiner, A.M., Lau, R.K., Jenkins, S., Baran, R. *et al.* (2018) Web of microbes (WoM): A curated microbial exometabolomics database for linking chemistry and microbes. *BMC Microbiology* 18(1), 115. DOI: 10.1186/s12866-018-1256-y.

Koza, N.A., Adedayo, A.A., Babalola, O.O. and Kappo, A.P. (2022) Microorganisms in plant growth and development: Roles in abiotic stress tolerance and secondary metabolites secretion. *Microorganisms* 10(8), 1528. DOI: 10.3390/microorganisms10081528.

Kumar, A., Nirmal, P., Kumar, M., Jose, A., Tomer, V. *et al.* (2023a) Major phytochemicals: Recent advances in health benefits and extraction method. *Molecules* 28(2), 887. DOI: 10.3390/molecules28020887.

Kumar, B., Lorusso, E., Fosso, B. and Pesole, G. (2024) A comprehensive overview of microbiome data in the light of machine learning applications: Categorization, accessibility, and future directions. *Frontiers in Microbiology* 15, 1343572. DOI: 10.3389/fmicb.2024.1343572.

Kumar, S., Mathew, S.O., Aharwal, R.P., Tulli, H.S., Mohan, C.D. *et al.* (2023b) Withaferin A: A pleiotropic anticancer agent from the Indian medicinal plant *Withania somnifera* (L.) Dunal. *Pharmaceuticals* 16(2), 160. DOI: 10.3390/ph16020160.

Kumari, P., Deepa, N., Trivedi, P.K., Singh, B.K., Srivastava, V. *et al.* (2023) Plants and endophytes interaction: A "secret wedlock" for sustainable biosynthesis of pharmaceutically important secondary metabolites. *Microbial Cell Factories* 22(1), 226. DOI: 10.1186/s12934-023-02234-8.

Kyrkou, I., Pusa, T., Ellegaard-Jensen, L., Sagot, M.F. and Hansen, L.H. (2018) Pierce's disease of grapevines: A review of control strategies and an outline of an epidemiological model. *Frontiers in Microbiology* 9, 2141. DOI: 10.3389/fmicb.2018.02141.

Liu, C., Ahmad, N., Tao, Y., Hussain, H., Chang, Y. *et al.* (2025a) Reprogramming hairy root cultures: A synthetic biology framework for precision metabolite biosynthesis. *Plants* 14(13), 1928. DOI: 10.3390/plants14131928.

Liu, H., Carvalhais, L.C., Crawford, M., Singh, E., Dennis, P.G. *et al.* (2017) Inner plant values: Diversity, colonization and benefits from endophytic bacteria. *Frontiers in Microbiology* 8, 2552. DOI: 10.3389/fmicb.2017.02552.

Liu, R., Bao, Z.X., Zhao, P.J. and Li, G.H. (2021) Advances in the study of metabolomics and metabolites in some species interactions. *Molecules* 26(11), 3311. DOI: 10.3390/molecules26113311.

Liu, Z., Liu, X., Wang, S., Liang, S., Li, S. *et al.* (2025b) Comparative transcriptome analysis of *Arabidopsis* seedlings under heat stress on whole plants, shoots, and roots reveals new HS-regulated genes, organ-specific responses, and shoots-roots communication. *International Journal of Molecular Sciences* 26(6), 2478. DOI: 10.3390/ijms26062478.

Lv, J., Yang, S., Zhou, W., Liu, Z., Tan, J. *et al.* (2024) Microbial regulation of plant secondary metabolites: Impact, mechanisms and prospects. *Microbiological Research* 283, 127688. DOI: 10.1016/j.micres.2024.127688.

Ma, W., Tang, S., Dengzeng, Z., Zhang, D., Zhang, T. *et al.* (2022) Root exudates contribute to belowground ecosystem hotspots: A review. *Frontiers in Microbiology* 13, 937940. DOI: 10.3389/fmicb.2022.937940.

Maitra, S., Brestic, M., Bhadra, P., Shankar, T., Praharaj, S. *et al.* (2021) Bioinoculants –natural biological resources for sustainable plant production. *Microorganisms* 10(1), 51. DOI: 10.3390/microorganisms10010051.

Malik, J., Naushad, Z., Singh, S., Shrivastav, D., Verma, V.V. *et al.* (2025) Emerging contaminants in healthcare effluents: Antibiotic resistance and virulence factor dissemination. *Folia Microbiologica* 70, 1201–1216. DOI: 10.1007/s12223-025-01356-3.

Masood, H.A., Qi, Y., Zahid, M.K., Li, Z., Ahmad, S. *et al.* (2024) Recent advances in nano-enabled immunomodulation for enhancing plant resilience against phytopathogens. *Frontiers in Plant Science* 15, 1445786. DOI: 10.3389/fpls.2024.1445786.

Medha, L.M., Islam, S. and Islam, T. (2025) Metagenomic insights into plant–probiotic bacterial interactions boosting crop growth and yield. *Stress-Resilient Crops: Coordinated Omics-CRISPR-Nanotechnology Strategies* 540–557. DOI: 10.1079/9781800627307.0023.

Meel, S. and Saharan, B.S. (2024) Enhancing crop resilience towards drought: By integrating nanotechnology, microbiomes, and growth-promoting rhizobacteria. *Discover Agriculture* 2(1), 112. DOI: 10.1007/s44279-024-00131-1.

Mehra, S., Tyagi, C., Dudha, N., Reddy, Y.P., Manjhi, J. *et al.* (2022) Molecular characterization and phylogenetic analysis of *Pseudomonas aeruginosa* obtained from wound infection. *Arab Gulf Journal of Scientific Research* 39(3), 221–235. DOI: 10.51758/AGJSR-03-2021-0027.

Mehra, S., Sharma, V.K., Tygai, C. and Tomar, L.K. (2023) Ethnomedicinal pertinence and anti-bacterial prospective of Himalayan medicinal plants of Uttarakhand in India. In: Mishra, M.K. and Kumari, N. (eds) *Plants for Immunity and Conservation Strategies*. Springer, Singapore, pp. 311–348. DOI: 10.1007/978-981-99-2824-8_15.

Mehra, S., Afoma, U.M., Dudha, N., Singh, S., Sharma, S. *et al.* (2024) Characterization and evaluation of antioxidant properties of leaf extracts from *Rhododendron arboreum*. *Journal of Bioscience and Biotechnology* 13(2), 147–153. Available at: https://editorial.uni-plovdiv.bg/index.php/JBB/article/view/545 (accessed 12 May 2026).

Miękus, N., Marszałek, K., Podlacha, M., Iqbal, A., Puchalski, C. *et al.* (2020) Health benefits of plant-derived sulfur compounds, glucosinolates, and organosulfur compounds. *Molecules* 25(17), 3804. DOI: 10.3390/molecules25173804.

Mipeshwaree Devi, A., Khedashwori Devi, K., Premi Devi, P., Lakshmipriyari Devi, M. and Das, S. (2023) Metabolic engineering of plant secondary metabolites: Prospects and its technological challenges. *Frontiers in Plant Science* 14, 1171154. DOI: 10.3389/fpls.2023.1171154.

Mishra, M.K. (2024) Phytochemical analysis and evaluation of antibacterial activity of various extracts from leaves, stems and roots of *Thalictrum foliolosum*. *Future Journal of Pharmaceutical Sciences* 10(1), 70. DOI: 10.1186/s43094-024-00643-z.

Mishra, M.K., Omar, R.A., Srivastava, V., Singh, S., Verma, V.V. *et al.* (2025) Advanced nanoparticle-based strategies for plant viral disease management: Mechanisms and innovations for crop protection. *3 Biotech* 15(12), 429. DOI: 10.1007/s13205-025-04617-7.

Misu, I.J., Kayess, M.O., Siddiqui, M.N., Gupta, D.R., Islam, M.N. *et al.* (2025) Microbiome engineering for sustainable rice production: Strategies for biofertilization, stress tolerance, and climate resilience. *Microorganisms* 13(2), 233. DOI: 10.3390/microorganisms13020233.

Mitra, S. and Murthy, G.S. (2022) Bioreactor control systems in the biopharmaceutical industry: A critical perspective. *Systems Microbiology and Biomanufacturing* 2(1), 91–112. DOI: 10.1007/s43393-021-00048-6.

Morgan, E.W., Perdew, G.H. and Patterson, A.D. (2022) Multi-omics strategies for investigating the microbiome in toxicology research. *Toxicological Sciences* 187(2), 189–213. DOI: 10.1093/toxsci/kfac029.

Muthu Narayanan, M., Ahmad, N., Shivanand, P. and Metali, F. (2022) The role of endophytes in combating fungal- and bacterial-induced stress in plants. *Molecules* 27(19), 6549. DOI: 10.3390/molecules27196549.

Nadarajah, K. and Abdul Rahman, N.S.N. (2023) The microbial connection to sustainable agriculture. *Plants* 12(12), 2307. DOI: 10.3390/plants12122307.

Najmi, A., Javed, S.A., Al Bratty, M. and Alhazmi, H.A. (2022) Modern approaches in the discovery and development of plant-based natural products and their analogues as potential therapeutic agents. *Molecules* 27(2), 349. DOI: 10.3390/molecules27020349.

Nam, N.N., Do, H.D.K., Loan Trinh, K.T. and Lee, N.Y. (2023) Metagenomics: An effective approach for exploring microbial diversity and functions. *Foods* 12(11), 2140. DOI: 10.3390/foods12112140.

Nazir, A., Hussain, F.H.N. and Raza, A. (2024) Advancing microbiota therapeutics: The role of synthetic biology in engineering microbial communities for precision medicine. *Frontiers in Bioengineering and Biotechnology* 12, 1511149. DOI: 10.3389/fbioe.2024.1511149.

Negre Rodríguez, M., Pioppi, A. and Kovács, Á.T. (2025) The role of plant host genetics in shaping the composition and functionality of rhizosphere microbiomes. *mSystems* 10(8), e0004124. DOI: 10.1128/msystems.00041-24.

Nguyen, Q.M., Iswanto, A.B.B., Son, G.H. and Kim, S.H. (2021) Recent advances in effector-triggered immunity in plants: New pieces in the puzzle create a different paradigm. *International Journal of Molecular Sciences* 22(9), 4709. DOI: 10.3390/ijms22094709.

Nimbulkar, P., Gupta, G., Virkhare, U., Althubiani, A.S., Dutta, A. *et al.* (2025) Bacterial endophytes and their secondary metabolites: Mechanisms of biosynthesis and applications in sustainable agriculture. *Journal of Umm Al-Qura University for Applied Sciences* 1–13. DOI: 10.1007/s43994-025-00233-6.

Ninkuu, V., Aluko, O.O., Yan, J., Zeng, H., Liu, G. *et al.* (2025) Phenylpropanoids metabolism: Recent insight into stress tolerance and plant development cues. *Frontiers in Plant Science* 16, 1571825. DOI: 10.3389/fpls.2025.1571825.

Olanrewaju, O.S., Glick, B.R. and Babalola, O.O. (2024) RETRACTED: Beyond correlation: Understanding the causal link between microbiome and plant health. *Heliyon* 10(23), e40517. DOI: 10.1016/j.heliyon.2024.e40517.

Oulas, A., Pavloudi, C., Polymenakou, P., Pavlopoulos, G.A., Papanikolaou, N. *et al.* (2015) Metagenomics: Tools and insights for analyzing next-generation sequencing data derived from biodiversity studies. *Bioinformatics and Biology Insights* 9, 75–88. DOI: 10.4137/BBI.S12462.

Ozyigit, I.I., Dogan, I., Hocaoglu-Ozyigit, A., Yalcin, B., Erdogan, A. *et al.* (2023) Production of secondary metabolites using tissue culture-based biotechnological applications. *Frontiers in Plant Science* 14, 1132555. DOI: 10.3389/fpls.2023.1132555.

Palladino, F., Marcelino, P.R.F., Schlogl, A.E., José, Á.H.M., Rodrigues, R.d.C.L.B. *et al.* (2024) Bioreactors: Applications and innovations for a sustainable and healthy future – a critical review. *Applied Sciences* 14(20), 9346. DOI: 10.3390/app14209346.

Panche, A.N., Diwan, A.D. and Chandra, S.R. (2016) Flavonoids: An overview. *Journal of Nutritional Science* 5, e47. DOI: 10.1017/jns.2016.41.

Pandey, P., Irulappan, V., Bagavathiannan, M.V. and Senthil-Kumar, M. (2017) Impact of combined abiotic and biotic stresses on plant growth and avenues for crop improvement by exploiting physio-morphological traits. *Frontiers in Plant Science* 8, 537. DOI: 10.3389/fpls.2017.00537.

Pandey, S.S., Singh, S., Pandey, H., Srivastava, M., Ray, T. *et al.* (2018) Endophytes of *Withania somnifera* modulate *in planta* content and the site of withanolide biosynthesis. *Scientific Reports* 8(1), 5450. DOI: 10.1038/s41598-018-23716-5.

Pang, Z., Chen, J., Wang, T., Gao, C., Li, Z. *et al.* (2021) Linking plant secondary metabolites and plant microbiomes: A review. *Frontiers in Plant Science* 12, 621276. DOI: 10.3389/fpls.2021.621276.

Pantigoso, H.A., Newberger, D. and Vivanco, J.M. (2022) The rhizosphere microbiome: Plant-microbial interactions for resource acquisition. *Journal of Applied Microbiology* 133(5), 2864–2876. DOI: 10.1111/jam.15686.

Pardo-Esté, C., Cortés, J., Castro-Severyn, J., Pérez, V., Henriquez-Aedo, K. *et al.* (2024) Secondary metabolites with antimicrobial activity produced by thermophilic bacteria from a high-altitude hydrothermal system. *Frontiers in Microbiology* 15, 1477458. DOI: 10.3389/fmicb.2024.1477458.

Peiffer, J.A., Spor, A., Koren, O., Jin, Z., Tringe, S.G. *et al.* (2013) Diversity and heritability of the maize rhizosphere microbiome under field conditions. *Proceedings of the National Academy of Sciences USA* 110(16), 6548–6553. DOI: 10.1073/pnas.1302837110.

Peng, M., Wang, C., Wang, Z., Huang, X., Zhou, F. *et al.* (2022) Differences between the effects of plant species and compartments on microbiome composition in two halophyte *Suaeda species*. *Bioengineered* 13(5), 12475–12488. DOI: 10.1080/21655979.2022.2076009.

Pérez-Cobas, A.E., Gomez-Valero, L. and Buchrieser, C. (2020) Metagenomic approaches in microbial ecology: An update on whole-genome and marker gene sequencing analyses. *Microbial Genomics* 6(8), mgen000409. DOI: 10.1099/mgen.0.000409.

Petrova, M., Miladinova-Georgieva, K. and Geneva, M. (2024) Influence of abiotic and biotic elicitors on organogenesis, biomass accumulation, and production of key secondary metabolites in *Asteraceae* plants. *International Journal of Molecular Sciences* 25(8), 4197. DOI: 10.3390/ijms25084197.

Pott, D.M., Osorio, S. and Vallarino, J.G. (2019) From central to specialized metabolism: An overview of some secondary compounds derived from the primary metabolism for their role in conferring nutritional and organoleptic characteristics to fruit. *Frontiers in Plant Science* 10, 835. DOI: 10.3389/fpls.2019.00835.

Praveen, A., Singh, S. and Sharma, V.K. (2023) Action of nanoparticles in the amelioration of heavy metal phytotoxicity. *Cereal Research Communications* 51(3), 537–544. DOI: 10.1007/s42976-022-00340-8.

Rocha, L.O. and Silva, N.C.C. (2023) Editorial: Application of omics-based technologies and the impact on food science. *Frontiers in Microbiology* 14, 1155757. DOI: 10.3389/fmicb.2023.1155757.

Rodriguez, P.A., Rothballer, M., Chowdhury, S.P., Nussbaumer, T., Gutjahr, C. *et al.* (2019) Systems biology of plant-microbiome interactions. *Molecular Plant* 12(6), 804–821. DOI: 10.1016/j.molp.2019.05.006.

Roszkowski, S. (2023) Application of polyphenols and flavonoids in oncological therapy. *Molecules* 28(10). DOI: 10.3390/molecules28104080.

Roy, A., Khan, A., Ahmad, I., Alghamdi, S., Rajab, B.S. *et al.* (2022) Flavonoids, a bioactive compound from medicinal plants and its therapeutic applications. *BioMed Research International* 2022(1), 5445291. DOI: 10.1155/2022/5445291.

Saeed, Q., Xiukang, W., Haider, F.U., Kučerik, J., Mumtaz, M.Z. *et al.* (2021) Rhizosphere bacteria in plant growth promotion, biocontrol, and bioremediation of contaminated sites: A comprehensive review of effects and mechanisms. *International Journal of Molecular Sciences* 22(19), 10529. DOI: 10.3390/ijms221910529.

Sagharyan, M., Ganjeali, A., Cheniany, M. and Mousavi Kouhi, S.M. (2020) Optimization of callus induction with enhancing production of phenolic compounds production and antioxidants activity in Callus cultures of *Nepeta binaloudensis* Jamzad (Lamiaceae). *Iranian Journal of Biotechnology* 18(4), e2621. DOI: 10.30498/IJB.2020.2621.

Sahoo, A., Yadav, G., Mehta, T., Meena, M. and Swapnil, P. (2025) Omics-driven insights into plant growth-promoting microorganisms for sustainable agriculture. *Discover Sustainability* 6(1), 659. DOI: 10.1007/s43621-025-01582-2.

Samantara, K., Shiv, A., de Sousa, L.L., Sandhu, K.S., Priyadarshini, P. *et al.* (2021) A comprehensive review on epigenetic mechanisms and application of epigenetic modifications for crop improvement. *Environmental and Experimental Botany* 188, 104479. DOI: 10.1016/j.envexpbot.2021.104479.

Sasidharan Pillai, S., Gagnon, C.A., Foster, C. and Ashraf, A.P. (2024) Exploring the gut microbiota: Key insights into its role in obesity, metabolic syndrome, and type 2 diabetes. *Journal of Clinical Endocrinology and Metabolism* 109(11), 2709–2719. DOI: 10.1210/clinem/dgae499.

Satam, H., Joshi, K., Mangrolia, U., Waghoo, S., Zaidi, G. *et al.* (2023) Next-generation sequencing technology: Current trends and advancements. *Biology* 12(7), 997. DOI: 10.3390/biology12070997.

Sessitsch, A., Wakelin, S., Schloter, M., Maguin, E., Cernava, T. *et al.* (2023) Microbiome interconnectedness throughout environments with major consequences for healthy people and a healthy planet. *Microbiology and Molecular Biology Reviews* 87(3), e0021222. DOI: 10.1128/mmbr.00212-22.

Sharma, I., Kashyap, S. and Agarwala, N. (2023) Biotic stress-induced changes in root exudation confer plant stress tolerance by altering rhizospheric microbial community. *Frontiers in Plant Science* 14, 1132824. DOI: 10.3389/fpls.2023.1132824.

Shen, Y., Fan, N., Ma, S.X., Cheng, X., Yang, X. *et al.* (2025) Gut microbiota dysbiosis: Pathogenesis, diseases, prevention, and therapy. *MedComm* 6(5), e70168. DOI: 10.1002/mco2.70168.

Sheshadri, S.A., Nishanth, M.J. and Simon, B. (2016) Stress-mediated *cis*-element transcription factor interactions interconnecting primary and specialized metabolism *in planta*. *Frontiers in Plant Science* 7, 1725. DOI: 10.3389/fpls.2016.01725.

Shrivastav, D., Mishra, J., Sharma, V.K., Singh, S., Khan, M.I. *et al.* (2026) Biochemical and physiological response during oxidative stress: A cross-species perspective. *Rejuvenation Research* 29, 29–39. DOI: 10.1177/15491684251386712.

Singh, S., Praveen, A., Dudha, N., Sharma, V.K. and Bhadrecha, P. (2024) Single-cell transcriptomics: A new frontier in plant biotechnology research. *Plant Cell Reports* 43(12), 294. DOI: 10.1007/s00299-024-03383-9.

Sivakumar, N., Sathishkumar, R., Selvakumar, G., Shyamkumar, R. and Arjunekumar, K. (2020) Phyllospheric microbiomes: Diversity, ecological significance, and biotechnological applications. *Plant Microbiomes for Sustainable Agriculture* 25, 113–172. DOI: 10.1007/978-3-030-38453-1_5.

Smith, A.M., Heisler, L.E., St. Onge, R.P., Farias-Hesson, E., Wallace, I.M. *et al.* (2010) Highly-multiplexed barcode sequencing: An efficient method for parallel analysis of pooled samples. *Nucleic Acids Research* 38(13), e142. DOI: 10.1093/nar/gkq368.

Sun, Y., Chen, L., Zhang, S., Miao, Y., Zhang, Y. *et al.* (2022) Plant interaction patterns shape the soil microbial community and nutrient cycling in different intercropping scenarios of aromatic plant species. *Frontiers in Microbiology* 13, 888789. DOI: 10.3389/fmicb.2022.888789.

Tabanelli, R., Brogi, S. and Calderone, V. (2021) Improving curcumin bioavailability: Current strategies and future perspectives. *Pharmaceutics* 13(10), 1715. DOI: 10.3390/pharmaceutics13101715.

Tan, B., Ng, C., Nshimyimana, J.P., Loh, L.L., Gin, K.Y.-H. *et al.* (2015) Next-generation sequencing (NGS) for assessment of microbial water quality: Current progress, challenges, and future opportunities. *Frontiers in Microbiology* 6, 1027. DOI: 10.3389/fmicb.2015.01027.

Thepbandit, W. and Athinuwat, D. (2024) Rhizosphere microorganisms supply availability of soil nutrients and induce plant defense. *Microorganisms* 12(3), 558. DOI: 10.3390/microorganisms12030558.

Tiwari, S.K., Manoj, G., Prasanth, K., Sivaram, G., Sharma, V.K. *et al.* (2009) Simplified and versatile method for bisulfite-based DNA methylation analysis of small amounts of DNA. *Journal of Clinical Laboratory Analysis* 23(3), 172–174. DOI: 10.1002/jcla.20314.

Tiwari, S.K., Sharma, V., Sharma, V.K., Gopi, M., Saikant, R. *et al.* (2011) Phylogenetic analysis, based on EPIYA repeats in the CagA gene of Indian *Helicobacter pylori*, and the implications of sequence variation in tyrosine phosphorylation motifs on determining the clinical outcome. *Genetics and Molecular Biology* 34(2), 280–285. DOI: 10.1590/S1415-47572011005000003.

Tripathi, A., Pandey, P., Tripathi, S.N. and Kalra, A. (2025) Plant growth-promoting endophytic consortium improved artemisinin biosynthesis via modulating antioxidants, gene expression, and transcriptional profile in *Artemisia annua* (L.) under stressed environments. *Plant Stress* 15, 100757. DOI: 10.1016/j.stress.2025.100757.

Tripathi, A.K., Ray, A.K. and Mishra, S.K. (2022) Molecular and pharmacological aspects of piperine as a potential molecule for disease prevention and management: Evidence from clinical trials. *Beni-Suef University Journal of Basic and Applied Sciences* 11(1), 16. DOI: 10.1186/s43088-022-00196-1.

Turner, T.R., James, E.K. and Poole, P.S. (2013) The plant microbiome. *Genome Biology* 14(6), 209. DOI: 10.1186/gb-2013-14-6-209.

Tuteja, N. and Sopory, S.K. (2008) Plant signaling in stress: G-protein coupled receptors, heterotrimeric G-proteins and signal coupling via phospholipases. *Plant Signaling and Behavior* 3(2), 79–86. DOI: 10.4161/psb.3.2.5303.

van Leeuwen, P.T., Brul, S., Zhang, J. and Wortel, M.T. (2023) Synthetic microbial communities (SyncComs) of the human gut: Design, assembly, and applications. *FEMS Microbiology Reviews* 47(2), fuad012. DOI: 10.1093/femsre/fuad012.

Verdú-Navarro, F., Moreno-Cid, J.A., Weiss, J. and Egea-Cortines, M. (2023) The advent of plant cells in bioreactors. *Frontiers in Plant Science* 14, 1310405. DOI: 10.3389/fpls.2023.1310405.

Verma, V.V., Vimal, S., Mishra, M.K. and Sharma, V.K. (2025) A comprehensive review on structural insights through molecular visualization: Tools, applications, and limitations. *Journal of Molecular Modeling* 31(6), 173. DOI: 10.1007/s00894-025-06402-y.

Wankhade, A., Wilkinson, E., Britt, D.W. and Kaundal, A. (2025) A review of plant–microbe interactions in the rhizosphere and the role of root exudates in microbiome engineering. *Applied Sciences* 15(13), 7127. DOI: 10.3390/app15137127.

White, P.J. and Brown, P.H. (2010) Plant nutrition for sustainable development and global health. *Annals of Botany* 105(7), 1073–1080. DOI: 10.1093/aob/mcq085.

Wijerathna-Yapa, A. and Hiti-Bandaralage, J. (2023) Tissue culture – A sustainable approach to explore plant stresses. *Life* 13(3), 780. DOI: 10.3390/life13030780.

Wijesekara, T. and Xu, B. (2023) Health-promoting effects of bioactive compounds from plant endophytic fungi. *Journal of Fungi* 9(10), 997. DOI: 10.3390/jof9100997.

Wishart, D.S., Cheng, L.L., Copié, V., Edison, A.S., Eghbalnia, H.R. *et al.* (2022) NMR and metabolomics – A roadmap for the future. *Metabolites* 12(8), 678. DOI: 10.3390/metabo12080678.

Xia, Y., Liu, J., Chen, C., Mo, X., Tan, Q. *et al.* (2022) The multifunctions and future prospects of endophytes and their metabolites in plant disease management. *Microorganisms* 10(5), 1072. DOI: 10.3390/microorganisms10051072.

Xiao, J.F., Zhou, B. and Ressom, H.W. (2012) Metabolite identification and quantitation in LC-MS/MS-based metabolomics. *Trends in Analytical Chemistry* 32, 1–14. DOI: 10.1016/j.trac.2011.08.009.

Xiao, X., Han, L., Chen, H., Wang, J., Zhang, Y. *et al.* (2023) Intercropping enhances microbial community diversity and ecosystem functioning in maize fields. *Frontiers in Microbiology* 13, 1084452. DOI: 10.3389/fmicb.2022.1084452.

Xiong, W., Abraham, P.E., Li, Z., Pan, C. and Hettich, R.L. (2015) Microbial metaproteomics for characterizing the range of metabolic functions and activities of human gut microbiota. *Proteomics* 15(20), 3424–3438. DOI: 10.1002/pmic.201400571.

Xu, W.J., Lin, Y., Mi, C.L., Pang, J.Y. and Wang, T.Y. (2023) Progress in fed-batch culture for recombinant protein production in CHO cells. *Applied Microbiology and Biotechnology* 107(4), 1063–1075. DOI: 10.1007/s00253-022-12342-x.

Yang, C.X., Chen, S.J., Hong, X.Y., Wang, L.Z., Wu, H.M. *et al.* (2025) Plant exudates-driven microbiome recruitment and assembly facilitates plant health management. *FEMS Microbiology Reviews* 49, fuaf008. DOI: 10.1093/femsre/fuaf008.

Yin, J.Y., Lai, M., Yu, X.Y., Su, D.D., Xiong, X.Y. *et al.* (2024) Comprehensive strategies for paclitaxel production: Insights from plant cell culture, endophytic microorganisms, and synthetic biology. *Horticulture Research* 12(3), uhae346. DOI: 10.1093/hr/uhae346.

Yusuf, A., Li, M., Zhang, S.Y., Odedishemi-Ajibade, F., Luo, R.F. *et al.* (2025) Harnessing plant-microbe interactions: Strategies for enhancing resilience and nutrient acquisition for sustainable agriculture. *Frontiers in Plant Science* 16, 1503730. DOI: 10.3389/fpls.2025.1503730.

Zagoskina, N.V., Zubova, M.Y., Nechaeva, T.L., Kazantseva, V.V., Goncharuk, E.A. *et al.* (2023) Polyphenols in plants: Structure, biosynthesis, abiotic stress regulation, and practical applications (review). *International Journal of Molecular Sciences* 24(18), 13874. DOI: 10.3390/ijms241813874.

Zam, W. (2018) Gut microbiota as a prospective therapeutic target for curcumin: A review of mutual influence. *Journal of Nutrition and Metabolism* 2018, 1367984. DOI: 10.1155/2018/1367984.

Zhu, J. and He, L. (2024) The modulatory effects of curcumin on the gut microbiota: A potential strategy for disease treatment and health promotion. *Microorganisms* 12(4), 642. DOI: 10.3390/microorganisms12040642.

11 Roles of Endophytes in Medicinal Plants: Synthesis of Bioactive Compounds and Plant Pathogen Resistance

Meenu Singh Sengar[1], Shalini Tiwari[2] and Manoj Kumar Mishra[3]*

[1]*Department of Agronomy, Horticulture and Plant Science, South Dakota State University, Brookings, USA*
[2]*Division of Natural Sciences and Mathematics, University of the Ozarks, Clarksville, USA*
[3]*Department of Biotechnology, S.R. Institute of Management and Technology, Lucknow, India*

11.1 Introduction

11.1.1 Definition and ecological context of endophytes

Endophytes are microorganisms, primarily bacteria and fungi, that reside asymptomatically within the tissues of living plants, including roots, stems, leaves, and reproductive organs, without causing symptoms to their host (Nazir *et al.*, 2024; Zotchev, 2024). These microbes have evolved sophisticated mechanisms to colonize their hosts and establish mutualistic or commensal relationships that often confer significant advantages to the plant. The term "endophyte" was first formally defined in the 1880s, but our understanding of these organisms has undergone revolutionary transformation in recent decades, particularly with the advent of molecular biology and genomic sequencing technologies (El-Zawawy *et al.*, 2023).

The ecological significance of endophytes cannot be overstated. It is estimated that most terrestrial plants harbor endophytic microbial communities, suggesting that these associations are not exceptional but rather fundamental to plant biology (Zotchev, 2024). For medicinal plants specifically, endophytes

*Corresponding author: manoj30biotech@gmail.com

 Medicinal Plants and the Rhizosphere (eds M.K. Mishra and S. Tiwari)
DOI: 10.1079/9781836990475.0011

have become the focus of intensive research due to their demonstrated capacity to synthesize bioactive compounds that rival or exceed those produced by the host plant itself (Nazir *et al.*, 2024).

11.1.2 Diversity of endophytic microorganisms

The endophytic microbiome of medicinal plants encompasses extraordinary diversity. The most identified endophytic bacteria belong to genera including *Bacillus*, *Pseudomonas*, *Streptomyces*, *Paenibacillus*, and *Chromobacterium*, among numerous others (Hnamte *et al.*, 2024). Of particular note, approximately 75% of characterized bacterial endophytes are *Streptomyces* species, which are filamentous, Gram-positive actinomycetes with remarkable potential for secondary metabolite biosynthesis (El-Zawawy *et al.*, 2023).

Fungal endophytes primarily comprise members of the ascomycetes, with notable genera including *Trichoderma*, *Curvularia*, *Colletotrichum*, and *Pestalotipora*. The presence of specific endophytic communities varies substantially based on plant species, geographic location, soil composition, and environmental conditions (Zotchev, 2024).

11.1.3 Significance for medicinal plants

Medicinal plants have been the foundation of traditional medicine systems across all cultures and remain the source of approximately 50% of modern pharmaceutical compounds (Nazir *et al.*, 2024). The discovery that many bioactive compounds in medicinal plants are partially or wholly synthesized by their endophytic partners has profound implications for pharmaceutical development, agricultural sustainability, and our understanding of plant evolution (Zotchev, 2024). This knowledge transforms medicinal plants from simple repositories of bioactive compounds into complex consortia of plant–microbial partnerships orchestrating sophisticated chemistry (Nazir *et al.*, 2024).

11.2 Endophytic Contribution to Secondary Metabolite Synthesis

11.2.1 Major classes of bioactive compounds synthesized by endophytes

Endophytic microorganisms produce an extraordinary array of secondary metabolites with diverse chemical structures and biological activities. The major classes are as follows (Nazir *et al.*, 2024; Zotchev, 2024).

- *Alkaloids*: Nitrogen-containing heterocyclic compounds with potent pharmacological activities, including antimicrobial, anticancer, and antiinflammatory properties.

- *Terpenoids*: Volatile and nonvolatile compounds derived from isoprene units, including monoterpenes, sesquiterpenes, and triterpenes with antimicrobial and immunomodulatory functions.
- *Flavonoids*: Polyphenolic compounds with antioxidant, antiinflammatory, and cardioprotective activities.
- *Anthraquinones*: Aromatic compounds with antimicrobial, laxative, and anticancer properties.
- *Polyketides*: Complex metabolites synthesized by polyketide synthase enzymes with diverse biological activities.
- *Phenols and polyphenols*: Aromatic compounds with substantial antioxidant potential.
- *Steroids and quinones*: Compounds with hormonal and antimicrobial activities.
- *Peptides and enzymatic proteins*: Including chitinases, proteases, and lipopeptides with biocontrol functions.

11.2.2 Mechanisms of endophytic metabolite production

11.2.2.1 Biosynthetic capacity and genome mining

Recent advances in genomic sequencing and bioinformatics have revealed that endophytes possess extensive biosynthetic gene clusters (BGCs) encoding the enzymatic machinery for secondary metabolite production (El-Zawawy *et al.*, 2023). Modern approaches such as genome mining enable researchers to predict the biosynthetic potential of endophytic strains by analyzing their genetic content prior to cultivation (El-Zawawy *et al.*, 2023). This computational approach has accelerated the discovery of novel bioactive compounds with pharmaceutical potential.

The biosynthetic capacity of endophytes appears substantially greater than their expressed phenotype in culture, suggesting that many metabolites remain cryptic under standard laboratory conditions (El-Zawawy *et al.*, 2023). Coculture systems, elicitation strategies, and in planta production represent emerging approaches to unlock this latent biosynthetic potential (Nazir *et al.*, 2024).

11.2.2.2 Molecular signaling and metabolite elicitation

Recent research demonstrates that endophytic elicitor compounds and signals from endophytes significantly enhance secondary metabolite accumulation in medicinal plants (Chowdhury *et al.*, 2024; Wang *et al.*, 2025). These elicitation effects operate through well-characterized molecular pathways involving the following.

- *Jasmonic acid (JA) signaling*: Activation of JA-responsive genes that upregulate secondary metabolite biosynthesis.
- *Salicylic acid (SA) signaling*: Enhancement of defense-related gene expression and metabolite production.

- *Abscisic acid (ABA) pathways*: Stress-responsive signaling that amplifies secondary metabolite accumulation.
- *Reactive oxygen species (ROS) generation*: Controlled ROS production that serves as a stress signal triggering metabolite synthesis.

11.2.3 Exemplary endophytes and their metabolite production

11.2.3.1 Fungal endophytes in medicinal plant chemistry

Curvularia papendorfii, an endophytic fungus, produces bioactive compounds including mannitol and kheiric acid, which demonstrate potent antiviral, antimicrobial, and antioxidant activities (Ribeiro *et al.*, 2025). Similarly, *Trichoderma* species, found extensively in medicinal plants, biosynthesize numerous secondary metabolites and simultaneously display antagonistic effects against phytopathogens through the production of cell wall-degrading enzymes (Guzmán-Guzmán *et al.*, 2025).

The endophytic fungus *Pestalotipora* has been isolated from plants traditionally used in ethnopharmacology and shown to produce compounds with immunosuppressive and anticancer properties (Zotchev, 2024).

11.2.3.2 Bacterial endophytes and bioactive production

Actinomycete endophytes, particularly *Streptomyces* species, represent perhaps the most prolific producers of bioactive compounds among all microorganisms. *Streptomyces roseoverticillatus* 63 produces carbazomycin B, an antimicrobial compound with potent activity against pathogenic bacteria and demonstrated therapeutic potential (Shi *et al.*, 2021). *Bacillus* species produce various antimicrobial lipopeptides including surfactin, fengycin, and iturin, compounds with pharmaceutical applications in infection control (Chowdhury *et al.*, 2024).

11.2.4 Implications for drug discovery and development

The endophytic route to bioactive compound acquisition offers several distinct advantages over traditional plant extraction (El-Zawawy *et al.*, 2023).

- *Scalability*: Endophytes can be cultured in bioreactors independent of host plant cultivation, enabling sustainable production of bioactive compounds.
- *Enhanced biosynthesis*: Metabolic engineering of endophytic strains can amplify metabolite production beyond levels achievable in nature.
- *Genetic engineering accessibility*: Microorganisms are more amenable to genetic modification than plants, enabling rational design of biosynthetic pathways.

- *Reduced environmental impact*: Microbial cultivation requires substantially less land, water, and resources compared to agricultural production of medicinal plants.
- *Novel compound generation*: Engineered endophytes can produce derivatives and analogues of natural products with potentially improved pharmacological properties.

11.3 Endophytes and Plant Pathogen Resistance: Mechanisms of Plant Defense

11.3.1 Overview of endophyte-mediated plant immunity

Endophytes enhance plant defense through multiple complementary mechanisms, functioning as sophisticated immune system partners that strengthen the plant's capacity to resist pathogens while maintaining tissue homeostasis (Chowdhury *et al.*, 2024). The defense mechanisms involve both direct antagonism of pathogens and indirect enhancement of the host plant's innate immune responses (Gore *et al.*, 2025).

11.3.2 Direct antagonistic mechanisms

11.3.2.1 Production of antimicrobial compounds

Endophytes produce an extensive arsenal of antimicrobial compounds that directly inhibit pathogen growth and virulence (Chowdhury *et al.*, 2024) (Table 11.1).

The lipopeptide antimicrobials represent a particularly important class. *Bacillus velezensis* strains produce lipopeptides that inhibit *Fusarium oxysporum* f. sp., the causative agent of *Fusarium* wilt, and *Colletotrichum gloeosporioides*, a major strawberry pathogen (Malviya *et al.*, 2020). These compounds

Table 11.1. Major antimicrobial compounds produced by endophytes and their mechanisms of action.

Antimicrobial class	Production mechanism	Target pathogens
Lipopeptides (surfactin, iturin, fengycin)	Ribosomal synthesis by NRPS enzymes	Fungi, Gram-positive bacteria
Antibiotics (hydrogen cyanide, HCN)	Enzymatic synthesis pathways	Bacteria, some fungi
Antifungal metabolites (β–1,3-glucanase)	Cell wall-degrading enzymes	Fungal cell walls
Siderophores	Iron-chelating compounds	Pathogenic bacteria
Secondary metabolites (carbazomycin B)	Polyketide biosynthesis	Gram-negative bacteria

HCN, hydrogen cyanide; NRPS, nonribosomal peptide synthetase.

function through multiple mechanisms including cell membrane disruption, protein denaturation, and inhibition of virulence factor expression.

11.3.2.2 Hydrolytic enzymes and cell wall degradation

Endophytic bacteria and fungi secrete hydrolytic enzymes that degrade the structural components of pathogen cell walls, rendering them vulnerable to lysis (Nimbulkar *et al.*, 2025).

- *Chitinases*: Degrade chitin in fungal cell walls, particularly effective against ascomycete and basidiomycete pathogens.
- *Proteases*: Disrupt pathogenic protein structures and inactivate virulence factors.
- *Cellulases and hemicellulases*: Degrade structural polysaccharides in pathogen cell walls.
- *Glucanases*: Disrupt *β*-glucan components of fungal and bacterial cell walls.

Bacillus thuringiensis SY33.3, for example, produces extracellular chitinase that provides effective biocontrol against multiple phytopathogens including *Aspergillus niger*, *Verticillium dahliae*, and *Fusarium oxysporum* f. sp. Niveum (Azizoglu *et al.*, 2021).

11.3.2.3 Competitive exclusion and resource limitation

Endophytes compete with pathogens for limited nutritional resources and ecological niches within plant tissues. Production of siderophores, iron-chelating compounds, creates iron-limiting conditions that suppress pathogenic bacterial and fungal growth (Ramírez *et al.*, 2024). This mechanism is particularly important because iron availability represents a critical limiting factor for microbial proliferation in plant tissues.

11.3.3 Indirect immunological enhancement mechanisms

11.3.3.1 Systemic resistance induction

Endophytes trigger systemic acquired resistance (SAR) and induced systemic resistance (ISR) pathways in host plants, priming the plant's immune system for rapid and enhanced response to pathogenic challenge (Kurokawa *et al.*, 2021).

- *Systemic acquired resistance*: Salicylic acid-dependent resistance mechanism that confers broad-spectrum protection against diverse pathogens.
- *Induced systemic resistance*: Jasmonic acid and ethylene-dependent pathway that enhances plant defensive capacity.
- *Priming*: Endophytes lower the threshold for activation of plant defense responses, enabling faster and more robust immune reactions upon pathogenic challenge.

These resistance mechanisms persist systemically throughout the plant, providing protection far exceeding the localized presence of endophytes.

11.3.3.2 Stimulation of host defense gene expression

Endophytes stimulate the expression of stress-responsive genes in their host plants, upregulating the biosynthesis of defensive phytochemicals (El-Zawawy *et al.*, 2023; Wang *et al.*, 2025). This includes enhanced production of:

- *phytoalexins*: Antimicrobial compounds synthesized de novo in response to pathogenic challenge
- *pathogenesis-related (PR) proteins*): Including chitinases, glucanases, and proteinase inhibitors
- *antioxidant enzymes*: Superoxide dismutase, catalase, and peroxidases that neutralize pathogenic ROS
- *phenolic compounds*: Accumulation of antimicrobial and antioxidant phenols and flavonoids.

The molecular basis of this stimulation involves the production of endophytic elicitors, conserved microbe-associated molecular patterns (MAMPs) that activate pattern recognition receptors (PRRs) on plant cell surfaces, triggering the initial immune cascade (El-Zawawy *et al.*, 2023).

11.3.3.3 Production of plant growth-promoting hormones

Endophytes enhance plant vigor and stress tolerance through the production of phytohormones (El-Zawawy *et al.*, 2023; Li *et al.*, 2024).

- *Auxins (indole-3-acetic acid, IAA)*: Promote root development, nutrient acquisition, and stress resilience.
- *Gibberellins*: Enhance plant growth, seed germination, and photosynthetic capacity.
- *Brassinosteroids*: Increase plant biomass accumulation and systemic defense capacity.

Bacillus circulans GN03, for example, enhances disease resistance in its host through the production of multiple growth-promoting hormones including brassinosteroids, IAA, and gibberellic acid, simultaneously promoting plant growth while increasing pathogen resistance (Qin *et al.*, 2021).

11.3.4 Quantified examples of endophyte-mediated pathogen control

11.3.4.1 Case study: Bacterial endophytes against fungal pathogens

Bacillus subtilis XZ18-3 exhibits potent antagonistic activity against *Rhizoctonia cerealis*, the causative agent of wheat rot and blight. The mechanism involves accumulation of ROS that damage pathogenic cell structures and suppress virulence factor expression (Yi *et al.*, 2022). Under field conditions, wheat plants colonized by this endophyte show approximately 60–70% reduction in disease severity compared to untreated controls.

Bacillus velezensis Bs006 produces antimicrobial cyclic lipopeptides including surfactins and fengycins, conferring protection against *Fusarium oxysporum* f. sp., which causes *Fusarium* wilt in goldenberries and numerous other crop species (Moreno-Velandia *et al.*, 2021). In controlled greenhouse experiments, disease incidence was reduced from approximately 85% in untreated plants to 15–20% in endophyte-colonized plants.

11.3.4.2 Case study: Streptomyces endophytes against bacterial pathogens

Streptomyces roseoverticillatus 63 demonstrates remarkable antagonistic activity against *Xanthomonas oryzae* pv. *oryzae*, a devastating pathogenic bacterium causing bacterial leaf blight in rice. The endophyte produces carbazomycin B, which suppresses bacterial protein expression and reduces pathogenic metabolic enzymes, including malate dehydrogenase (Shi *et al.*, 2021). This represents a particularly important example because it demonstrates endophyte efficacy against difficult-to-control bacterial pathogens that currently require extensive chemical pesticide applications.

11.3.5 Clinical context: Approximately 70% of plant infections are attributed to fungal and bacterial pathogens

It is well established that approximately 70% of documented plant infections are caused by fungal and bacterial pathogens, resulting in substantial global crop losses (Hnamte *et al.*, 2024). The major pathogenic genera include *Xanthomonas*, *Ralstonia*, *Pseudomonas*, *Clavibacter*, *Agrobacterium*, *Xylella*, *Burkholderia*, *Erwinia*, and various fungal genera including *Ustilago*, *Verticillium*, *Botrytis*, and *Phytophthora* (Hnamte *et al.*, 2024).

These pathogens cause characteristic disease phenotypes including blights, leaf spots, cankers, wilts, scabs, and soft rots, collectively responsible for approximately 40% of global crop losses in vulnerable commodities (www.fao.org/plant-production-protection/about/en). Traditional management approaches rely heavily on chemical pesticides but the emergence of pesticide-resistant pathogenic strains, combined with concerns regarding pesticide residual toxicity and environmental persistence, necessitates development of sustainable alternatives (Anand *et al.*, 2023). Endophytic microorganisms represent promising biological alternatives to chemical pesticides, offering the potential for sustained, environmentally compatible pathogen management (Anand *et al.*, 2023).

11.4 Representative Endophytes Identified from Medicinal Plants with Bioactive Compounds

A comprehensive survey of endophytes isolated from medicinal plants reveals remarkable diversity in secondary metabolite production and biological activities. Table 11.2 presents representative examples of characterized endophytes from various medicinal plants, their bioactive compounds, and corresponding biological activities.

11.4.1 Mechanisms of endophytic colonization

Successful endophytic colonization requires sophisticated mechanisms enabling the microbe to breach plant defenses, establish residence within plant tissues, and avoid triggering excessive host immune responses (Ahmad *et al.*, 2025).

- *Cell wall modification*: Endophytes secrete cell wall-degrading enzymes and polysaccharidases that facilitate penetration through plant tissue barriers while simultaneously avoiding recognition as pathogens.
- *Modulation of phytohormones*: Endophytes manipulate auxin, gibberellin, and other hormone levels to facilitate host acceptance and systemic colonization.
- *Secretion of effector proteins*: Microbe-derived effector proteins suppress or modulate plant immune signaling cascades.
- *Production of antioxidants*: Endophytes produce superoxide dismutase, catalase, and other antioxidant enzymes that neutralize plant-derived ROS generated during immune activation.
- *Biofilm formation*: Many endophytes form biofilms within intercellular and intracellular spaces, protecting themselves from antimicrobial host defense compounds.

11.4.2 Metabolic crosstalk and synthetic symbiosis

The interaction between medicinal plants and their endophytic partners involves extensive metabolic crosstalk (El-Zawawy *et al.*, 2023; Nazir *et al.*, 2024). Endophytes can:

- *obtain resources*: Access photosynthetically derived carbohydrates and other metabolic resources from the host plant
- *provide benefits*: Supply fixed nitrogen (for N-fixing endophytes), bioavailable phosphate, and bioactive compounds that enhance host fitness
- *regulate host metabolism*: Modulate host gene expression and metabolic flux, often steering metabolism toward secondary metabolite production

Table 11.2. Representative endophytes isolated from medicinal plants with their bioactive compounds, types, and biological activities.

Medicinal plant (host)	Endophyte (genus/ species)	Metabolite/bioactivity	Type	Key functions	Reference
Taxus brevifolia (Pacific yew)	*Taxomyces andreanae*	Taxol-like diterpenoids	Fungus	Anticancer, antiproliferative	(Zotchev, 2024)
Artemisia annua	*Colletotrichum gloeosporioides*	Artemisinin	Fungus	Antimalarial, elicitation	(Nazir *et al.*, 2024)
Maesa japonica	*Streptomyces olivaceus* JB1	Lobophorins	Actinomycete	Antiinflammatory, cytotoxic	(Um *et al.*, 2022; Zotchev, 2024)
Cinnamomum cassia	*Streptomyces* sp. HBQ95	Lydiamycins A–H	Actinomycete	Antimycobacterial, antimetastatic	(Wang *et al.*, 2023; Zotchev, 2024)
Bruguiera gymnorrhiza	*Streptomyces* sp. GT2002/1503	Xiamycins	Actinomycete	Anti-HIV, antibacterial	(Ding *et al.*, 2010; Zotchev, 2024)
Kandelia candel	*Streptomyces* sp. SCSIO 02999	Sespenins	Actinomycete	Antibacterial, cytotoxic	(Ding *et al.*, 2011; Zotchev, 2024)
Withania somnifera	*Pseudomonas* spp., *Bacillus* spp.	Plant growth promotion, withanolide enhancement	Bacteria	Biocontrol, PGP, immunomodulation	(Zotchev, 2024)
Dendrobium huoshanense	*Streptomyces* sp. HS-3-L-1	Huoshanmycins A–C	Actinomycete	Cytotoxic, polyketide dimers	(Zhu *et al.*, 2022; Zotchev, 2024)
Opuntia ficus-indica	*Rhizopus oryzae*	L-tyrosine	Fungus	Antibacterial, antibiofilm	(El-Zawawy *et al.*, 2023)
Olea europaea	*Fusarium redolens*	Fumaric acid, chrysophanol	Fungus	Antimicrobial, antifungal	(Nazir *et al.*, 2024; Zotchev, 2024)
Lycoris aurea	*Burkholderia gladioli*	Toxoflavin	Bacterium	Antifungal, cytotoxic	(Li *et al.*, 2019; Zotchev, 2024)
Vochysia divergens	*Aeromicrobium ponti*	Indole alkaloids, brevianamide F	Actinomycete	Anti-MRSA, antibacterial	(Gos *et al.*, 2017; Zotchev, 2024)

Bacopa monnieri	*Aspergillus*, *Penicillium* spp.	Alkaloids, flavonoids, terpenoids	Fungi	PGP, stress tolerance, bioaccumulation	(Barthwal *et al.*, 2025)
Bergenia pacumbis	*Helotiales* sp.	Cochlioquinone B, isofusidienol A	Fungus	Biocontrol, diverse secondary metabolites	(Tao *et al.*, 2022)
Calendula officinalis	*Bacillus halotolerans* Cal.l.30	Mojavensin A	Bacterium	Antifungal, cytotoxic, biocontrol	(Tsalgatidou *et al.*, 2022; Zotchev, 2024)
Camellia sinensis (*Tea plant*)	*Pestalotipora fici*	Polyketides, peptides, terpenoids	Fungus	74 biosynthetic gene clusters (PKS, NRPS)	(Wang *et al.*, 2014; Zotchev, 2024)
Anisomeles malabarica	Diverse endophytic fungi	Cryptotanshinone-like diterpenes	Fungi	Secondary metabolite elicitation, biocontrol	(Toppo *et al.*, 2024)
Atractylodes lancea	*Pseudomonas fluorescens* ALEB7B	Oxygenous sesquiterpenoids	Bacterium	Bioactive enhancement, plant growth promotion	(Zhou *et al.*, 2018; Zotchev, 2024)
Papaver somniferum	*Acinetobacter* sp.	Benzylisoquinoline alkaloid gene upregulation	Bacterium	Enhanced secondary metabolite production	(Pandey *et al.*, 2016; Zotchev, 2024)
Ligusticum chuanxiong	*Bacillus subtilis* LB5	Tetramethylpyrazine	Bacterium	Antiinflammatory, nootropic, blood circulation	(Yin *et al.*, 2018; Zotchev, 2024)

PGP, plant growth-promoting; PKS, polyketide synthase; NRPS, nonribosomal peptide synthetase.

- *manage stress responses*: Buffer host stress responses, enabling the plant to tolerate otherwise deleterious environmental conditions.

This represents a form of synthetic symbiosis where plant and microbe function as an integrated metabolic unit, exceeding the capacity of either component alone.

11.4.3 Environmental responsiveness and stress tolerance

Endophytes significantly enhance their host plant's tolerance to diverse environmental stresses including:

- *biotic stress*: Pathogen infection, herbivory, and parasitic attack
- *abiotic stress*: Drought, temperature extremes, salinity, and nutrient limitation
- *xenobiotic stress*: Heavy metals, pesticide residues, and other contaminants.

Medicinal plants colonized by endophytes often display enhanced accumulation of secondary metabolites under stress conditions, likely to reflect the dual effects of stress signaling and endophyte-mediated metabolic stimulation. (El-Zawawy *et al.*, 2023; Nazir *et al.*, 2024).

11.5 Biotechnological Applications and Future Perspectives

11.5.1 Metabolic engineering of endophytes for enhanced bioactive production

The ease of genetic manipulation and growth of endophytic microorganisms, combined with their biosynthetic capacity, make them exceptional candidates for metabolic engineering (El-Zawawy *et al.*, 2023).

- *Pathway amplification*: Overexpression of biosynthetic genes increases target metabolite accumulation.
- *Heterologous pathway integration*: Introduction of biosynthetic genes from unrelated organisms enables production of novel compounds.
- *Precursor feeding*: Supplementation of biosynthetic precursors increases product yields.
- *Rational design*: Computational modeling guides targeted modifications to biosynthetic enzymes, improving catalytic efficiency.

These approaches enable production of valuable pharmaceuticals at scales and costs unattainable through conventional plant cultivation.

11.5.2 Coculture systems and in planta metabolite expression

Emerging research demonstrates that endophytes express substantially different metabolic profiles when cultured in association with their host plants compared to axenic cultures (El-Zawawy *et al.*, 2023). Coculture systems and in planta cultivation unlock cryptic biosynthetic capacity, enabling discovery of compounds not produced under standard laboratory conditions (El-Zawawy *et al.*, 2023). This represents a critical frontier in endophyte-based drug discovery.

11.5.3 Biocontrol formulation and agricultural application

Endophytic biocontrol agents represent sustainable alternatives to chemical pesticides (Anand *et al.*, 2023; Hnamte *et al.*, 2024).

- *Inoculant formulation*: Endophytic strains are formulated as seed treatments, root inoculants, or foliar applications.
- *Integrated pest management*: Endophytes function as components of comprehensive IPM strategies.
- *Resistance management*: Rotation of endophytic strains prevents emergence of resistant pathogen populations.
- *Environmental sustainability*: Biocontrol eliminates chemical pesticide residues, benefiting soil microbiota, beneficial insects, and human health.

11.5.4 Genomic mining and computational discovery

Advanced genomic sequencing technologies combined with bioinformatic analysis enable systematic identification of biosynthetic gene clusters and prediction of novel metabolite structures (El-Zawawy *et al.*, 2023). These computational approaches accelerate discovery of pharmaceutically relevant compounds without requiring cultivation or chemical isolation.

11.5.5 Personalized and precision medicine applications

Understanding the endophytic contribution to medicinal plant bioactivity has implications for quality control and standardization of herbal medicines (Nazir *et al.*, 2024). Future medicinal preparations may be formulated to include specific endophytic strains alongside plant material, ensuring consistent bioactive compound delivery and enhanced therapeutic efficacy.

11.6 Conclusion

Endophytes represent transformative partners in medicinal plant biology, orchestrating sophisticated secondary metabolism and providing critical defense against pathogens. Their involvement in bioactive compound synthesis, including alkaloids, terpenoids, flavonoids, and anthraquinones, expands our understanding of pharmaceutical compound origins and opens new frontiers in drug discovery and development. The multiple mechanisms through which endophytes enhance plant pathogen resistance, from direct antimicrobial production to induced systemic immunity, demonstrate the remarkable evolutionary sophistication of plant–microbe partnerships.

The comprehensive survey of endophytes from medicinal plants (Table 11.2) demonstrates the extraordinary chemical diversity these microorganisms can generate. From actinomycete endophytes producing complex spirotetronate macrolides and indolosesquiterpenes to bacterial endophytes synthesizing polyketides and peptides, the pharmaceutical potential of endophytic microorganisms is becoming increasingly evident (Zotchev, 2024). The correlation between endophytic secondary metabolite production and traditional medicinal uses of host plants evidenced by cases such as lobophorin production by *Streptomyces olivaceus* in *Maesa japonica*, and lydiamycin production by *Streptomyces* sp. HBQ95 in *Cinnamomum cassia* (Um *et al.*, 2022) suggests that endophytes may be fundamental contributors to the therapeutic efficacy of traditional herbal medicines (Wang *et al.*, 2023).

The convergence of genomic sequencing, metabolic engineering, and agricultural biotechnology enables unprecedented opportunities for harnessing endophytic potential. Future research should prioritize the following.

- *Mechanistic understanding*: Elucidating the molecular basis of endophyte–plant–secondary metabolite interactions.
- *Strain optimization*: Identifying and engineering endophytic strains with enhanced bioactive production.
- *Scalable production*: Developing cost-effective, sustainable methods for endophyte cultivation and bioactive compound synthesis.
- *Field validation*: Rigorous evaluation of endophyte-based biocontrol agents under agricultural field conditions.
- *Standardization*: Establishing regulatory frameworks and quality control standards for endophyte-containing pharmaceuticals and biocontrol products.
- *Ecosystem dynamics*: Understanding the long-term ecological impacts of endophyte inoculation on agricultural and natural plant communities.

The endophytes inhabiting medicinal plants are not passive components of traditional medicine systems but active architects of pharmaceutical chemistry and plant immunity. As we transition toward sustainable agriculture and personalized medicine, understanding and leveraging endophytic partnerships will prove increasingly central to achieving both agricultural security and pharmaceutical innovation.

References

Ahmad, F., Wang, P., Sun, P., Liu, Y., Ge, J. *et al.* (2025) Colonization mechanism of endophytes with plants and their role in pesticides degradation. *Journal of Agricultural and Food Chemistry* 73(36), 22133–22149. DOI: 10.1021/acs.jafc.5c06182.

Anand, U., Pal, T., Yadav, N., Singh, V.K., Tripathi, V. *et al.* (2023) Current scenario and future prospects of endophytic microbes: Promising candidates for abiotic and biotic stress management for agricultural and environmental sustainability. *Microbial Ecology* 86(3), 1455–1486. DOI: 10.1007/s00248-023-02190-1.

Azizoglu, Z.B., Yilmaz, S., Azizoglu, U., Karabörklü, S., Temizgul, R. *et al.* (2021) Molecular characterization of the chitinase genes of native *Bacillus thuringiensis* isolates and their antagonistic activity against three important phytopathogenic fungi. *Biologia* 76(9), 2745–2755. DOI: 10.1007/s11756-021-00802-0.

Barthwal, H., Sharma, C., Kumar, V., Rajput, V., Naik, B. *et al.* (2025) Bacopa monnieri, a wonder plant in the backyard: Emphasizing the role of the microbiome in increasing its potential. *Phytomedicine Plus* 5(4), 100865. DOI: 10.1016/j.phyplu.2025.100865.

Chowdhury, T., Shishir, Md.A. and Fakruddin, M. (2024) Endophytic bacteria are potential source of medicinal plant therapeutics and bioactive compound synthesis. *Advances in Herbal Research* 7(1), 1–12. DOI: 10.25163/ahi.719985.

Ding, L., Münch, J., Goerls, H., Maier, A., Fiebig, H.H. *et al.* (2010) Xiamycin, a pentacyclic indolosesquiterpene with selective anti-HIV activity from a bacterial mangrove endophyte. *Bioorganic and Medicinal Chemistry Letters* 20(22), 6685–6687. DOI: 10.1016/j.bmcl.2010.09.010.

Ding, L., Maier, A., Fiebig, H.H., Lin, W.H. and Hertweck, C. (2011) A family of multicyclic indolosesquiterpenes from a bacterial endophyte. *Organic and Biomolecular Chemistry* 9(11), 4029. DOI: 10.1039/c1ob05283g.

El-Zawawy, N.A., Ali, S.S. and Nouh, H.S. (2023) Exploring the potential of *Rhizopus oryzae* AUMC14899 as a novel endophytic fungus for the production of L-tyrosine and its biomedical applications. *Microbial Cell Factories* 22(1), 31. DOI: 10.1186/s12934-023-02041-1.

Gore, S., Singh, S., Swarnkar, P. and Rathore, K.S. (2025) Endophyte mediated plant defence responses and their potential against pathogenic bacteria and fungi. *Discover Plants* 2(1), 331. DOI: 10.1007/s44372-025-00429-4.

Gos, F.M.W.R., Savi, D.C., Shaaban, K.A., Thorson, J.S., Aluizio, R. *et al.* (2017) Antibacterial activity of endophytic actinomycetes isolated from the medicinal plant *Vochysia divergens* (Pantanal, Brazil). *Frontiers in Microbiology* 8, 1642. DOI: 10.3389/fmicb.2017.01642.

Guzmán-Guzmán, P., Etesami, H. and Santoyo, G. (2025) Trichoderma: A multifunctional agent in plant health and microbiome interactions. *BMC Microbiology* 25(1), 434. DOI: 10.1186/s12866-025-04158-2.

Hnamte, L., Kumar, A., Yadav, M.K. and Singh, P.K. (2024) An updated view of bacterial endophytes as antimicrobial agents against plant and human pathogens. *Current Research in Microbial Sciences* 7, 100241. DOI: 10.1016/j.crmicr.2024.100241.

Kurokawa, M., Nakano, M., Kitahata, N., Kuchitsu, K. and Furuya, T. (2021) An efficient direct screening system for microorganisms that activate plant immune responses based on plant-microbe interactions using cultured plant cells. *Scientific Reports* 11(1), 7396. DOI: 10.1038/s41598-021-86560-0.

Li, X., Li, Y., Wang, R., Wang, Q. and Lu, L. (2019) Toxoflavin produced by *Burkholderia gladioli* from *Lycoris aurea* is a new broad-spectrum fungicide. *Applied and Environmental Microbiology* 85(2), e00106–19. DOI: 10.1128/AEM.00106-19.

Li, M., Liu, X., Wu, F., Shi, X., Kong, D. *et al.* (2024) Fermentation broth of a novel endophytic fungus enhanced maize salt tolerance by regulating sugar metabolism and phytohormone biosynthesis or signaling. *Plant Physiology and Biochemistry* 216, 109125. DOI: 10.1016/j.plaphy.2024.109125.

Malviya, D., Sahu, P.K., Singh, U.B., Paul, S., Gupta, A. *et al.* (2020) Lesson from ecotoxicity: Revisiting the microbial lipopeptides for the management of emerging diseases for crop protection. *International Journal of Environmental Research and Public Health* 17(4), 1434. DOI: 10.3390/ijerph17041434.

Moreno-Velandia, C.A., Ongena, M. and Cotes, A.M. (2021) Effects of fengycins and iturins on *Fusarium oxysporum* f. sp. *physali* and root colonization by *Bacillus velezensis* Bs006 protect golden berry against vascular wilt. *Phytopathology* 111(12), 2227–2237. DOI: 10.1094/PHYTO-01-21-0001-R.

Nazir, A., Puthuveettil, A.R., Hussain, F.H.N., Hamed, K.E. and Munawar, N. (2024) Endophytic fungi: Nature's solution for antimicrobial resistance and sustainable agriculture. *Frontiers in Microbiology* 15, 1461504. DOI: 10.3389/fmicb.2024.1461504.

Nimbulkar, P., Gupta, G., Virkhare, U., Althubiani, A.S., Dutta, A. *et al.* (2025) Bacterial endophytes and their secondary metabolites: Mechanisms of biosynthesis and applications in sustainable agriculture. *Journal of Umm Al-Qura University for Applied Sciences* 1–13. DOI: 10.1007/s43994-025-00233-6.

Pandey, S.S., Singh, S., Babu, C.S.V., Shanker, K., Srivastava, N.K. *et al.* (2016) Endophytes of opium poppy differentially modulate host plant productivity and genes for the biosynthetic pathway of *Benzylisoquinoline alkaloids*. *Planta* 243(5), 1097–1114. DOI: 10.1007/s00425-016-2467-9.

Qin, L., Tian, P., Cui, Q., Hu, S., Jian, W. *et al.* (2021) *Bacillus circulans* GN03 alters the microbiota, promotes cotton seedling growth and disease resistance, and increases the expression of phytohormone synthesis and disease resistance-related genes. *Frontiers in Plant Science* 12, 644597. DOI: 10.3389/fpls.2021.644597.

Ramírez, C., Cardozo, M., López Gastón, M., Galdeano, E. and Collavino, M.M. (2024) Plant growth promoting activities of endophytic bacteria from *Melia azedarach* (Meliaceae) and their influence on plant growth under gnotobiotic conditions. *Heliyon* 10(15), e35814. DOI: 10.1016/j.heliyon.2024.e35814.

Ribeiro, V.P., Bajsa-Hirschel, J., Tamang, P., Harries, M.D. and Meepagala, K.M. (2025) Bioactive secondary metabolites from *Curvularia* spp.: Natural alternatives for pest management in agriculture. *Journal of Natural Pesticide Research* 12, 100117. DOI: 10.1016/j.napere.2025.100117.

Shi, T., Guo, X., Zhu, J., Hu, L., He, Z. *et al.* (2021) Inhibitory effects of carbazomycin B produced by *Streptomyces roseoverticillatus* 63 against *Xanthomonas oryzae* pv. *oryzae*. *Frontiers in Microbiology* 12, 616937.

Tao, J., Bai, X., Zeng, M., Li, M., Hu, Z. *et al.* (2022) Whole-genome sequence analysis of an endophytic fungus *Helotiales* sp. and its biosynthetic potential of bioactive secondary metabolites. *Microorganisms* 10(9), 1789. DOI: 10.3390/microorganisms10091789.

Toppo, P., Jangir, P., Mehra, N., Kapoor, R. and Mathur, p (2024) Bioprospecting of endophytic fungi from medicinal plant *Anisomeles indica* L. for their diverse role in agricultural and industrial sectors. *Scientific Reports* 14, 588. DOI: 10.1038/s41598-023-51057-5.

Tsalgatidou, P.C., Thomloudi, E.-E., Baira, E., Papadimitriou, K., Skagia, A. *et al.* (2022) Integrated genomic and metabolomic analysis illuminates key secreted metabolites produced by the novel endophyte *Bacillus halotolerans* Cal.l.30 involved in diverse biological control activities. *Microorganisms* 10(2), 399. DOI: 10.3390/microorganisms10020399.

Um, S., Lee, J. and Kim, S.H. (2022) Lobophorin producing endophytic *Streptomyces olivaceus* JB1 associated with *Maesa japonica* (Thunb.) Moritzi & Zoll. *Frontiers in Microbiology* 13, 881253. DOI: 10.3389/fmicb.2022.881253.

Wang, X., Wang, H., Liu, T. and Xin, Z. (2014) A PKS I gene-based screening approach for the discovery of a new polyketide from *Penicillium citrinum* salicorn 46. *Applied Microbiology and Biotechnology* 98(11), 4875–4885. DOI: 10.1007/s00253-014-5572-3.

Wang, W., Kim, S., Vu, T.H.N., Quach, N.T., Oh, E. *et al.* (2023) Bioactive piperazic acid-bearing cyclodepsipeptides, lydiamycins E–H, from an endophytic *Streptomyces* sp. associated with *Cinnamomum cassia*. *Journal of Natural Products* 86(4), 751–758. DOI: 10.1021/acs.jnatprod.2c00902.

Wang, Y., Chen, N., Deng, K., Zhong, X., Li, Z. *et al.* (2025) Effects and molecular mechanisms of endophytic elicitors on secondary metabolite accumulation in medicinal plants. *Applied Microbiology and Biotechnology* 109(5), 2156–2178. DOI: 10.1007/s00253-025-13456-x.

Yi, Y., Luan, P., Liu, S., Shan, Y., Hou, Z. *et al.* (2022) Efficacy of *Bacillus subtilis* XZ18-3 as a biocontrol agent against *Rhizoctonia cerealis* on wheat. *Agriculture* 12(2), 258. DOI: 10.3390/agriculture12020258.

Yin, D.D., Yang, M., Wang, Y.L., Yin, D.K., Liu, H.K. *et al.* (2018) High tetramethylpyrazine production by the endophytic bacterial *Bacillus subtilis* isolated from the traditional medicinal plant *Ligusticum chuanxiong* Hort. *AMB Express* 8(1), 193. DOI: 10.1186/s13568-018-0721-1.

Zhou, J.-Y., Sun, K., Chen, F., Yuan, J., Li, X. *et al.* (2018) Endophytic *Pseudomonas* induces metabolic flux changes that enhance medicinal sesquiterpenoid accumulation in *Atractylodes lancea*. *Plant Physiology and Biochemistry* 130, 473–481. DOI: 10.1016/j.plaphy.2018.07.016.

Zhu, Y., Kong, Y., Hong, Y., Zhang, L., Li, S. *et al.* (2022) Huoshanmycins A–C, new polyketide dimers produced by endophytic *Streptomyces* sp. HS-3-L-1 from *Dendrobium huoshanense*. *Frontiers in Chemistry* 9, 807508. DOI: 10.3389/fchem.2021.807508.

Zotchev, S.B. (2024) Unlocking the potential of bacterial endophytes from medicinal plants for drug discovery. *Microbial Biotechnology* 17(2), e14382. DOI: 10.1111/1751-7915.14382.

12 Induction of Specific Secondary Metabolite Pathways in Medicinal Plants, Addressing Their Nutritional Needs

Archana Prasad[1,2*], Shiv Rattan[3] and Gauri Saxena[2]

[1]Grape Physiology and Breeding Lab, Department of Fruit Tree Sciences, Institute of Plant Sciences, Agricultural Research Organization (ARO) – Volcani Institute, Rishon Lezion, Israel
[2]Department of Botany, University of Lucknow, Lucknow, India
[3]Department of Ornamental Plants and Agricultural Biotechnology, Agricultural Research Organization (ARO) – Volcani Institute, Rishon Lezion, Israel

12.1 Introduction

Medicinal plants are recognized for their ability to synthesize a wide spectrum of bioactive secondary metabolites that play pivotal roles in plant defense, stress adaptation, and ecological interactions. These low molecular weight compounds, including terpenoids, coumarins, flavonoids, phenolics, and alkaloids, are not essential for basic physiological functions but are crucial for survival under adverse conditions and also possess immense therapeutic and commercial value. The rising global demands for natural therapeutics and sustainable agriculture mean that we need better understanding of regulatory factors of the biosynthesis of these metabolites.

The synthesis of these metabolites is governed by complex biochemical pathways that are influenced by various environmental stimuli and plant physiological conditions. Plant mineral nutrients are not only the major contributors for primary and secondary organic compounds but are also crucial components as enzyme cofactors involved in major metabolic fluxes.

*Corresponding author: archanaambt@gmail.com

DOI: 10.1079/9781836990475.0012

The minerals required by the plant for growth and metabolism are macronutrients. The major macronutrients that underpin plant metabolism include nitrogen (N), phosphorus (P), potassium (K), sulfur (S), and magnesium (Mg). N, P, and S are important constituents of primary and secondary organic compounds while K acts as a cofactor for many enzymatic reactions and as a counter-ion for metabolite transport (Bhat *et al.*, 2020). Nitrogen is a key component for synthesis of the backbone of nucleic acids, amino acids, phenolics, and flavonoids (Robertson and Vitousek, 2009). Potassium is a precursor for many secondary metabolites and also regulates the synthesis of sugars and organic acids (Xin *et al.*, 2016). Phosphorus activates various biosynthetic enzymes that regulate metabolic fluxes in various plants. It is also important for signal transduction and energy transfer (Dong *et al.*, 2004). Sulfur promotes the synthesis of sulfur-containing amino acids like cysteine and methionine. It supports the formation of key secondary metabolites such as glucosinolates, thiols, and certain flavonoids, and enhances enzymatic activity and redox regulation through compounds like glutathione, influencing stress tolerance and metabolic efficiency (Xue *et al.*, 2022). Magnesium (Mg), as the central atom in chlorophyll and a cofactor for numerous enzymes, is indispensable for photosynthesis and carbohydrate metabolism, indirectly modulating the synthesis of flavonoids and terpenoids.

Together, these macronutrients orchestrate a complex network of metabolic pathways that determine the type and abundance of secondary metabolites, underscoring the importance of balanced nutrition in enhancing the phytochemical profile of medicinal plants.

Micronutrients, though required in smaller quantities, are equally vital for secondary metabolism. Elements such as boron (B), copper (Cu), iron (Fe), manganese (Mn), molybdenum (Mo), and zinc (Zn) are involved in enzymatic reactions, redox processes, and signal transduction pathways that facilitate the biosynthesis of various secondary compounds. For instance, zinc has been shown to influence the production of glucosinolates and organic acids, while iron plays a role in chlorophyll synthesis and electron transport, indirectly affecting metabolite accumulation. The availability, form, and concentration of these nutrients in the soil can significantly alter the metabolic levels of a plant, thereby impacting the quantity and quality of its secondary metabolites.

In this chapter, we explain the role of important macro- and micronutrients in regulating the biosynthesis of secondary metabolites in medicinal plants. We review the key findings from recent research that emphasize nutrient management for targeting the enhancement of various metabolites in different plants. By bridging insights from plant physiology, soil science, and metabolomics, this chapter aims to provide a comprehensive understanding of nutritional requirements to regulate the important metabolites that can be further exploited for pharmaceutical and agricultural innovation.

12.2 Effect of Macronutrients on the Biosynthesis of Secondary Metabolites

Macronutrients are essential for plant growth and development, serving as fundamental building blocks for structural components, energy transfer, and enzymatic functions. Beyond their roles in primary metabolism, these nutrients significantly influence the biosynthesis of secondary metabolites—bioactive compounds that contribute to plant defense, stress resilience, and medicinal value.

Deficiency of macronutrients such as nitrogen, phosphorus, potassium, sulfur, and magnesium can significantly impair the biosynthesis of secondary metabolites in medicinal plants. Nitrogen deficiency reduces the availability of amino acid precursors, leading to diminished production of alkaloids and phenolics. Phosphorus scarcity disrupts energy transfer and enzyme activation, limiting the synthesis of flavonoids and terpenoids. Potassium deficiency affects carbohydrate metabolism and osmotic balance, reducing the accumulation of sugars and organic acids essential for antioxidant production. Sulfur deficiency impairs the formation of sulfur-containing compounds like glucosinolates and thiols, weakening plant defense mechanisms. Magnesium deficiency compromises chlorophyll synthesis and photosynthetic efficiency, indirectly reducing the metabolic flux toward secondary metabolite pathways.

Overall, nutrient deficiencies not only hinder plant growth but also diminish the concentration and diversity of bioactive compounds, thereby affecting the therapeutic potential of medicinal plants.

12.2.1 Nitrogen (N)

Nitrogen is a major macronutrient essential to plants for growth and development, serving as a key component for the synthesis of chlorophyll, amino acids, nucleic acids, proteins, hormones, vitamins, and enzymes (Robertson and Vitousek, 2009; Li *et al.*, 2017). Plants mainly use nitrogen in the form of nitrate (NO_3^-) or ammonium (NH_4^+). Its deficiency can result in a series of physiological and metabolic changes that may lead to a shift of N-based metabolites to C-based metabolites such as flavonoids, phenolics, and anthocyanins that are involved in regulating various phytohormones, accumulation of stress-related metabolites, and root architecture (Wang *et al.*, 2000; Wang, 2003; Lv *et al.*, 2020). Extensive studies have been performed on *Oryza sativa*, *Triticum aestivum*, *Zea mays*, *Hordeum vulgare*, *Solanum lycopersicum*, *Glycine max*, and *Brassica napus* showing the limitation of total N content regulating differential accumulation of various classes of their metabolites (Schlüter *et al.*, 2012; Sung *et al.*, 2015; Tawaraya *et al.*, 2018; Xin *et al.*, 2019; Liu *et al.*, 2020; Wang *et al.*, 2021; Shen *et al.*, 2022). Table 12.1 summarizes metabolomic studies under N stress.

Nitrogen deficiency leads to a significant reduction in photosynthetic rate accompanied by downregulation of photosynthesis-related genes and proteins, which correlates with the accumulation of carbohydrate under N-limiting conditions (Xin *et al.*, 2019; Sun *et al.*, 2021). Scheible *et al.* (2004) studied metabolic regulation as well as the transcripts involved in *Arabidopsis* seedlings under N deprivation and N supply. They identified a cascade of genes involved in the regulation of metabolites under N deficiency and reported a decrease in the level of amino acids, mainly glutamine, and chlorophyll content and increase in specific flavonoids, starch, and phenylpropanoids. Another study by Zhao *et al.* (2021) showed that limited N markedly increased the levels of sugars such as glucose, maltose, sucrose, and galactose in barley leaves. Similar results have also been found in apple that showed that N deficiency has a strong correlation with carbon metabolism (Sun *et al.*, 2021). The lower carbohydrate is found to be correlated with a reduced photosynthetic rate during N limitation. The tricarboxylic acid (TCA) cycle also showed reduced synthesis of many intermediate compounds such as malate, fumarate, isocitrate, succinate, and 2-oxaloglutarate in tomato leaves under N deficiency (Urbanczyk-Wochniak and Fernie, 2005).

Nitrogen deficiency disrupts nitrogen metabolism, leading to reduced amino acid concentrations across multiple crops. In rice, free amino acids declined by 12.5% under low-N treatment (Xin *et al.*, 2019), while maize and barley showed decreases in glutamate, asparagine, alanine, serine, glycine, and other amino acids (Schlüter *et al.*, 2012; Zhao *et al.*, 2021). This reduction is linked to the downregulation of key enzymes such as glutamine synthetase (GS), glutamine aminotransferase (GOGAT), and NADH-dependent glutamate synthase (OsGLT1) (Xin *et al.*, 2019; Sun *et al.*, 2021). Similar declines in amino acid levels have been observed in soybean and tomato leaves under N deficiency (Liu *et al.*, 2020), reflecting an energy-conserving strategy during low-N stress.

Nitrogen deficiency promotes the generation of reactive oxygen species (ROS), leading to lipid peroxidation and oxidative stress. For instance, H_2O_2 levels rise significantly in apple leaves under low-N treatment (Sun *et al.*, 2021). Plants counteract this stress by activating antioxidant enzymes and accumulating protective metabolites such as galactinol, raffinose, sugar alcohols, ascorbic acid, and polyamines (Shi and Chan, 2014). In barley, ascorbic acid, putrescine, and 5-hydroxytryptamine accumulate under N deficiency, enhancing stress tolerance (Zhao *et al.*, 2021). Secondary metabolites, including flavonoids like cinnamic acid, dihydroquercetin, pelargonidin-3-O-glucoside, and cyanidin-3-O-glucoside, also increase in apple leaves (Sun *et al.*, 2021), providing additional ROS protection. Moreover, osmoprotectants such as β-alanine, proline, and γ-aminobutyric acid (GABA) rise under N limitation (Ashraf and Foolad, 2007), contributing to stress tolerance, though their precise roles vary among crops.

12.2.2 Phosphorus (P)

Phosphorus is an essential structural component of protein, nucleic acids, and membrane lipids which are required for various physiological and metabolic

Table 12.1. Summary of metabolic responses studied under macronutrient (N, P, K) stress 2015–2023.

Plant species	Metabolite studied	Metabolic response under stress	References
K stress			
Solanum lycopersicum, *Hordeum vulgare*	Glucose, sucrose, fructose	Sugar increase under K deficiency	(Sung *et al*., 2015, Zeng *et al*., 2018)
Hordeum vulgare, *Arachis hypogaea*	Proline	Tolerant cultivars showed enhanced proline	(Zeng *et al*., 2018, Patel *et al*., 2022)
Hordeum vulgare	L-phenylalanine Ascorbic acid	Elevated in tolerant barley Increased in roots and leaves	(Zeng *et al*., 2018)
Helianthus annuus	Glycerol-3-phosphate, fructose-6-phosphate Citrate, putrescine, malate, fumarate, and aconitate	Decrease under K stress Increased	(Cui *et al*., 2019)
Triticum aestivum	Asparagine, alanine, ornithine, histidine Citric acid, GABA, glutamic acid Glutathione	Increased under K deficiency Decreased accumulation in roots Elevated in tolerant wheat	(Zhao *et al*. 2020)
Solanum lycopersicum	Citric acid, α-ketoglutaric acid Succinic, cis-aconitic acids, malic Citramalic acid, dehydroascorbic acid	Increased under high K supply Increased under high K supply Decreased under high K supply	(Weinert *et al*. 2021)
Brassica napus	Citric acid, asparagine, and arginine	Increase in leaves under K deficiency	(Hu *et al*., 2021)
Arachis hypogaea	Abscisic acid (ABA) Jasmonic acid (JA) Salicylic acid (SA	Increased in leaves and roots Increased in leaves Increased in leaves, decreased in roots	(Patel *et al*., 2022)
P stress			
Solanum lycopersicum	Soluble sugar, organic acids, and amino acids	Decreased in roots and leaves	(Sung *et al*., 2015)
Lupinus albus	Fructose, sucrose, glucose, and glycerol-3-P 1-Kestose, raffinose Aspartate, glutamine, and alanine	Decreasing in roots Increased in roots Increased in leaves	(Müller *et al*., 2015)
Oryza sativa	L-phenylalanine, L-aspartate, GABA, adenine, guanosine, and cytidine T-zeatin, citrate, and D-glucosamine	Decreased in shoots Decreased in roots	(Tawaraya *et al*., 2018)

Avena sativa	Malic and citric acid and phosphorylated metabolites	Decreased in roots under P deprivation	(Wang *et al.*, 2018)
Triticum aestivum	Glycerol-3-P, glutamine, aspartate, and alanine	Decreased in leaves	(Nguyen *et al.*, 2019)
	Raffinose, 1-kestose	Increased in leaves	
Glycine max	Nucleic acids and phosphorylated lipids	Decreased in roots	(Mo *et al.*, 2022)
Chenopodium quinoa	Glucose-1-phosphate, uridine diphospho-D-glucose, 3-phospho-D-glyceric acid	Decreased in shoots	(Xiong *et al.*, 2022)
Artemisia argyi	Phenolic and flavonoid	Increased in leaves	(Wang *et al.*, 2023)
N stress			
Solanum lycopersicum	Amino acids and organic acids	Decrease in leaves and roots	(Sung *et al.*, 2015)
	Fructose-6-phosphate, glucose-6-phosphate, and sedoheptulose-7-P	Increase in leaves and roots	
Oryza sativa	L-phenylalanine, GABA, L-aspartate, guanosine, adenine, and cytidine	Decrease in shoots	(Xin *et al.*, 2019)
	Nicotinamide glycerol-3-P, sorbitol-6-P, L-phenylalanine, GABA, acetylserine, citrulline, histidinol	Decrease in roots	
	Ketoglutarate	Increase in leaves	
Glycine max	Sugars and organic acids	Increased in roots	(Li *et al.*, 2018)
Hordeum vulgare	Amino acids (lysine, threonine, tyrosine, glutamine, and ornithine)	Decrease in roots	(Zhao *et al.*, 2021)
Triticum aestivum	Sugars, ornithine, cysteine, tyrosine, aspartate	Increased in grains	(Wang *et al.*, 2021)
Malus pumila	Ornithine, asparagine, and arginine	Increased in leaves	(Sun *et al.*, 2021)
	Cinnamic acid, pelargonidin-3-O-glucoside and cyanidin-3-O-glucoside	Decreased in roots	
Brassica napus	Aspartic acid	Decrease in leaves and increase in roots	(Shen *et al.*, 2022)
Cannabis sativa	Cannabinoids and terpenoids, sugar	Increase under N deficiency	(Song *et al.*, 2023)

GABA, g-aminobutyric acid.

processes such as cell division, phospholipid biosynthesis, respiration, and DNA synthesis (Wang *et al.*, 2023). In plants, phosphorus is primarily available as phosphate or its ester form. P deficiency results in reduced crop production by affecting fruit and root development and various traits during vegetative growth (Lopez *et al.*, 2023).

In recent years, extensive research has been done to investigate the various physiological and biochemical mechanisms adopted by different plants to manage phosphorus deficiency. In one study performed by Hermans *et al.* (2006), P deficiency lead to an increase in phenolics and organic acids which delayed plant development. Another study performed in rice shoots and roots revealed the differential expression of two micro RNAs (miR399 and miR169) depicting their potential role in plant nutrient homeostasis under P deprivation (Cai *et al.*, 2011). *Lolium perenne* manages P limitation through bypassing via the glycolytic pathway and exchange of phospholipids with sulfolipids (Byrne *et al.*, 2011). P deficiency also affects the photosynthetic rate in *Ginko biloba* (Chen *et al.*, 2021). Metabolomics studies performed in *Phaseolus vulgaris*, *Avena sativa*, *Chenopodium quinoa*, and *Solanum lycopersicum* identified many metabolites involved in coping with P deficiency (Hernández *et al.*, 2009; Sung *et al.*, 2015; Wang *et al.*, 2018; Mo *et al.*, 2022). A summary of metabolomics studies under P stress is presented in Table 12.1.

Plants have adopted various adaptive mechanisms to enhance phosphate (Pi) under low P conditions. Under Pi deficiency, plants enhance Pi acquisition efficiency by reallocating carbon to roots, increasing the root-to-shoot ratio, and modifying root architecture to explore more soil (Mo *et al.*, 2022). Transcriptomic and proteomic studies reveal activation of phosphate starvation-induced (PSI) genes, while metabolomic analyses show sugar accumulation (maltose, sucrose, raffinose, kestose) that both fuels root growth and signals PSI gene expression (Huang *et al.*, 2008; Sung *et al.*, 2015; Wang *et al.*, 2018; Mo *et al.*, 2022). Nitrogen metabolism is also reprogrammed, with amino acids such as asparagine, lysine, histidine, ornithine, isoleucine, leucine, and arginine accumulating due to increased protein degradation and their reduced synthesis (Wu *et al.*, 2003; Misson *et al.*, 2005; Mo *et al.*, 2022). In parallel, organic acids are secreted to mobilize external Pi, and phospholipid remodeling conserves internal P by replacing phospholipids with non-P lipids. Together, these coordinated changes in root morphology, carbon and nitrogen metabolism, and metabolite regulation optimize Pi uptake and internal reutilization under low-P stress.

In addition to regulating root growth, (Dong *et al.*, 2004) reported the exudation of organic acid under Pi deficiency in two contrasting soybean genotypes. The role of organic acids under Pi limitation has also been recognized in different plants. Metabolomic studies show that rice roots release organic acids such as 3-dehydroshikimate, 2,6-diaminopimelate, hypoxanthine, fumarate, and D-galacturonate under P deficiency, aiding mobilization of insoluble soil P (Tawaraya *et al.*, 2018). Similarly, the P-efficient wheat genotype RAC875 exhibits increased exudation of organic acids (malic, succinic, and oxalic acids) (Nguyen *et al.*, 2019). Conversely, internal organic acids in roots often decline under low-P stress, as seen in common bean (tartaric acid, 2,4-dihydroxybutanoic acid) (Hernández *et al.*, 2009) and barley (α-ketoglutarate, succinate, fumarate, malate) (Huang *et al.*, 2008). This reduction may reflect their secretion into the rhizosphere, highlighting organic acid exudation as a key physiological mechanism for improving Pi utilization.

Phosphorus deprivation strongly influences various metabolic pathways that are linked to the synthesis of flavonoids, phenolic acids, and other phenylalanine-derived compounds. In *Salvia miltiorrhiza*, P deficiency upregulates the phosphate starvation response gene (SPX gene family: SYG1, Pho81, and XPR1), enhancing phenolic acid production (Liu *et al.*, 2016). Similarly, P deficiency increases flavonoid accumulation in *Cyclocarya paliurus* (Zhang *et al.*, 2021) and alters tissue-specific flavonoid distribution in white lupin roots, fine-tuning cluster root development (Xiong *et al.*, 2022). In lavender, volatile oil content rises with higher P application rates (Peçanha *et al.*, 2021).

12.2.3 Potassium (K)

Potassium (K) acts as a major inorganic cation or cofactor for various important enzymes involved in physiological and biochemical processes in plants (White and Karley, 2010). K is involved in the activation of enzymes that regulate protein synthesis, metabolism of sugar, nitrogen and carbon, osmoregulation, and photosynthesis (Hafsi *et al.*, 2014; Oosterhuis *et al.*, 2014). Deficiency of K in soil is one of the major limiting factors for sustainable crop production and metabolite accumulation. Metabolomic studies under K deficiency are presented in Table 12.1.

Armengaud *et al.* (2009) studied the molecular regulation of metabolites in *A. thaliana* under K deficiency. They performed tissue-specific regulation of metabolites under K stress. In roots, limited K resulted in a spike in the concentration of sugars along with specific N-rich amino acids while the concentration of pyruvate, nitrate, and other organic acid decreased. In shoots, amino acids and sugars changed dramatically in the same manner as the roots. However, organic acids did not show any variation. Both tissues showed an increase in total protein and amino acid.

In recent years extensive research has lead to the identification of various metabolites correlated with K deficiency in many important plants, including *Solanum lycopersicum*, *Helianthus annuus*, *Hordeum vulgare*, *Brassica napus*, and *Arachis hypogaea* (Sung *et al.*, 2015; Cui *et al.*, 2019; Hu *et al.*, 2021; Zhao *et al.*, 2021; Patel *et al.*, 2022).

Carbohydrate metabolism provides energy and supports protein and lipid metabolism in plants (Rolland *et al.*, 2006). Under K deficiency, sugars such as sucrose, glucose, and fructose accumulate in leaves and roots of barley and tomato, where sucrose acts as both an energy source and a signaling molecule regulating root growth (Hernández *et al.*, 2009; Rosa *et al.*, 2009; Schlüter *et al.*, 2012; Sung *et al.*, 2015; Zeng *et al.*, 2018; Zhao *et al.*, 2021). The low-K-tolerant genotype of barley showed higher accumulation of sugar than the sensitive genotype, highlighting its role in K stress adaptation (Zeng *et al.*, 2018). K is required for the transport of sucrose into the phloem, and its deficiency disrupts transport from leaves to roots, making it a key indicator of K tolerance (Armengaud *et al.*, 2009; White and Karley, 2010). Metabolomic analysis showed K limitation also alters nitrogen metabolism, while amino acids like tryptophan, alanine, asparagine, histidine, and ornithine increased in K-deficient wheat root, and citric acid, glutamic acid, and GABA also increased (Zeng *et al.*, 2018; Zhao *et al.*, 2021). In barley leaves and roots, most of the amino acids which are positively charged tend to rise, whereas negatively charged ones decline, reflecting the need to maintain cellular charge balance which is essential to overcoming low K stress (Zeng *et al.*, 2018). Additionally, regulation of the phenylpropanoid pathway, with elevated L-phenylalanine, was observed in low-K-tolerant barley genotypes, which supports synthesis of flavonoids, lignin, and alkaloids, contributing to low-K stress resilience in barley (Dixon *et al.*, 2002; Zhang and Liu, 2015; Zeng *et al.*, 2018).

Potassium deficiency also leads to the production of ROS and oxidative stress in plants (Hernandez *et al.*, 2012). Compatible solutes such as proline, amino acids, sugars, and polyols play important roles in osmotic adjustment (Khan *et al.*, 2020). Proline serve as a major antioxidant; its concentration rises in barley and peanut under K deprivation, especially in tolerant cultivars like XZ153 (Patel *et al.*, 2022). Ascorbic acid also increases in roots and leaves of tolerant barley, protecting membrane integrity, but declines in sensitive genotypes (Shalata and Neumann, 2001). Similarly, glutathione, crucial in ROS scavenging via the ascorbic acid–glutathione (AsA–GSH) cycle, accumulates in roots of tolerant wheat (KN9204) but not in sensitive lines (Zhao *et al.*, 2021). Collectively, proline, ascorbic acid, and glutathione represent vital metabolites for crop adaptation to K deficiency.

Phytohormones are key signaling molecules regulating plant growth and stress responses, and their metabolites are strongly influenced by K deficiency. Abscisic acid (ABA) acts as a stress signal, maintaining water balance through stomatal regulation and metabolic control (Peuke *et al.*, 2002; Kim *et al.*, 2014). Jasmonic acid (JA) supports abiotic stress tolerance by activating antioxidant systems, enhancing amino acid and sugar synthesis, and modulating stomatal function (Wang *et al.*, 2020). Salicylic acid (SA) protects membrane integrity and influences proteins linked to secondary metabolism (Patel *et al.*, 2022). Under K deficiency, ABA and JA levels rise in peanut leaves (and ABA also in roots), while SA increases in leaves but decreases in roots (Patel *et al.*, 2022). Collectively, ABA, JA, SA, and related phytohormones play vital roles in plant adaptation to low-K stress.

12.3 Effect of Micronutrients on Secondary Metabolites

The enormous importance of micronutrients in plant sciences cannot be overstated because plants rely heavily on micronutrients, which have a profound influence on a wide range of plant functions. Although micronutrients are abundant in soil, plants often acquire them in trace amounts, thereby classifying them as trace elements. Boron, copper, iron, manganese, molybdenum, and zinc are micronutrients that plants require in minute amounts yet they play a significant role in plant growth and development. Micronutrients play crucial roles in plant metabolism, nutrient management, reproductive growth, chlorophyll synthesis, carbohydrate generation, fruit and seed development, and other functions (Tripathi *et al.*, 2015; Rattan *et al.*, 2022). However, deficiencies can cause a variety of ailments in plants, ultimately reducing the quality and amount of food. Thus, the role of micronutrients in plants has sparked intense interest and is a significant topic among researchers. It has been well shown that micronutrients perform a variety of roles, and their proper availability enhances plant growth and yield, consequently shielding plants from the harmful impacts of numerous biotic and abiotic challenges (Tripathi *et al.*, 2015). Below, we summarize the overall benefits and significance of certain micronutrients in plants.

12.3.1 Boron (B)

Boron shortage is frequently reported in citrus orchards and is a primary cause of production and quality loss. The structural and morphological reactions of roots to B deficiency have been documented in several plants. Wu *et al.* (2020) studied B-deficient-induced cellular damage and target secondary metabolites in the shikimate pathway. Fluorescent vital staining, paraffin slice, transmission electron microscopy (TEM), and target metabolomics were used to study cell viability and structure, as well as target aromatic metabolites in the shikimate pathway in B-deficient trifoliate orange roots. Boron deprivation-induced ROS build-up exacerbated membrane peroxidation, weakening cell viability and causing cell rupture in roots. Furthermore, B shortage raised phenylalanine and tyrosine levels in roots, facilitating the production of salicylic acid, caffeic acid, and ferulic acid. Physiological studies showed that roots exposed to extended B starvation may stop dividing their cells, which is most likely a side-effect of B insufficiency (Dell and Huang, 1997; Wu *et al.*, 2020). When B was administered exogenously at 2 kg ha^{-1}, the amount of secondary metabolites in soybean plants increased significantly. Under extreme salinity conditions, the rise was substantial. Although salt stress and 2 kg ha^{-1} and B therapy yielded the most significant results, high B treatment created further increases (Tanveer *et al.*, 2025).

Nutrient deficits typically increase the amount of phenols in plant tissue, either by concentrating the existing pool or by altering the metabolic pathways that result in de novo synthesis of the molecules. When boric acid is sufficiently accessible in the cytoplasm, it forms cis-dihydroxyl-borate complexes with phenolic acids that have a cis-dihydroxyl structure. These substances increase in free forms when there is a B deficit because cis-dihydroxyl borate complexes are less likely to form. Additionally, B deprivation has been demonstrated to increase phenylalanine ammonia-lyase (PAL) activity in plant cells, which causes phenols to accumulate (Kohli *et al.*, 2023). Higher amounts were found for a number of metabolites that were differentially expressed and implicated in the formation of flavonoids in mulberry leaves, including prunin, chlorogenic acid, 5-O-caffeoyl shikimic acid, quercetin, kaempferol, and epigallocatechin (Zhang *et al.*, 2023).

Boron is frequently found in plants alongside nitrogen, phosphorus, potassium, and calcium, making it necessary for excellent agricultural yields. Its participation in plant hormone metabolism is vital and plants require a constant supply (Singh and Dwivedi, 2019). Boron has a role in numerous activities, including sugar transport, cell wall formation and maintenance, membrane integrity and RNA, indole-3-acetic acid (IAA), and phenol metabolism (Dordas and Brown, 2001). In iris plants, higher levels of B (20 ppm) resulted in maximum carbohydrate, chlorophyll, and oil content. Boron's primary purpose is to produce cell walls and cell membranes, germinate pollen, divide cells, transport carbohydrates, and metabolize calcium, IAA, and RNA (Singh and Dwivedi, 2019). Foliar application of boron may be more effective than soil application, where these minerals are adsorbed to soil particles and less available to the rooting medium (Sarkar *et al.*, 2007; Wissuwa *et al.*, 2008).

12.3.2 Copper (Cu)

Copper is a vital element for plants, acting as a cofactor in various enzymes. However, it becomes poisonous when consumed in excess, necessitating the use of homeostatic mechanisms (Akhtar *et al.*, 2025). Plant secondary metabolites, though not essential for basic metabolic functions like growth or reproduction, play a crucial role in plant survival and ecological interactions. These compounds include terpenoids, phenolics, and nitrogen-containing metabolites. In forest ecosystems, they serve as chemical defenses against herbivores and pathogens, with examples like tannins in oaks and oleoresins in pines. They also help plants adapt to abiotic stressors; importantly, Cu availability significantly influences the biosynthesis of these compounds, as Cu acts as a cofactor for key enzymes like laccases and polyphenol oxidases. Thus, secondary metabolites are indispensable for plant resilience, integrating environmental cues and nutrient status to support defense, competition, and stress tolerance (Akhtar *et al.*, 2025).

Medicago sativa (alfalfa) is a highly valued herbaceous plant in agriculture. The chemical composition of the substrate influences mineral availability as well as the growth, development, and metabolism of plants (Veljković *et al.*, 2024). Experiments to assess the effect of copper complex (Cu Cl_2) on lucerne germination and growth, as well as the amount of phenolic chemicals, were undertaken on alfalfa. Alfalfa seeds were planted in three independent groups and treated with the aforementioned complexes; the fourth group served as a control. After days 10, 20, and 30, the vegetative portion's growth was noted. For every experimental group, the germination percentage was noted. Following treatment, each group's aerial parts were dried and made ready for methanol extraction. Liquid chromatography was then used to determine the quantitative composition of phenolic chemicals. The results show that Cu Cl_2 had the best effect on the germination percentage, and chromatographic analysis showed that Cu Cl_2 improved the synthesis of lucerne phenolic compounds, with a significant content of catechins in the treatment with the copper complex compared to the control group. Based on the collected results, it was concluded that the complex compounds have a substantial effect on the germination, growth, and phenolic component content of lucerne, with varied effects on each of the analyzed processes (Veljković *et al.*, 2024).

In another study, Ibrahim *et al.* (2017) examined how copper and cadmium affected *Gyanura procumbens* growth, metabolite content, and the bioaccumulation of the two heavy metals. The combined Cd and Cu treatment was found to decrease the production of total phenolics, flavonoids, 1,1-diphenyl-2-picrylhydrazyl (DPPH), ferric reducing antioxidant potential (FRAP), and saponin. These effects were linked to a reduction in the activity of PAL, an enzyme crucial for secondary metabolite production. Overall, the study showed that contamination with copper and cadmium reduces the therapeutic qualities of *G. procumbens*. Additionally, it was shown that the accumulation of heavy metals was greater than the WHO recommended safety threshold for both individual and combined Cd and Cu treatments. These findings suggest that *G. procumbens* exposure to soil contaminated with Cd and Cu may be harmful to consumers because of metal bioaccumulation and decreased herbal product efficacy (Ibrahim *et al.*, 2017)

12.3.3 Iron (Fe)

Iron is essential for plant growth, but both deficiency and excess can cause stress. In cucumber plants, low and high Fe levels in hydroponics reduced growth, biomass, chlorophyll content, photosynthesis, and transpiration, with Fe deficiency having a more severe impact. Exogenous melatonin (N-acetyl-5-methoxytryptamine) alleviated these effects by boosting endogenous melatonin, reducing oxidative damage, and enhancing antioxidant enzyme activity, secondary metabolism, and levels of phenols and flavonoids. These findings underscore melatonin's role in mitigating Fe-induced stress and its potential in managing crops in Fe-affected soils (Ahammed *et al.*, 2020). Iron toxicity, caused by excessive Fe uptake, is a major issue in flooded soils of Asia and West Africa. Under such conditions, Fe^{2+} catalyzes the formation of ROS, notably hydroxyl radicals via the Fenton reaction, leading to oxidative stress (Wu *et al.*, 2017; Li *et al.*, 2019; Mahender *et al.*, 2019). This results in symptoms like leaf bronzing (especially in rice), chlorosis, growth inhibition, nutrient imbalances, and secondary water stress.

Plants counteract Fe toxicity through root-based mechanisms—such as forming iron plaques and compartmentalizing Fe—and shoot-based strategies that enhance antioxidant defenses, including glutathione, ascorbate, phenolics, and enzymes like superoxide dismutase and ascorbate peroxidase (Wu *et al.*, 2017; Mahender *et al.*, 2019). A study by Ahammed *et al.* (2020) showed that cucumber plants exposed to both low and high Fe circumstances displayed elevated levels of H_2O_2 and $O_2\bullet^-$, as well as increased electrolyte leakage and malondialdehyde (MDA), which are indicators of oxidative stress and membrane lipid peroxidation. Under Fe stress, antioxidant enzyme activity was reduced, indicating that ROS overload may affect their function or production. Similarly, the application of exogenous melatonin enhanced membrane stability, raised endogenous melatonin content, and dramatically decreased ROS and MDA levels. Under both Fe-deficient and Fe-excess conditions, it also improved the activity and gene expression of antioxidant enzymes like SOD, POD, and CAT.

Secondary metabolism enhances plant stress tolerance, with the phenylpropanoid pathway playing a key role via the PAL enzyme. Iron stress (both deficiency and excess) suppressed PAL activity and reduced phenol and flavonoid levels in cucumber leaves. Melatonin treatment reversed these effects, boosting PAL and polyphenol oxidase (PPO) activity and increasing phenolic and flavonoid accumulation. These compounds likely help scavenge ROS and mitigate oxidative stress. Melatonin-induced secondary metabolism may thus aid in cucumber adaptation to iron stress (Wu *et al.*, 2017; Ahammed *et al.*, 2020). Iron and other minerals may be more concrete when applied topically rather than when applied to soil, where they are adsorbed to soil particles and less accessible to the rooting medium (Wissuwa *et al.*, 2008).

12.3.4 Zinc (Zn)

Micronutrient deficiency or inappropriate presence or absence disrupts metabolic activities in plants at various stages of development, resulting in internal imbalance and symptoms such as chlorosis, necrosis, stunted growth, mottled leaves, and many other affected morphological parameters. Zinc is a crucial micronutrient for healthy plants and the accretion of metabolites. Zinc is an essential component of numerous enzymes, including alcoholic dehydrogenase, superoxide dismutase, and carbonic anhydrase. Furthermore, the synthesis of IAA is influenced by the shortage of Zn, which is claimed to be required for optimum tryptophan levels (Singh and Dwivedi, 2019). Zinc has a significant impact on protein synthesis and metabolism since it is an important component of both the ribosome and RNA polymerase (Singh and Dwivedi, 2019).

Foliar spraying with Zn as zinc sulfate considerably boosted growth, yield, and yield components, and enhanced the nutrient content of leaves and flowers and plant chemical constituents such as pigments, carbohydrates, and floral oil concentration in iris plants (Khalifa *et al.*, 2011). Also, oil percentage was dramatically raised by the application of ascorbic acid and zinc together, compared to control plants, in *Cupressus sempervirens* (Farahat *et al.*, 2007). Previous research on foliar application of Zn in *Allium cepa* demonstrated an increase in vital oil, growth, and yield of the plants (El-Tohamy *et al.*, 2009). Singh and Dwivedi (2019) found that foliar spraying of zinc sulfate promoted stevia growth by varying morphological characteristics such as size and color. It was observed that following treatment, the leaves appeared to have more area since they were bigger and greener. The leaves of plants serve as a source for light absorption and assimilation. Furthermore, when zinc sulfate was sprayed foliarly, the amount of stevioside was considerably higher than in the control.

The highest leaf number/plant values were obtained from iris plants sprayed with zinc at a higher rate (4.5 g/l) (Khalifa *et al.*, 2011). The maximum levels of carbohydrate, carotenoids, and floral oil content are obtained with a high concentration of zinc (4.5g/l). Using a three-factor, three-level orthogonal experiment, Zhou *et al.* (2025) examined the impact of zinc on the growth and therapeutic potential of *Stauntonia leucantha* leaves. They discovered that the content of leaf area (LA), leaf thickness (LT), chlorophyll, oleanolic acid (OA), rutin, and calceolarioside B was considerably raised by appropriate fertilizer concentrations of zinc. The study observed that the optimal Zn fertilizer content for the overall medicinal value of *Stauntonia leucantha* was zinc sulfate 6 g/plant. Furthermore, the study provides a reference for the artificial cultivation and fertilization of *Stauntonia leucantha*, which fills a research vacuum for this species.

Nekoukhou *et al.* (2022) investigated the effects of foliar application of zinc oxide nanoparticles (ZnONPs) and zinc suspension (ZnS) on dragonhead (*Dracocephalum moldavica*) under semiarid circumstances. ZnONPs considerably increased zinc bioaccumulation, chlorophyll content, biomass, and essential oil yield compared to ZnS and the control, with the best results obtained at 160 mg/l. Secondary metabolites such as neral, geraniol, and geranial have also increased. However, greater doses (400 mg/l) resulted in cellular stress. ZnONPs represent a promising, long-term method for increasing medicinal plant yield and economic rewards in arid settings.

12.3.5 Manganese (Mn)

Manganese may be more concrete when applied topically rather than when it is adsorbed to soil particles and less accessible to the rooting medium (Sarkar *et al.*, 2007; Wissuwa *et al.*, 2008). Mn is necessary for plant growth and is involved in several metabolic activities, including respiration, photosynthesis, secondary metabolism, and protein production (Jia *et al.*, 2020). Many enzymes, including those involved in the tricarboxylic acid cycle and SOD, require Mn as a cofactor. Additionally, Mn is involved in the formation of lignin and flavonoids (Lidon *et al.*, 2004; Jia *et al.*, 2020). Since Mn is a trace element, most plants only need 20–40 mg of it per kilogram of dry weight (Jia *et al.*, 2020). However, when Mn levels in plants approach 150 mg per kg dry weight, it is regarded as a heavy metal that might result in phytotoxicity (Millaleo *et al.*, 2010). Mn solubility is mostly influenced by pH levels and redox conditions, and accessible Mn levels in soils range from 450 to 4000 mg per kg (Sparrow and Uren, 2014).

In order to reduce Mn toxicity by lowering Mn availability, soil amelioration techniques such as lime treatment are commonly used; however, these methods are expensive from an economic and environmental standpoint (Jia *et al.*, 2020). These factors make breeding crop varieties with superior Mn tolerance a viable alternative agronomic approach but this necessitates a deeper comprehension of the responses of plants to Mn toxicity. The appearance of

Mn toxicity in most plants is typically characterized by leaf chlorosis, brown spots, crinkled leaves and brown roots, and eventually plant growth inhibition, although morphological changes in plants grown under Mn toxicity vary among plant species (Li *et al.*, 2019; Jia *et al.*, 2020). Mn toxicity has also been shown to have negative physiological effects on plant cells, including oxidative stress, lipid peroxidation, enzyme inhibition, impaired photosynthesis and chlorophyll biosynthesis, and disruption of the uptake and translocation of other mineral elements (Millaleo *et al.*, 2010; Jia *et al.*, 2020).

An experiment was undertaken to examine the effects of various Mn concentrations on the levels and correlations of many primary and secondary metabolites in *Mentha aquatica* (Jia *et al.*, 2020). Four different Mn concentrations were used: 40, 80, and 160 μM of Mn given as $MnSO_4.H_2O$, and basic Hoagland's solution (control). The findings showed that the plants treated with moderate amounts of Mn (40 and 80 μM) had higher biomass, photosynthetic pigment levels, and soluble carbohydrate contents than the control and 160 μM-treated plants. However, as the concentration of Mn in the nutrient solution increased, so did the levels of flavonoids, anthocyanins, hydrogen peroxide (H_2O_2), MDA, and antioxidant enzymes such as SOD, CAT, and POD. Furthermore, the chemical composition and amount of essential oils varied significantly across the treatment and control groups. The most prevalent components of essential oils in the control and Mn-treated plants in this investigation were 1,8-cineole, menthofuran, and β-caryophyllene. When the Mn concentration in the nutritional solution was higher, the correlation analysis between pairs of primary and secondary metabolites revealed both positive and negative correlations between the variables. These results unequivocally show that Mn up to 80 μM in the nutrient solution has a beneficial effect on *M. aquatica* growth.

Furthermore, Jia *et al.* (2020) found that high Mn levels reduced *Stylosanthes* growth due to oxidative stress. Essentially, transcriptome analysis revealed that *Stylosanthes* reaction to Mn toxicity may be influenced by a group of differently expressed genes (DEGs) that control ROS scavenging and defense responses. Additionally, the discovery of DEGs linked to secondary metabolic pathways indicated that *Stylosanthes* may be adapting to Mn toxicity by controlling secondary metabolism. Additionally, through transcriptional regulation, the activation of different transcription factors may aid *Stylosanthes* in tolerating Mn toxicity. When combined, these findings not only offer fresh perspectives on the molecular processes that underlie *Stylosanthes*' reaction to Mn toxicity, but also suggest candidate genes that could be helpful in the genetic engineering of crop lines with Mn tolerance.

12.3.6 Molybdenum (Mo)

In higher plants, Mo is a necessary trace element that is involved in numerous physiological and biochemical processes (Sun *et al.*, 2009; Wang *et al.*, 2013). The primary physiological role of Mo in higher plants is as an important component of mononuclear Mo enzymes which are essential for several metabolic activities in plants, including the C, N, and S cycles (Wang *et al.*, 2013). Mo treatment boosts crop and medicinal plant development, yields, and quality, as well as improving characteristics such as ascorbic acid and soluble sugar, and reducing nitrate concentrations (Hristozkova *et al.*, 2007). Furthermore, Mo may boost levels of active components in medicinal plants and promote the formation of secondary metabolites (He *et al.*, 2011; Yu *et al.*, 2012). According to Wang (2003), applying Mo to *Glycyrrhiza uralensis* could enhance root growth, increase root diameter and the number of lateral roots, and, most importantly, increase the amount of glycyrrhizic acid (GA). Liu (2005) conducted an additional experiment to confirm whether applying Mo might benefit *G. uralensis* growth, boost root biomass, and greatly improve GA accumulation. The impact of Mo on *Stauntonia leucantha* leaf growth and medicinal potential was investigated. The results showed that Mo considerably raised the levels of calceolarioside B and LA as well as rutin, OA, and chlorophyll a (Zhou *et al.*, 2025).

The effects of Mo on the secondary metabolic process of GA were examined using a pot experiment in *G. uralensis*. Sand and cleaned vermiculite were used to cultivate 1-year-old seedlings in pots. Four concentrations of Hoagland feeding solution—0, 0.52, 5.2, and 10.4 mg/l—were applied (Wang *et al.*, 2013). On days 35, 70, and 105, the accumulations of GA and its biosynthesis precursors (squalene and b-amyrin) were examined, followed by the expression of the main synthase (b-amyrin synthase, or b-AS). The b-AS gene expression and squalene and GA concentrations under 0.52 and 5.2 mg/l Mo were considerably higher than those under 0 and 10.4 mg/l Mo in the early stage, on days 35 and 70. The effect of b-amyrin was the opposite. On the other hand, 105 days had the highest squalene content under 0 mg/l Mo. Accordingly, it was proposed that by altering the GA biosynthesis process at a specific moment, a suitable concentration of Mo would encourage the accumulation of GA (Wang *et al.*, 2013).

Soybean seedlings are highly resistant to excess Mo, although the underlying metabolic pathways are not fully known. Researchers used ultra-performance liquid chromatography-tandem mass spectrometry (UPLC-MS/MS) to identify 42 changed metabolites in roots and 19 in leaves exposed to increased Mo. The amounts of lipids, hormones (salicylic and jasmonic acid), and organic acids (gluconic acid, D-glucarate, and citric acid) in the roots rose dramatically. While leaves mostly displayed alterations in lipid metabolism and salicylic acid, metabolites associated with ascorbate–glutathione metabolism and flavonoid/isoflavone production also accumulated in roots. Antioxidants and signaling molecules aid in the detoxification of ROS, whereas organic acids probably chelate Mo to lessen toxicity. Compared to leaves, roots showed higher metabolic alterations, suggesting a key role for roots in Mo tolerance (Xu *et al.*, 2018).

By upregulating important genes involved in Mo metabolism and transport, increasing enzyme activity, and increasing biomass accumulation, Mo promotes tobacco growth and cell wall biosynthesis. It changes metabolic pathways, including the creation of galactose and arginine/proline, which raises the amounts of stress-related chemicals, flavonoids, and carbohydrates.

Additionally, Mo enriches cellulose, hemicellulose, lignin, and pectin, which thicken the walls and increase their ability to withstand abiotic stress. These results demonstrate how Mo plays a crucial part in enhancing plant resilience and structure through metabolic and physiological regulation (Zhao *et al.*, 2021).

12.4 Conclusion and Future Prospects

Nutrient deficiency is always a limiting factor not only for plant growth and development but also for the production of commercially important metabolites. During recent years, various advances in the field of plant biotechnology have lead to improvements in detection and analytical techniques using metabolomics and bioinformatic tools, which have resulted in the identification of metabolite profiling associated with the nutrient deficiencies in various plants. The present review helps in compiling the systematic and advanced knowledge of metabolic response mechanisms adopted by different plants under the influence of macro and micronutrient stress conditions. The defense mechanism response against various nutritional deficiencies varies depending on the plant species. However, significant advances in detection platforms have helped researchers to unveil metabolite profiles under limiting nutritional scenarios. The continuous advancements in analytical techniques for metabolite profiling along with other tools such as microscopy, microarrays, and bioinformatics will greatly improve multilevel studies in many important agricultural crops. The need for sustainable production and high-quality crops encourages us to gain in-depth knowledge of the regulatory mechanisms adopted by different plants for metabolite synthesis under limiting nutrient conditions.

Acknowledgments

The authors are grateful to the Agricultural Research Organization (ARO) – Volcani Institute, Israel, and the University of Lucknow, India, for research facilities and funding.

References

Ahammed, G.J., Wu, M., Wang, Y., Yan, Y., Mao, Q. *et al.* (2020) Melatonin alleviates iron stress by improving iron homeostasis, antioxidant defense and secondary metabolism in cucumber. *Scientia Horticulturae* 265, 109205. DOI: 10.1016/j.scienta.2020.109205.

Akhtar, M.T., Lu, Z., Ren, S.X., Zou, H.L., Noor, I. *et al.* (2025) Copper homeostasis: Crosstalk with plant secondary metabolism and stress responses. *Plant Science* 362, 112795. DOI: 10.1016/j.plantsci.2025.112795.

Armengaud, P., Sulpice, R., Miller, A.J., Stitt, M., Amtmann, A. *et al.* (2009) Multilevel analysis of primary metabolism provides new insights into the role of potassium nutrition for glycolysis and nitrogen assimilation in Arabidopsis roots. *Plant Physiology* 150(2), 772–785. DOI: 10.1104/pp.108.133629.

Ashraf, M. and Foolad, M.R. (2007) Roles of glycine betaine and proline in improving plant abiotic stress resistance. *Environmental and Experimental Botany* 59(2), 206–216. DOI: 10.1016/j.envexpbot.2005.12.006.

Bhat, B.A., Islam, S.T., Ali, A., Sheikh, B.A., Tariq, L. *et al.* (2020) Role of micronutrients in secondary metabolism of plants. In: Aftab, T. and Hakeem, K. (eds) *Plant Micronutrients: Deficiency and Toxicity Management*. Springer International Publishing, Cham, Switzerland, pp. 311–329.

Byrne, S.L., Foito, A., Hedley, P.E., Morris, J.A., Stewart, D. *et al.* (2011) Early response mechanisms of perennial ryegrass (*Lolium perenne*) to phosphorus deficiency. *Annals of Botany* 107(2), 243–254. DOI: 10.1093/aob/mcq234.

Cai, H., Xie, W., Zhu, T. and Lian, X. (2011) Transcriptome response to phosphorus starvation in rice. *Acta Physiologiae Plantarum* 34(1), 327–341. DOI: 10.1007/s11738-011-0832-7.

Chen, C.J., Luo, D.D., Miao, Y.H., Kang, L.P., Guo, L.P. *et al.* (2021) Analysis and evaluation of volatile oil content in leaves of different *Artemisia argyi* germplasm resources. *Zhongguo Zhong Yao Za Zhi* 46(15), 3814–3823. DOI: 10.19540/j.cnki.cjcmm.20210523.102.

Cui, J., Abadie, C., Carroll, A., Lamade, E. and Tcherkez, G. (2019) Responses to K deficiency and waterlogging interact via respiratory and nitrogen metabolism. *Plant, Cell and Environment* 42(2), 647–658. DOI: 10.1111/pce.13450.

Dell, B. and Huang, L. (1997) Physiological response of plants to low boron. *Plant and Soil* 193(1–2), 103–120. DOI: 10.1023/A:1004264009230.

Dixon, R.A., Achnine, L., Kota, P., Liu, C.-J., Reddy, M.S.S. *et al.* (2002) The phenylpropanoid pathway and plant defence – a genomics perspective. *Molecular Plant Pathology* 3(5), 371–390. DOI: 10.1046/j.1364-3703.2002.00131.x.

Dong, D., Peng, X. and Yan, X. (2004) Organic acid exudation induced by phosphorus deficiency and/or aluminium toxicity in two contrasting soybean genotypes. *Physiologia Plantarum* 122(2), 190–199. DOI: 10.1111/j.1399-3054.2004.00373.x.

Dordas, C. and Brown, P.H. (2001) Permeability and the mechanism of transport of boric acid across the plasma membrane of *Xenopus laevis* oocytes. *Biological Trace Element Research* 81(2), 127–139. DOI: 10.1385/BTER:81:2:127.

El-Tohamy, W.A., Khalid, A.K., El-Abagy, H.M. and Abou-Hussein, S.D. (2009) Essential oil, growth and yield of onion (*Allium cepa* L.) in response to foliar application of some micronutrients. *Australian Journal of Basic and Applied Sciences* 3(1), 201–205.

Farahat, M.M., Ibrahim, M.S., Taha, L.S. and El-Quesni, E.F. (2007) Response of vegetative growth and some chemical constituents of *Cupressus sempervirens* L. to foliar application of ascorbic acid and zinc at Nubaria. *World Journal of Agricultural Sciences* 3(4), 496–502.

Hafsi, C., Debez, A. and Abdelly, C. (2014) Potassium deficiency in plants: Effects and signaling cascades. *Acta Physiologiae Plantarum* 36(5), 1055–1070. DOI: 10.1007/s11738-014-1491-2.

He, Z.J., Liang, S.W., Zeng, B., Guan, K.Y. and Xiong, J.F. (2011) Effects of Mo fertilization on growth, nutrient and total saponin contents of *Paris polyphylla* var. *yunnanensis*. *Journal of Plant Nutrition and Fertilizers* 17(6), 1481–1486. DOI: 10.11674/zwyf.2011.0499.

Hermans, C., Hammond, J.P., White, P.J. and Verbruggen, N. (2006) How do plants respond to nutrient shortage by biomass allocation? *Trends in Plant Science* 11(12), 610–617. DOI: 10.1016/j.tplants.2006.10.007.

Hernández, G., Valdés-López, O., Ramírez, M., Goffard, N., Weiller, G. *et al.* (2009) Global changes in the transcript and metabolic profiles during symbiotic nitrogen fixation in phosphorus-stressed common bean plants. *Plant Physiology* 151(3), 1221–1238. DOI: 10.1104/pp.109.143842.

Hernandez, M., Fernandez-Garcia, N., Garcia-Garma, J., Rubio-Asensio, J.S., Rubio, F. *et al.* (2012) Potassium starvation induces oxidative stress in *Solanum lycopersicum* L. roots. *Journal of Plant Physiology* 169(14), 1366–1374. DOI: 10.1016/j.jplph.2012.05.015.

Hristozkova, M., Geneva, M., Stancheva, I. and Georgiev, G. (2007) Nitrogen assimilatory enzymes and amino acid content in inoculated foliar fertilized pea plants grown at reduced molybdenum concentration. *Journal of Plant Nutrition* 30(9), 1409–1419. DOI: 10.1080/01904160701555838.

Hu, W., Lu, Z., Meng, F., Li, X., Cong, R. *et al.* (2021) Potassium modulates central carbon metabolism to participate in regulating CO_2 transport and assimilation in *Brassica napus* leaves. *Plant Science* 307, 110891. DOI: 10.1016/j.plantsci.2021.110891.

Huang, C.Y., Roessner, U., Eickmeier, I., Genc, Y., Callahan, D.L. *et al.* (2008) Metabolite profiling reveals distinct changes in carbon and nitrogen metabolism in phosphate-deficient barley plants (*Hordeum vulgare* L.). *Plant and Cell Physiology* 49(5), 691–703. DOI: 10.1093/pcp/pcn044.

Ibrahim, M.H., Chee Kong, Y. and Mohd Zain, N.A. (2017) Effect of cadmium and copper exposure on growth, secondary metabolites and antioxidant activity in the medicinal plant Sambung nyawa (*Gynura procumbens* (Lour.) Merr). *Molecules* 22(10), 1623. DOI: 10.3390/molecules22101623.

Jia, Y., Li, X., Liu, Q., Hu, X., Li, J. *et al.* (2020) Physiological and transcriptomic analyses reveal the roles of secondary metabolism in the adaptive responses of *Stylosanthes* to manganese toxicity. *BMC Genomics* 21(1), 861. DOI: 10.1186/s12864-020-07279-2.

Khalifa, R.K.H.M., Shaaban, S.H.A. and Rawia, A. (2011) Effect of foliar application of zinc sulfate and boric acid on growth, yield and chemical constituents of iris plants. *Ozean Journal of Applied Sciences* 4(2), 129–144.

Khan, M.N., AlSolami, M.A., Basahi, R.A., Siddiqui, M.H., Al-Huqail, A.A. *et al.* (2020) Nitric oxide is involved in nano-titanium dioxide-induced activation of antioxidant defense system and accumulation of osmolytes under water-deficit stress in *Vicia faba* L. *Ecotoxicology and Environmental Safety* 190, 110152. DOI: 10.1016/j.ecoenv.2019.110152.

Kim, Y.H., Khan, A.L., Waqas, M., Shim, J.K., Kim, D.H. *et al.* (2014) Silicon application to rice root zone influenced the phytohormonal and antioxidant responses under salinity stress. *Journal of Plant Growth Regulation* 33(2), 137–149. DOI: 10.1007/s00344-013-9356-2.

Kohli, S.K., Kaur, H., Khanna, K., Handa, N., Bhardwaj, R. *et al.* (2023) Boron in plants: Uptake, deficiency and biological potential. *Plant Growth Regulation* 100(2), 267–282. DOI: 10.1007/s10725-022-00844-7.

Li, H., Hu, B. and Chu, C. (2017) Nitrogen use efficiency in crops: Lessons from Arabidopsis and rice. *Journal of Experimental Botany* 68(10), 2477–2488. DOI: 10.1093/jxb/erx101.

Li, J., Jia, Y., Dong, R., Huang, R., Liu, P. *et al.* (2019) Advances in the mechanisms of plant tolerance to manganese toxicity. *International Journal of Molecular Sciences* 20(20), 5096. DOI: 10.3390/ijms20205096.

Lidon, F.C., Barreiro, M.G. and Ramalho, J.C. (2004) Manganese accumulation in rice: Implications for photosynthetic functioning. *Journal of Plant Physiology* 161(11), 1235–1244. DOI: 10.1016/j.jplph.2004.02.003.

Liu, C.L. (2005) *Regulatory Mechanism of Glycyrrhizic Acid Accumulation in Glycyrrhiza Uralensis*. Beijing University of Chinese Medicine, Beijing, China.

Liu, L., Yang, D., Liang, T., Zhang, H., He, Z. *et al.* (2016) Phosphate starvation promoted the accumulation of phenolic acids by inducing the key enzyme genes in *Salvia miltiorrhiza* hairy roots. *Plant Cell Reports* 35(9), 1933–1942. DOI: 10.1007/s00299-016-2007-x.

Li, M.X., Xu, J.S., Wang, X.X., Fu, H., Zhao, M.L. *et al.* (2018) Photosynthetic characteristics and metabolic analyses of two soybean genotypes revealed adaptive strategies to low-nitrogen stress. *Journal of Plant Physiology* 229, 132–141. DOI: 10.1016/j.jplph.2018.07.009.

Liu, Y., Xue, Y., Xie, B., Zhu, S., Lu, X. *et al.* (2020) Complex gene regulation between young and old soybean leaves in responses to manganese toxicity. *Plant Physiology and Biochemistry* 155, 231–242. DOI: 10.1016/j.plaphy.2020.07.002.

Lopez, G., Ahmadi, S.H., Amelung, W., Athmann, M., Ewert, F. *et al.* (2023) Nutrient deficiency effects on root architecture and root-to-shoot ratio in arable crops. *Frontiers in Plant Science* 13, 1067498. DOI: 10.3389/fpls.2022.1067498.

Lv, X., Zhang, Y., Hu, L., Zhang, Y., Zhang, B. *et al.* (2020) Low-nitrogen stress stimulates lateral root initiation and nitrogen assimilation in wheat: Roles of phytohormone signaling. *Journal of Plant Growth Regulation* 40(1), 436–450. DOI: 10.1007/s00344-020-10112-5.

Mahender, A., Swamy, B.P.M., Anandan, A. and Ali, J. (2019) Tolerance of iron-deficient and -toxic soil conditions in rice. *Plants* 8(2), 31. DOI: 10.3390/plants8020031.

Millaleo, R., Reyes- Diaz, M., Ivanov, A.G., Mora, M.L. and Alberdi, M. (2010) Manganese as essential and toxic element for plants: Transport, accumulation and resistance mechanisms. *Journal of Soil Science and Plant Nutrition* 10(4), 470–481. DOI: 10.4067/S0718-95162010000200008.

Misson, J., Raghothama, K.G., Jain, A., Jouhet, J., Block, M.A. *et al.* (2005) A genome-wide transcriptional analysis using *Arabidopsis thaliana* Affymetrix gene chips determined plant responses to phosphate deprivation. *Proceedings of the National Academy of Sciences USA* 102(33), 11934–11939. DOI: 10.1073/pnas.0505266102.

Mo, X., Liu, G., Zhang, Z., Lu, X., Liang, C. *et al.* (2022) Mechanisms underlying soybean response to phosphorus deficiency through integration of omics analysis. *International Journal of Molecular Sciences* 23(9), 4592. DOI: 10.3390/ijms23094592.

Müller, J., Gödde, V., Niehaus, K., Zörb, C. (2015) Metabolic adaptations of white Lupin roots and shoots under phosphorus deficiency. *Frontiers in Plant Science*. 6, 1014. DOI: 10.3389/fpls.2015.01014.

Nekoukhou, M., Fallah, S., Abbasi-Surki, A., Pokhrel, L.R. and Rostamnejadi, A. (2022) Improved efficacy of foliar application of zinc oxide nanoparticles on zinc biofortification, primary productivity and secondary metabolite production in dragonhead. *Journal of Cleaner Production* 379, 134803. DOI: 10.1016/j.jclepro.2022.134803.

Nguyen, V.L., Palmer, L., Roessner, U. and Stangoulis, J. (2019) Genotypic variation in the root and shoot metabolite profiles of wheat (*Triticum aestivum* L.) indicate sustained, preferential carbon allocation as a potential mechanism in phosphorus efficiency. *Frontiers in Plant Science* 10, 995. DOI: 10.3389/fpls.2019.00995.

Oosterhuis, D., Loka, D., Kawakami, E. and Pettigrew, W. (2014) The physiology of potassium in crop production. *Advances in Agronomy* 126, 203–234. DOI: 10.1016/B978-0-12-800132-5.00003-1.

Patel, M., Fatnani, D. and Parida, A.K. (2022) Potassium deficiency stress tolerance in peanut (*Arachis hypogaea*) through ion homeostasis, activation of antioxidant defense, and metabolic dynamics: Alleviatory role of silicon supplementation. *Plant Physiology and Biochemistry* 182, 55–75. DOI: 10.1016/j.plaphy.2022.04.013.

Peçanha, D.A., Freitas, M.S.M., Vieira, M.E., Cunha, J.M. and de Jesus, A.C. (2021) Phosphorus fertilization affects growth, essential oil yield and quality of true lavender in Brazil. *Industrial Crops and Products* 170, 113803. DOI: 10.1016/j.indcrop.2021.113803.

Peuke, A.D., Jeschke, W.D. and Hartung, W. (2002) Flows of elements, ions and abscisic acid in *Ricinus communis* and site of nitrate reduction under potassium limitation. *Journal of Experimental Botany* 53(367), 241–250. DOI: 10.1093/jexbot/53.367.241.

Rattan, S., Partap, M., Kanika Kumar, S. and Warghat, A.R. (2022) Nutrient feeding approach enhances the vegetative growth biomass, volatile oil composition, and myristicin content in hydroponically cultivated *Petroselinum crispum* (Mill.) Nyman. *Journal of Applied Research on Medicinal and Aromatic Plants* 26, 100359. DOI: 10.1016/j.jarmap.2021.100359.

Robertson, G.P. and Vitousek, P.M. (2009) Nitrogen in agriculture: Balancing the cost of an essential resource. *Annual Review of Environment and Resources* 34(1), 97–125. DOI: 10.1146/annurev.environ.032108.105046.

Rolland, F., Baena-Gonzalez, E. and Sheen, J. (2006) Sugar sensing and signaling in plants: Conserved and novel mechanisms. *Annual Review of Plant Biology* 57, 675–709. DOI: 10.1146/annurev.arplant.57.032905.105441.

Rosa, M., Prado, C., Podazza, G., Interdonato, R., González, J.A. *et al.* (2009) Soluble sugars—metabolism, sensing and abiotic stress: A complex network in the life of plants. *Plant Signaling and Behavior* 4, 388–393. DOI: 10.4161/psb.4.5.8294.

Sarkar, D., Mandal, B. and Kundu, M.C. (2007) Increasing use efficiency of boron fertilisers by rescheduling the time and methods of application for crops in India. *Plant and Soil* 301(1–2), 77–85. DOI: 10.1007/s11104-007-9423-1.

Scheible, W.R., Morcuende, R., Czechowski, T., Fritz, C., Osuna, D. *et al.* (2004) Genome-wide reprogramming of primary and secondary metabolism, protein synthesis, cellular growth processes, and the regulatory infrastructure of Arabidopsis in response to nitrogen. *Plant Physiology* 136(1), 2483–2499. DOI: 10.1104/pp.104.047019.

Schlüter, U., Mascher, M., Colmsee, C., Scholz, U., Bräutigam, A. *et al.* (2012) Maize source leaf adaptation to nitrogen deficiency affects not only nitrogen and carbon metabolism but also control of phosphate homeostasis. *Plant Physiology* 160, 1384–1406. DOI: 10.1104/pp.112.204420.

Shalata, A. and Neumann, P.M. (2001) Exogenous ascorbic acid (vitamin C) increases resistance to salt stress and reduces lipid peroxidation. *Journal of Experimental Botany* 52(364), 2207–2211. DOI: 10.1093/jexbot/52.364.2207.

Shen, X., Yang, L., Han, P., Gu, C., Li, Y. *et al.* (2022) Metabolic profiles reveal changes in the leaves and roots of rapeseed (*Brassica napus* L.) seedlings under nitrogen deficiency. *International Journal of Molecular Sciences* 23(10), 5784. DOI: 10.3390/ijms23105784.

Shi, H. and Chan, Z. (2014) Improvement of plant abiotic stress tolerance through modulation of the polyamine pathway. *Journal of Integrative Plant Biology* 56(2), 114–121. DOI: 10.1111/jipb.12128.

Singh, P. and Dwivedi, P. (2019) Micronutrients zinc and boron enhance stevioside content in *Stevia rebaudiana* plants while maintaining genetic fidelity. *Industrial Crops and Products* 140, 111646. DOI: 10.1016/j.indcrop.2019.111646.

Song, C., Saloner, A., Fait, A., Bernstein, N. (2023) Nitrogen deficiency stimulates cannabinoid biosynthesis in medical cannabis plants by inducing a metabolic shift towards production of low-N metabolites. *Industrial Crops and Products* 202, 116969. DOI: 10.1016/j.indcrop.2023.116969.

Sparrow, L.A. and Uren, N.C. (2014) Manganese oxidation and reduction in soils: Effects of temperature, water potential, pH and their interactions. *Soil Research* 52(5), 483–494. DOI: 10.1071/SR13159.

Sun, X., Hu, C., Tan, Q., Liu, J. and Liu, H. (2009) Effects of molybdenum on expression of cold-responsive genes in abscisic acid (ABA)-dependent and ABA-independent pathways in winter wheat under low-temperature stress. *Annals of Botany* 104(2), 345–356. DOI: 10.1093/aob/mcp133.

Sun, T., Zhang, J., Zhang, Q., Li, X., Li, M. *et al.* (2021) Integrative physiological, transcriptome, and metabolome analysis reveals the effects of nitrogen sufficiency and deficiency conditions in apple leaves and roots. *Environmental and Experimental Botany* 192, 104633. DOI: 10.1016/j.envexpbot.2021.104633.

Sung, J., Lee, S., Lee, Y., Ha, S., Song, B. *et al.* (2015) Metabolomic profiling from leaves and roots of tomato (*Solanum lycopersicum* L.) plants grown under nitrogen, phosphorus or potassium-deficient condition. *Plant Science* 241, 55–64. DOI: 10.1016/j.plantsci.2015.09.027.

Tanveer, M., Mahmood, A., Alawadi, H.F.N., Adiba, A., Javaid, M.M. *et al.* (2025) Impact of boron on *Glycine max* L. to mitigate salt stress by modulating the morpho-physiological and biochemical responses. *BMC Plant Biology* 25(1), 286. DOI: 10.1186/s12870-024-06037-w.

Tawaraya, K., Horie, R., Wagatsuma, T., Saito, K. and Oikawa, A. (2018) Metabolite profiling of shoot extract, root extract, and root exudate of rice under nitrogen and phosphorus deficiency. *Soil Science and Plant Nutrition* 64(3), 312–322. DOI: 10.1080/00380768.2018.1476828.

Tripathi, D.K., Singh, S., Singh, S., Mishra, S., Chauhan, D.K. *et al.* (2015) Micronutrients and their diverse role in agricultural crops: Advances and future prospective. *Acta Physiologiae Plantarum* 37(7), 139. DOI: 10.1007/s11738-015-1870-3.

Urbanczyk-Wochniak, E. and Fernie, A.R. (2005) Metabolic profiling reveals altered nitrogen nutrient regimes have diverse effects on the metabolism of hydroponically-grown tomato (*Solanum lycopersicum*) plants. *Journal of Experimental Botany* 56(410), 309–321. DOI: 10.1093/jxb/eri059.

Veljković, B., Ćirić, A. and Soldatović, T. (2024) The influence of complex compounds of zinc, copper and platinum on germination, growth and amount of secondary metabolites of alfalfa (*Medicago sativa* L.). *Phytochemistry Letters* 62, 68–72. DOI: 10.1016/j.phytol.2024.07.002.

Wang, J.Y. (2003) *Research on Cultivation Nutrition of Glycyrrhiza Uralensis*. Beijing Forestry University, Beijing, China.

Wang, R., Okamoto, M., Xing, X. and Crawford, N.M. (2000) Microarray analysis of the nitrate response in Arabidopsis roots and shoots: Expression patterns and novel metabolic and potential regulatory genes induced by nitrate. *Plant Cell* 12(8), 1491–1509. DOI: 10.1105/tpc.12.8.1491.

Wang, D., Pang, Y.-X., Wang, W.-Q., Wan, C.-Y., Hou, J.-L. *et al.* (2013) Effect of molybdenum on secondary metabolic process of glycyrrhizic acid in *Glycyrrhiza uralensis* Fisch. *Biochemical Systematics and Ecology* 50, 93–100. DOI: 10.1016/j.bse.2013.03.045.

Wang, Y., Lysøe, E., Armarego-Marriott, T., Erban, A., Paruch, L. *et al.* (2018) Transcriptome and metabolome analyses provide insights into root and root-released organic anion responses to phosphorus deficiency in oat. *Journal of Experimental Botany* 69(15), 3759–3771. DOI: 10.1093/jxb/ery176.

Wang, J., Song, L., Gong, X., Xu, J. and Li, M. (2020) Functions of jasmonic acid in plant regulation and response to abiotic stress. *International Journal of Molecular Sciences* 21(4), 1446. DOI: 10.3390/ijms21041446.

Wang, Y., Wang, D., Tao, Z., Yang, Y., Gao, Z. *et al.* (2021) Impacts of nitrogen deficiency on wheat (*Triticum aestivum* L.) grain during the medium filling stage: Transcriptomic and metabolomic comparisons. *Frontiers in Plant Science* 12, 674433. DOI: 10.3389/fpls.2021.674433.

Wang, Z., Ma, L., Chen, C., Guo, L., Guo, L. *et al.* (2023) Effects of phosphorus stress on the growth and secondary metabolism of *Artemisia argyi*. *Journal of Plant Research* 136(6), 879–889. DOI: 10.1007/s10265-023-01479-z.

Weinert, C.H., Sonntag, F., Egert, B., Pawelzik, E., Kulling, S.E. *et al.* (2021) The effect of potassium fertilization on the metabolite profile of tomato fruit (*Solanum lycopersicum* L.). *Plant Physiology and Biochemistry* 159, 89-99. DOI: 10.1016/j.plaphy.2020.12.010.

White, P.J. and Karley, A.J. (2010) Potassium. In: Hell, R. and Mendel, R. (eds) *Cell Biology of Metals and Nutrients*. Springer, Berlin, Germany, pp. 199–224.

Wissuwa, M., Ismail, A.M. and Graham, R.D. (2008) Rice grain zinc concentrations as affected by genotype, native soil-zinc availability, and zinc fertilization. *Plant and Soil* 306(1–2), 37–48. DOI: 10.1007/s11104-007-9368-4.

Wu, P., Ma, L., Hou, X., Wang, M., Wu, Y. *et al.* (2003) Phosphate starvation triggers distinct alterations of genome expression in Arabidopsis roots and leaves. *Plant Physiology* 132(3), 1260–1271. DOI: 10.1104/pp.103.021022.

Wu, L.B., Ueda, Y., Lai, S.K. and Frei, M. (2017) Shoot tolerance mechanisms to iron toxicity in rice (*Oryza sativa* L.). *Plant, Cell and Environment* 40(4), 570–584. DOI: 10.1111/pce.12733.

Wu, X., Riaz, M., Yan, L., Zhang, Z. and Jiang, C. (2020) How the cells were injured and the secondary metabolites in the shikimate pathway were changed by boron deficiency in trifoliate orange root. *Plant Physiology and Biochemistry* 151, 630–639. DOI: 10.1016/j.plaphy.2020.04.009.

Xin, W., Zhang, L., Zhang, W., Gao, J., Yi, J. *et al.* (2016) An integrated analysis of hydroponically grown rice reveals adaptive strategies in response to low nitrogen stress. *Journal of Plant Physiology* 206, 59–67.

Xin, W., Zhang, L., Zhang, W., Gao, J., Yi, J. *et al.* (2019) An integrated analysis of the rice transcriptome and metabolome reveals differential regulation of carbon and nitrogen metabolism in response to nitrogen availability. *International Journal of Molecular Sciences* 20(9), 2349. DOI: 10.3390/ijms20092349.

Xiong, C., Li, X., Wang, X., Wang, J., Lambers, H. *et al.* (2022) Flavonoids are involved in phosphorus-deficiency-induced cluster-root formation in white lupin. *Annals of Botany* 129(1), 101–112. DOI: 10.1093/aob/mcab131.

Xu, S., Hu, C., Hussain, S., Tan, Q., Wu, S. *et al.* (2018) Metabolomics analysis reveals potential mechanisms of tolerance to excess molybdenum in soybean seedlings. *Ecotoxicology and Environmental Safety* 164, 589–596. DOI: 10.1016/j.ecoenv.2018.08.062.

Xue, Y., Zhu, S., Schultze-Kraft, R., Liu, G. and Chen, Z. (2022) Dissection of crop metabolome responses to nitrogen, phosphorus, potassium, and other nutrient deficiencies. *International Journal of Molecular Sciences* 23(16), 9079. DOI: 10.3390/ijms23169079.

Yu, W., Cao, F. and Wu, G. (2012) Leaf growth and medicinal quality of ginkgo seedling with spraying Mg, Zn, Mo fertilizers. *Journal of Plant Nutrition and Fertilizers* 18, 981–999. DOI: 10.11674/zwyf.2012.11385.

Zeng, J., Quan, X., He, X., Cai, S., Ye, Z. *et al.* (2018) Root and leaf metabolite profiles analysis reveals the adaptive strategies to low potassium stress in barley. *BMC Plant Biology* 18(1), 187. DOI: 10.1186/s12870-018-1404-4.

Zhao, Y., Sun, R., Liu, H., Liu, X., Xu, K. *et al.* (2020) Multi-omics analyses reveal the molecular mechanisms underlying the adaptation of wheat (*Triticum aestivum* L.) to potassium deprivation. *Frontiers in Plant Science* 11, 588994. DOI: 10.3389/fpls.2020.588994.

Zhang, X. and Liu, C.J. (2015) Multifaceted regulations of gateway enzyme phenylalanine ammonia-lyase in the biosynthesis of phenylpropanoids. *Molecular Plant* 8(1), 17–27. DOI: 10.1016/j.molp.2014.11.001.

Zhang, X., Zhang, M., Zhao, T. and Deng, B. (2021) Phosphate availability regulates flavonoid accumulation associated with photosynthetic carbon partitioning in *Cyclocarya paliurus*. *Physiologia Plantarum* 173(4), 1956–1966. DOI: 10.1111/ppl.13539.

Zhang, Q., Ackah, M., Wang, M., Amoako, F.K., Shi, Y. *et al.* (2023) The impact of boron nutrient supply in mulberry (*Morus alba*) response to metabolomics, enzyme activities, and physiological parameters. *Plant Physiology and Biochemistry* 200, 107649. DOI: 10.1016/j.plaphy.2023.107649.

Zhao, H., Ni, S., Cai, S. and Zhang, G. (2021) Comprehensive dissection of primary metabolites in response to diverse abiotic stress in barley at seedling stage. *Plant Physiology and Biochemistry* 161, 54–64. DOI: 10.1016/j.plaphy.2021.01.048.

Zhou, Y., Zhou, J., Chen, J., Chang, Y., Lin, X. *et al.* (2025) Effects of zinc, manganese, and molybdenum fertilizers on growth and main medicinal metabolites of *Stauntonia leucantha* leaves. *Horticulturae* 11(2), 123. DOI: 10.3390/horticulturae11020123.

13 Role of Soil Microorganisms and Mycorrhizal Fungi in the Synthesis of Bioactive Compounds in Medicinal Plants

RAHUL KUMAR[1], SOURAV CHATTARAJI[2], PERIYASAMY PANNEERSELVAM[3], SERGIO DE LOS SANTOS VILLALOBOS[4] AND DEBASIS MITRA[1*]

[1]*Department of Microbiology, Graphic Era (Deemed to be University), Dehradun, India*
[2]*Centre for Industrial Biotechnology Research, School of Pharmaceutical Sciences, Siksha 'O' Anusandhan (Deemed to be University), Bhubaneswar, India*
[3]*Crop Production Division, ICAR Central Rice Research Institute, Cuttack, India*
[4]*Instituto Tecnológico de Sonora, Ciudad Obregón, Mexico*

13.1 Introduction

Medicinal plants have been a part of human health and traditional medicine systems for millennia (Davis and Choisy, 2024). They form the foundation of many traditional therapeutic practices, such as Ayurveda, Traditional Chinese Medicine, and indigenous herbal practices, and provide preventive and curative solutions for a broad array of diseases (Goyal and Chauhan, 2024). In addition to conventional applications, medicinal plants are a vital reservoir of contemporary drugs, and numerous modern medications either come directly from plant metabolites or are patterned after plant chemical structures. The global demand for herbal medicines and plant-based drugs has grown tremendously because of their perceived safety and curative efficacy, and the increasing incidence of chronic diseases (Khare and Khare,

*Corresponding author: debasismitra3@gmail.com

 Medicinal Plants and the Rhizosphere
(eds M.K. Mishra and S. Tiwari)
DOI: 10.1079/9781836990475.0013

2025; Singh *et al.*, 2026). This highlights the need to maximize the cultivation and biochemical potential of medicinal plants to achieve a consistent quality and strength of bioactive compounds.

Bioactive compounds are secondary metabolites which are synthesized by plants and have a biological impact on humans and other living organisms (Twaij and Hasan, 2022). Phytochemicals and essential oils are some examples of bioactive compounds (El-Saadony *et al.*, 2025). Alkaloid compounds such as vincristine and morphine have broad applications in anticancer and analgesic research. Phenolic and flavonoid compounds have antioxidant, antiinflammatory, and cardioprotective properties (Kumar *et al.*, 2025e).

Biosynthesis of these compounds is directly influenced by environmental conditions, genetic composition, and biotic interactions, particularly with soil microbes and mycorrhizal fungi (Ehlers *et al.*, 2020). These interactions can modulate plant metabolic pathways and enhance secondary metabolite production. Plant–microbe interactions are critical for plant growth, health, and secondary metabolite production (Korenblum *et al.*, 2022). Soil microbes such as bacteria, fungi, and actinomycetes interact with plant roots through symbiotic relationships. Plant growth-promoting rhizobacteria and other beneficial microbes enhance nutrient uptake, stimulate phytohormone production, and improve plant resistance to biotic and abiotic stressors (Hasan *et al.*, 2024). Symbiotic relationships between mycorrhizal fungi and plant roots improve stress tolerance, nutrient uptake, and secondary metabolism. Many microbes function as natural elicitors by modulating metabolic pathways and gene expression, thereby stimulating the synthesis of bioactive compounds. These enhanced interactions not only improve plant growth and resilience but also increase the therapeutic quality of medicinal plants (Wang *et al.*, 2022).

Understanding the mechanisms by which soil microbes and mycorrhizal fungi influence metabolite synthesis is essential for developing effective cultivation strategies aimed at enhancing metabolite yields and advancing biotechnological applications. The application of microbial inoculants in medicinal plant farming offers a sustainable approach to meet the global demand for high-quality herbal products, while promoting resilient agriculture (Díaz-Rodríguez *et al.*, 2025).

13.2 Metabolite Production and Plant Growth Soil Microorganisms

13.2.1 Diversity of soil microorganisms

Bacteria are an extensive and metabolic group involved in nutrient cycling and decomposition of organic matter, nitrogen fixation, and synthesis of phytohormones (Chauhan *et al.*, 2023). Actinomycete bacteria are filamentous fungi that decompose complicated organic compounds such as cellulose, chitin, and antibiotics production (Javed *et al.*, 2021). Some fungi are saprophytic and symbiotic and are responsible for nutrients, organic matter, mineralization, and stable decomposition in soil. Arbuscular mycorrhizal fungi (AMF) symbiotic relationships with higher terrestrial plants promote nutrient acquisition, specifically phosphorus, and biotic and abiotic stress tolerance in plants (Wahab *et al.*, 2023). Microbial communities, including nematode-borne microbes and protozoa, indirectly affect plant growth by controlling the growth of bacteria and fungi. The ecological functions of these microbes are diverse, including nutrient solubilization and inhibition of pathogen induction of plants as secondary metabolites, which determine the bioactive synthesis of molecules in medicinal plants (Koza *et al.*, 2022).

In this group, microorganisms are broadly classified as either beneficial or pathogenic. Beneficial microorganisms, such as plant growth-promoting rhizobacteria (PGPR), nitrogen-fixing bacteria, and mycorrhizal fungi, enhance development, stress tolerance, and metabolite synthesis in plants (Khan *et al.*, 2020; Kamboj *et al.*, 2025). Pathogenic microorganisms, such as *Fusarium*, *Phytophthora*, and some bacterial species, can induce diseases, inhibit growth, and adversely influence the quality and yield of bioactive metabolites (Heredia-Bátiz *et al.*, 2025). A delicate balance between beneficial and pathogenic microorganisms in the rhizosphere is essential for the best plant health and productivity, and knowledge of such diversity is fundamental for exploring the microbial potential for sustainable cultivation and value addition in medicinal plant metabolites (Chen *et al.*, 2024).

13.3 Mechanisms by Which Microorganisms Influence Bioactive Compound Production

13.3.1 Nutrient solubilization

Nutrient availability is a significant determinant of growth and secondary metabolite production in medicinal plants. Most major nutrients in soils, such as nitrogen (N), phosphorus (P), and potassium (K), and micronutrients, such as iron, zinc, and manganese, are in insoluble or unavailable biological forms. Soil microorganisms, especially PGPR, actinomycetes, and some fungi, increase nutrient availability via solubilization and mobilization mechanisms (Nagrale *et al.*, 2023).

Nitrogen-fixing bacteria, including *Azospirillum*, *Rhizobium*, and *Bacillus* species, reduce atmospheric nitrogen to ammonia or nitrate and give plants a form of nitrogen that can be used for the synthesis of amino acids, proteins, and secondary metabolites. Phosphate-solubilizing microorganisms produce organic acids that chelate fixed phosphates to make them available for plant uptake and enhance enzymatic activities of the shikimate and phenylpropanoid biosynthetic pathways. Potassium-solubilizing microorganisms and fungi mobilize potassium from insoluble mineral forms, maintaining enzymatic activity for metabolite biosynthesis of metabolites (Verma *et al.*, 2017). Likewise, siderophore-producing microorganisms sequester trace elements, such as iron, thus increasing their solubility and uptake. Enhanced availability of nutrients promotes not only vegetative growth but also the bioengineering of biosynthetic pathways to increase the accumulation of alkaloids, flavonoids, terpenoids, and phenolic compounds.

Through effective nutrient recycling, soil microbes serve as natural biofertilizers, providing supportive conditions for the optimal growth of medicinal crops and production of bioactive compounds (Vaghela and Gohel, 2023; Gupta *et al.*, 2025; Kumar *et al.*, 2025a).

13.3.2 Production of phytohormones

The soil microbiota strongly affects plant growth and the biosynthesis of secondary metabolites through the production of phytohormones. Endophytic bacteria, PGPR, and biocontrol fungi synthesize important phytohormones such as indole-3-acetic acid (IAA), gibberellins (GA), and cytokinins, which control several physiological and biochemical processes in medicinal plants (Kumar *et al.*, 2022). IAA promotes elongation of the roots and root branching, enhances the surface area for nutrient uptake, and increases the ability of the plant to capture elements required for the biosynthesis of secondary metabolites (Kumar *et al.*, 2025b; Samal *et al.*, 2025). Microbial production of gibberellins induces cell division and growth, promotes vegetative development, and indirectly promotes the flow of metabolism toward the biosynthesis of bioactive compounds. Cytokinins control cell differentiation, shoot development, and nutrient mobilization, and hence secondary metabolic processes.

Microbial phytohormones also regulate plant stress responses, which tend to induce elevated synthesis of defense-related secondary metabolites, such as alkaloids, flavonoids, and phenolics (Fig. 13.1). For example, inoculation with *Bacillus subtilis* and *Pseudomonas fluorescens* elevates IAA and cytokinin concentrations, leading to increased growth and greater accumulation of bioactive compounds in plants such as *Withania somnifera* and *Ocimum sanctum* (Rizvi *et al.*, 2022). Through the production of these phytohormones, soil microbes function as natural bioregulators to maximize plant physiology and provide ideal conditions for increased metabolite production and medicinal activity.

13.3.3 Induction of plant defense pathways

Microorganisms in the soil play a major part in inducing plant defense responses, which have a direct effect on the production of secondary metabolites in medicinal plants. Useful microbes such as PGPR and endophytic fungi release signaling molecules, volatile organic compounds, and microbe-associated molecular patterns (MAMPs) that are detected by plant receptors, initiating defense-related pathways (Jalmi and Sinha, 2022). This elicitation frequently simulates stressful conditions, resulting in increased

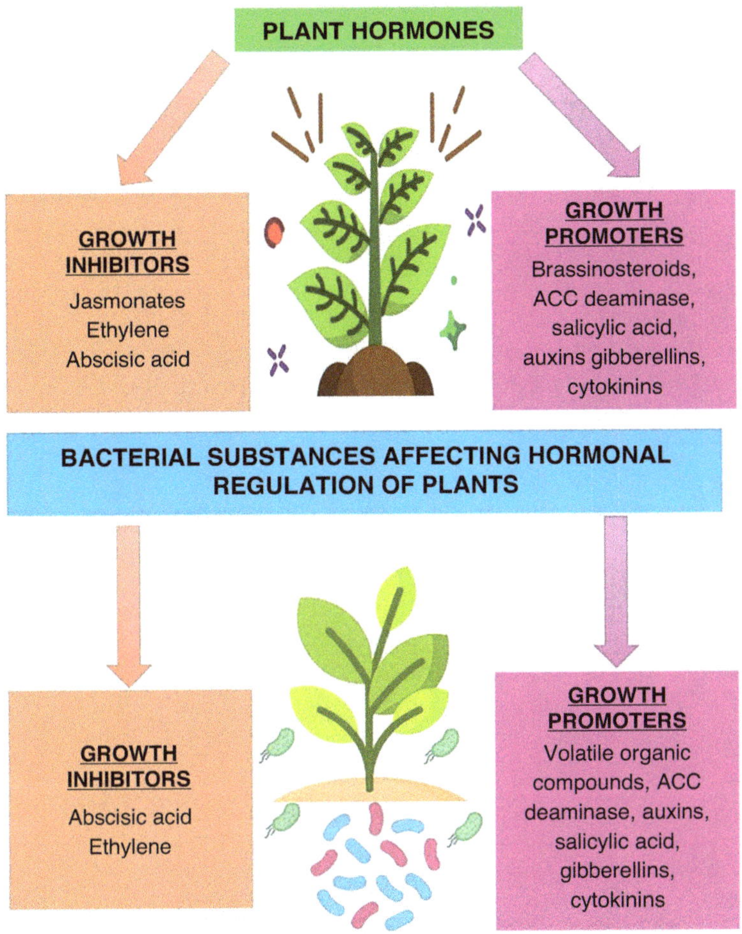

Fig. 13.1. Types and production of phytohormones.

synthesis of secondary metabolites, such as alkaloids, flavonoids, phenolics, and terpenoids, which act as protective agents against pathogens, herbivores, and environmental stresses. The process of elicitation includes the upregulation of primary biosynthetic enzymes, such as phenylalanine ammonia lyase (PAL), chalcone synthase, and strictosidine synthase, which are essential for the shikimate, phenylpropanoid, and terpenoid pathways (Cardillo *et al.*, 2021). For instance, inoculation of *Withania somnifera* with *Bacillus subtilis* enhanced the accumulation of withanolides, whereas endophytic fungi in *Hypericum perforatum* induced hypericin synthesis (Castronovo *et al.*, 2021).

Microbial elicitation improves not only plant defense but also the quality and therapeutics of medicinal plants. The regulation of plant metabolic networks via defense signaling and soil microbiota functions as a biological bio-elicitor, providing a greener method for enhancing the yield and pharmacological activity of bioactive substances (Bhaskar *et al.*, 2022).

13.3.4 Microbial synthesis of bioactive compounds

In addition to promoting plant secondary metabolism, some soil microbes can directly synthesize bioactive compounds, which may supplement the plant's native metabolite content. Endophytic fungi and bacteria are of special interest for the production of pharmacologically valuable molecules, such as alkaloids, flavonoids, terpenoids, and phenolics (Singh and Kumar, 2023). These microbial metabolites may be directly absorbed by the host plant or may act synergistically with plant biosynthetic routes, resulting in greater overall bioactive compound accumulation. Actinomycetes, particularly *Streptomyces* species, are known to produce antibiotics, phenolic compounds, and other bioactive compounds that may defend the plant against pathogens while concurrently enhancing its metabolite pool (Al-Quwaie, 2024). Concomitantly, endophytic fungi from medicinal plants such as *Taxus*, *Hypericum*, and *Withania* have been demonstrated to produce taxol, hypericin, and withanolides, respectively (Singh *et al.*, 2021).

Microbial production not only increases plant resistance but also provides a valuable avenue for biotechnological production of value-added compounds, irrespective of plant growth. By integrating plant–microbe associations with the inherent biosynthetic potential of microbes, scientists can utilize these microbes as natural biofactories. This strategy has great potential for sustainable scale and environmental production of medicines, and is important for secondary metabolites and improving the quality of drug activity of medicinal plants (Motolinía-Alcántara *et al.*, 2021).

13.3.5 Beneficial rhizobia in plant growth and secondary metabolite enhancement

Plant growth-promoting rhizobia microbes have been well studied for their potential to promote production of secondary metabolites in medicinal plants (Jabborova *et al.*, 2024). Some species, such as *Bacillus*, are famous for the production of phytohormone solubilization of nutrients and mechanisms of plant defense. *Catharanthus roseus* and *Bacillus* strains have been improved to produce alkaloids such as vinblastine and vincristine by activating important enzymes in biosynthetic pathways (Mistry *et al.*, 2022). *Pseudomonas fluorescens* improves the production of phenolic compounds and essential oils in *Ocimum sanctum* by promoting nutrients and inducing secondary metabolism, leading to enhanced antimicrobial activities and antioxidants. *Azospirillum* species are known to fix nitrogen to enhance the synthesis of metabolites in medicinal plants (Fukami *et al.*, 2018). Coinoculation of PGPR with other phytobeneficial microbes can produce a synergistic accumulation of flavonoids, alkaloids, and terpenoids (Mashabela *et al.*, 2022). Through the regulation of nutrient availability, defense mechanisms, and phytohormones, these microbes are used to enhance the yield and quality of bioactive metabolites in medicinal plants (Pandey *et al.*, 2023).

13.4 Symbiotic Role of Mycorrhizal Fungi

13.4.1 Mechanisms of interaction

Medicinal plants interact with mycorrhizal fungi, which together promote stress, growth, and secondary metabolite production. Enhanced nutrient achievement is one of the major roles of the mycorrhizal relationships. Hyphal fungi, which are extensive beyond the root zone, expand the root surface area for absorption of nutrients, especially nitrogen, phosphorus, and micronutrients (Bhantana *et al.*, 2021). This facilitates the critical enzymatic and metabolic pathways involved in the biosynthesis of bioactive molecules. Colonization of mycorrhizal plants affects water requirements and allows plants to tolerate drought, abiotic stress, and salinity, which tend to constrain growth and secondary metabolites (Diagne *et al.*, 2020). The nutritional assistance of mycorrhizal fungi plays an active role in the control of plant secondary metabolism via signaling interactions. It induces colonization and release of plant hormones and molecules that regulate biosynthetic pathways, leading to greater build-up of pharmacological compounds (Jamiołkowska, 2020). Inoculation of AMF in *Withania somnifera* increases with anolide, whereas in *Hypericum perforatum*, the level of hypericin and flavonoids increases with fungal induction of metabolic pathways (Johny *et al.*, 2021).

At the molecular level, mycorrhizal fungi regulate the expression of genes involved in secondary metabolite production. The important genes encoding enzymes such as phenylalanine ammonia lyase (PAL), terpene, and chalcone synthase are induced upon colonization, which enhances flux through the shikimate, terpenoid, and phenylpropanoid pathways (Tohge *et al.*, 2013). This regulation of genes, in addition to better nutrient and water status in medicinal plants, maximizes the production of bioactive compounds, even under optimal environmental conditions. These mechanisms demonstrate the complex functions of mycorrhizal fungi in maintaining the health of plants, their metabolic processes, and the production of therapeutic secondary metabolites (Table 13.1).

13.4.2 Influence on specific bioactive compounds

Mycorrhizal fungi significantly influence the production of major bioactive compounds such as terpenoids, alkaloids, phenolics, and flavonoids in medicinal plants. These secondary metabolites are essential for the medicinal value of plants and tend to be produced in response to nutrient levels, stress, and signal-mediated interactions through fungal colonization (Modolon *et al.*, 2020). Terpenoids have anticancer, antiinflammatory, and antimicrobial properties that are enhanced by mycorrhizal plants (Khoza *et al.*, 2025; Kumar *et al.*, 2025c). Examples include *Panax ginseng* infected by AMF, which produces elevated ginsenoside levels and a dominant terpenoid with adaptogenic and immunomodulatory properties. Alkaloids, consisting

Table 13.1. Types of mycorrhizal fungi.

Type of mycorrhizal fungi	Host plants	Colonization pattern	Functions
Arbuscular mycorrhizal fungi (AMF)	Most herbaceous plants, including many medicinal plants	Penetrates cortical cells, forming arbuscules and vesicles	Enhances nutrient (especially P) and water uptake, improves stress tolerance, stimulates secondary metabolite synthesis (Ran *et al.*, 2022)
Ectomycorrhizal fungi	Mostly woody plants, trees (e.g. Pinaceae, Fagaceae)	Forms a mantle around roots and Hartig net between cortical cells	Improves nutrient uptake (N, P), provides protection against pathogens, modulates secondary metabolism (Khullar and Reddy, 2019)
Ericoid mycorrhiza	Ericaceae family (heathers, blueberries)	Hyphal coils in epidermal cells of fine roots	Facilitates nutrient acquisition in acidic, nutrient-poor soils, enhances secondary metabolite production (Daghino *et al.*, 2025)
Orchid mycorrhiza	Orchid species	Penetrates root cortical cells, forming pelotons	Supports seed germination, nutrient acquisition, and synthesis of bioactive compounds during early development (Kaur and Deb, 2025)

of powerful compounds such as vincristine, morphine, and withanolides, are also beneficial to mycorrhizal symbiosis. In *Glycyrrhiza glabra*, AMF inoculation has been shown to increase the accumulation of glycyrrhizin, a triterpenoid saponin that displays antiviral and antiinflammatory activities (Yuan *et al.*, 2023). Mycorrhizal associations commonly enhance phenolic and flavonoid contents with antioxidant, antiinflammatory, and cardioprotective activities. In *Hypericum perforatum*, colonization with AMF elevates flavonoid and hypericin levels, thereby increasing the pharmacological activity of the plant (Agnolucci *et al.*, 2020).

Mycorrhizal fungi generally act as natural bioenhancers to both quantitatively and qualitatively enhance therapeutically useful compounds (Sangwan and Prasanna, 2022). Their capacity to induce various classes of secondary metabolites in medicinal plants is reflective of their potential in sustainable agricultural practices, providing a promising approach to the production of high-quality plant-based pharmaceuticals on an ecologically balanced basis (Shukla *et al.*, 2025).

13.4.3 Synergistic effects with other soil microbes

The synergistic effect that results from the interaction of mycorrhizal fungi with other beneficial soil microbes, including PGPR, is a promising approach to improve the production of secondary metabolites in medicinal plants. Coinoculation with mycorrhizal fungi and bacteria, such as *Bacillus*, *Pseudomonas*, or *Azospirillum*, tends to yield complementary effects that are greater than those from single inoculants (Ribeiro *et al.*, 2022). Mycorrhizal fungi mainly enhance nutrient acquisition and stress tolerance, but PGPR are also involved in phytohormone production, nutrient solubilization, and triggering plant defense mechanisms. The synergism resulting from the action of these microbes in concert generates a maximized rhizosphere environment that enhances plant growth, root structure, and metabolic processes (Mohanram and Kumar, 2019; Kumar *et al.*, 2025d).

It has been shown experimentally that coinoculation technologies have the potential to significantly enhance the pharmaceutical production value of secondary metabolites. For instance, *Withania somnifera* plants inoculated with AMF and *Bacillus subtilis* had greater withanolide content than plants treated with either microbe in isolation. Likewise, coinoculation in *Ocimum sanctum* increases the essential oil composition and flavonoid content, and in *Hypericum perforatum*, hypericin and phenolic content increases (Sahu *et al.*, 2020). These synergistic interactions are usually contributed by the microbiome-mediated regulation of plant metabolic processes, such as the upregulation of critical biosynthetic enzymes and enhanced nutrient and water availability. Apart from nutrient and metabolic promotion, microbial communities can enhance plant tolerance to abiotic and biotic stresses, thereby enhancing secondary metabolite accumulation (Anand *et al.*, 2023).

The combination of mycorrhizal fungi with cooperative soil microbes shows that understanding plant–microbiome interactions is of the utmost importance (Fig. 13.2). Using these synergistic relationships, it is possible to design sustainable practices for cultivating high-quality medicinal plants with optimized bioactive compound profiles, thereby decreasing reliance on chemical fertilizers and increasing ecological sustainability.

13.5 Molecular and Biochemical Insights

13.5.1 Metabolic pathways modulated by microbes and fungi

Soil microorganisms and mycorrhizal fungi play an essential role in regulating major metabolic pathways in medicinal plants, resulting in the stimulation

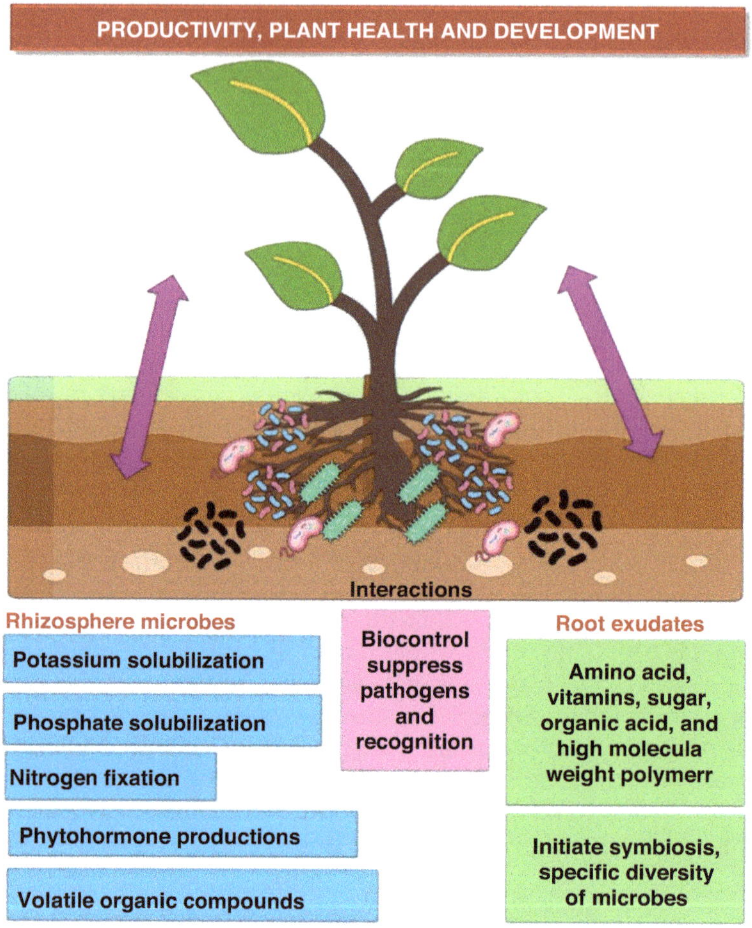

Fig. 13.2. Interaction and effect of microbes on plant development.

of bioactive compound synthesis. The shikimate, mevalonate, and phenylpropanoid pathways have pivotal roles in the synthesis of alkaloids, terpenoids, flavonoids, and phenolics (Bhatla and Lal, 2023). The shikimate pathway, which is involved in the synthesis of aromatic amino acids such as phenylalanine, tyrosine, and tryptophan, is a precursor for many secondary metabolites, such as flavonoids, alkaloids, and lignins. The colonization of microbial enzymes can induce this pathway to enhance efflux metabolite production (Dai *et al.*, 2020). The precursor pathway is regulated by microbial interactions to boost the production of terpenoids, such as ginsenosides in *Panax ginseng* and withanolides in *Withania somnifera*. The phenylpropanoid pathway is a major pathway for phenolics, flavonoids, and other defense-related metabolites that are heavily impacted by microbial elicitors and mycorrhizal symbiosis (Kaur and Suseela, 2020). Beneficial microbes will activate important enzymes such as PAL, cinnamate-4-hydroxylase (C_4H), and chalcone synthase to increase the accumulation of antioxidant and pharmacologically active metabolites.

Through the enhanced utilization of nutrients, production of phytohormones, and induction of plant defense mechanisms, these metabolic pathways increase the yield concentration of therapeutic metabolites. These pathways controlled by microbial interactions form the basis of formulations to enhance the quality and effectiveness of medicinal plants in perfect cultivation systems (Vishwakarma *et al.*, 2020).

13.5.2 Enzyme activity in gene expression

Mycorrhizal microbes control the production of secondary metabolites in medicinal plants not only by hormonal and nutrient modulation but also by controlling gene expression and enzyme activity.

Microbe colonization and molecules can increase the expression of genes that encode pivotal enzymes in biosynthetic pathways, leading to the enhanced synthesis of terpenoids, alkaloids, phenolics, and flavonoids (Huang *et al.*, 2022). These enzymes play essential roles in the terpenoid, shikimate, and phenylpropanoid pathways, which comprise the backbone of bioactive compounds (Yu *et al.*, 2025). They are detected by plant receptors, which induce intricate transcriptional cascades that activate genes involved in secondary metabolite biosynthesis (Meraj *et al.*, 2020). For example, mycorrhizal fungi secrete signaling compounds that increase the expression of genes linked to the production of terpenoids and flavonoids in *Hypercum perforatum* and *Panax ginseng*. PGPR, such as *Bacillus* and *Pseudomonas*, can cause systemic responses, leading to the upregulation of alkaloid biosynthetic genes in *Catharanthus roseus* and withanolide biosynthesis genes in *Withania somnifera* (Jain *et al.*, 2024).

Increased enzyme activity is a direct result of gene upregulation, which improves catalytic efficiency in the critical steps of metabolite biosynthesis. This regulation of coordination enables plants to respond quickly by accumulating bioactive compounds in response to microbial signals, thereby improving their growth and therapeutic utility (George and Brandl, 2021).

Understanding the molecular mechanisms of microbe-mediated gene expression provides a basis for bio-inoculant design strategies to optimize secondary metabolite production and approaches for the high-quality cultivation of medicinal plants (Suman *et al.*, 2022; Kumar *et al.*, 2025c).

13.5.3 Omics approaches

Omics technology, such as metabolomics, genomics, and transcriptomics, has transformed our data on plant–microbe interactions and their secondary metabolite synthesis in medicinal plants. Plant genomics studies, along with microorganisms, enable the identification of genes involved in nutrient uptake, secondary metabolite biosynthesis, and phytohormone synthesis

(Lal *et al.*, 2024). Transcriptomics also identifies dynamic regulation of gene expression following colonization of microbes and upregulation of biosynthetic genes such as chalcone synthase, PAL, strictosidine synthase, and terpene synthase in response to microbes (Ji *et al.*, 2023). Through the profile of actual secondary metabolites produced in plant tissues, these strategies detect shifts in metabolite quantity and diversity resulting from interactions between microbes (Shen *et al.*, 2023). Techniques such as gas chromotography mass spectrometry (GC-MS), liquid chromotography (LC)-MS, and nuclear magnetic resonance (NMR) enable the accurate quantification of bioactive molecule correlations between gene expression and metabolite build-up (Marchev *et al.*, 2021).

These omics platforms provide an integrated understanding of plant–microbe interactions by elucidating the signaling processes that mediate microbial colonization. The omics systems are great tools for deconstructing biochemical networks and for the design of high-value production practices (Amer and Baidoo, 2021).

13.6 Agronomic and Environmental Factors

The soil activity of microbes promotes medicinal plant growth, and secondary metabolite yields are strongly dependent on soil properties and environmental factors. Types of soil control, such as aeration, texture control, and water-holding capacity, influence the colonization of microbes in root–fungi relationships (Pinto *et al.*, 2024). The pH of soil is important because optimal growth of microbes and fungi is promoted by slightly acidic to neutral conditions, and excessive acidity or alkalinity can lower microbial counts and impair symbiotic efficacy (Naz *et al.*, 2022). The availability of nutrients, especially nitrogen, phosphorus, potassium, and trace elements, affects enzyme activity in the metabolic pathways of plants and has a direct effect on the biosynthesis of alkaloids, flavonoids, terpenoids, and phenolics. Soil water is another critical parameter, with sufficient water favoring microbial metabolism, hyphal elongation, and transport of nutrients, whereas drought or waterlogging can damage microbial activity and decrease metabolite concentration (Zhang *et al.*, 2025). Temperature affects microbial growth and enzyme kinetics, with high or low temperatures breaking up symbiotic relationships and plant metabolic processes.

Optimization of environmental conditions is essential for maximizing microbial inoculant benefits, uniformity, and the production of high-quality bioactive metabolites in medicinal plants (Kavusi *et al.*, 2023; Mitra *et al.*, 2024).

13.6.1 Impact of agricultural practices on microbial populations

Microbial soil communities are greatly influenced by agricultural practices and this has an effect on the growth and production of secondary metabolites in medicinal plants. The application of chemical fertilizers and pesticides disrupts the microbial population, leading to low soil biodiversity, and symbiotic relationships, such as colonization by mycorrhiza (Meena *et al.*, 2020). These types of disruptions result in reduced availability of nutrients, decreased biosynthesis of phytohormones, and decreased secondary metabolites, ultimately reducing the quality and quantity of bioactive compounds. However, beneficial agricultural practices such as organic farming, intercropping, crop rotation, and use of compost or green manure build a healthy microbial environment (Diacono *et al.*, 2021). Mycorrhizal inoculants and PGPR have been found to cause strong accumulation of metabolites in medicinal plants (Shokati and Poudineh, 2017). Organic matter and minimum tillage maintenance also support microbial communities for long-term soil fertility and plant health (Srour *et al.*, 2020).

13.6.2 Microbial diversity and organic versus conventional cultivation

Cultivation methods affect the microbial diversity of the soil and biosynthesis of bioactive metabolites in medicinal plants. Organic farming, which uses compost manure and biofertilizers, avoids the excessive use of chemical pesticides and fertilizers to promote the development of microbial ecosystems in the rhizosphere (Vurukonda *et al.*, 2024). Improved microbial diversity in organic systems enhances the cycling of nutrients, phytohormone biosynthesis, and efficient mycorrhizal colonization, which lead to enhanced plant growth and improved accumulation of secondary metabolites. Experiments have shown that when grown organically, medicinal plants such as *Withania somnifera* and *Ocimum sanctum* have more bioactive compounds than their conventionally grown equivalents, mainly because of the increased activity of disease-suppressing bacteria and fungi in organic soil (Greff *et al.*, 2023).

Intensive conventional cultivation based on massive use of chemical fertilizers and pesticides may decrease microbial diversity and break symbiotic associations. The overuse of chemicals can promote opportunistic, quick-growing microbial species at the expense of beneficial microbes such as PGPR and mycorrhizal fungi, thus reducing their beneficial impact on plant metabolism (Qasim, 2023). The synergistic use of organic approaches combined with specific microbial inoculants provides an effective means of enhancing the quality and activity of medicinal plants while minimizing their environmental footprint (Kumar *et al.*, 2021).

13.7 Biotechnological Applications

13.7.1 Microbial inoculants to enhance the quality of medicinal plants

Inoculation of microbes, such as PGPR, nitrogen-fixing bacteria, phosphate-solubilizing microbes, and AMF, has been shown to improve the growth of the bioactive metabolite content of medicinal plants (Ansabayeva *et al.*,

2025). These microbial inoculants increase the nutrients and enhance their metabolites. For example, inoculation of *Withania somnifera* with AMF and *Bacillus subtilis* has been found to increase the withanolide content, whereas PGPR treatment in *Ocimum sanctum* increases the composition of essential oils and flavonoids (Raj *et al.*, 2022). Microbial inoculation also enhances plant tolerance to abiotic stresses, such as salinity, drought, and nutrient stress, which indirectly contribute to metabolite production. To enhance the yield of concentrated metabolites, the application of bio-inoculants to chemical pesticides and fertilizers supports eco-friendly agricultural practices (Santos *et al.*, 2021). Synergistic interactions between beneficial microbes and plant inoculant microbes present a practical approach to obtain high-quality medicinal plants with improved therapeutic activity and are central to biotechnological applications and herbal agriculture (Tripathi *et al.*, 2022).

13.7.2 Microbial consortia and development of biofertilizers

The development of microbial consortia used as biofertilizers promotes the growth and production of bioactive compounds in medicinal plants. Biofertilizer is a product of microbial benefits, including AMF, nitrogen-fixing bacteria, PGPR, and phosphate-solubilizing bacteria, which improve nutrient availability and induce the production of phytohormones and metabolic activity in plants (Nosheen *et al.*, 2021; Pandit *et al.*, 2025). Attention has recently focused on the development of microbial consortia by pairing two or more complementary species to create synergistic effects. Development strategies for effective microbial consortia include the selection of compatible stains, optimization of inoculum ratio, and formulation methods for microbial viability in field conditions (Chaudhary *et al.*, 2023). Consortia can be developed for medicinal plants or cultivation. Microbial consortia and biofertilizer use in agricultural practices improve soil health and support the cultivation of high-quality medicinal plants (Folina *et al.*, 2025).

13.7.3 Bioactive compound production through plant–microbe interactions

Plant–microbe culture systems are highly promising biotechnological techniques for the large-scale production of bioactive molecules from medicinal plants. Culturing is associated with microbes such as AMF, endophyte bacteria, and fungi to improve metabolite production (Zhao *et al.*, 2022), triggering the production of biosynthetic enzymes and metabolic pathways related to the biosynthesis of phenolics, terpenoids, alkaloids, and flavonoids. Coculture methods have a much higher yield of bioactive compounds than solo-cultured plants (Vinale *et al.*, 2017). In coculture, choice of microbial stain, culture medium components, inoculum density, and environmental parameters are optimized to further enhance metabolite yield (Liu *et al.*, 2022).

13.7.4 Cultivation of medicinal plants

Beneficial microbes improve water uptake, stress resistance, and nutrient availability, allowing the plant to survive stressful environmental conditions such as poor soil, drought, and heat (Enebe and Babalola, 2018). The symbiosis of mycorrhizal fungi improves root structure and soil organization, whereas microbial consortia enhance phytohormone production and mechanisms of plant defense, with the result that growth becomes steady and bioactive molecules accumulate even under changing climate conditions. These interactions minimize the need for chemical pesticides and fertilizers, encourage beneficial practices, and maintain soil health for extended periods of time (Tahat *et al.*, 2020). The microbial inoculum and coculture system are customized for the production of secondary metabolites from individual medicinal plants, even under field stress conditions. The growing plant resistance and productivity of metabolites profit from smart climate agriculture and production maintenance of herbal drugs (Vishnoi and Goel, 2024). Microbial-based methods ensure high-quality bioactive compound production and promote biodiversity, soil integrity, and resources. These innovations in agricultural systems are a feasible method to develop economically competitive medicinal plant systems to meet the growing global demand (Fei *et al.*, 2025).

13.8 Challenges and Future Perspectives

Despite the potential of soil microorganisms and mycorrhizal fungi as biofertilizers and biotic elicitors of growth and secondary metabolites of medicinal plants, several constraints must be overcome in order to achieve their complete biotechnological utilization. Standardization of microbial inoculants is of great concern because differences in strain viability, formulation, and field performance may result in variable effects (Sammauria *et al.*, 2020). Recognition of plant microbiome specificity is also essential, as the activity of microbial symbionts relies on compatible plant–host interactions, root architecture, and environmental conditions. Omics technologies together with artificial intelligence and machine learning provide tools to unravel intricate plant–microbe interactions and maximize metabolite production, but their translation into large-scale cultivation systems remains limited (Satrio *et al.*, 2024). Regulatory environments and economic factors also affect the uptake of microbial-based approaches, as biofertilizers and coculture systems must conform to safety, quality, and commercial feasibility requirements.

13.9 Conclusion

The use of medicinal plants for the production of secondary metabolites represents a sustainable alternative to conventional approaches. However, beneficial soil microbes such as mycorrhizal fungi can further enhance plant growth, health, and metabolite accumulation. In this context, microbial consortia and coinoculation strategies offer promising opportunities to improve both plant performance and bioactive compound yield.

Our understanding of plant–microbe interactions has expanded due to developments in omics and molecular biology, which have also made it feasible to precisely customize microbial inoculum and growing methods. The coculture system of plant–microbe methods for pharmaceutical development and the synthesis of biotechnological high-value metabolites are less reliant on field cultivation. There are still issues with host microbe-specific considerations. These constraints require multidisciplinary research that integrates agronomy, plant biotechnology, microbiology, and computational resources. The utilization of mycorrhizal fungi and soil microbes is a groundbreaking technique for growing valuable, climate-resistant, and sustainable medicinal plants.

References

Agnolucci, M., Avio, L., Palla, M., Sbrana, C., Turrini, A. *et al.* (2020) Health-promoting properties of plant products: The role of mycorrhizal fungi and associated bacteria. *Agronomy* 10(12), 1864.

Al-Quwaie, D.A. (2024) The role of streptomyces species in controlling plant diseases: A comprehensive review. *Australasian Plant Pathology* 53(1), 1–14. DOI: 10.1007/s13313-023-00959-z.

Amer, B. and Baidoo, E.E. (2021) Omics-driven biotechnology for industrial applications. *Frontiers in Bioengineering and Biotechnology* 9, 613307.

Anand, U., Pal, T., Yadav, N., Singh, V.K., Tripathi, V. *et al.* (2023) Current scenario and future prospects of endophytic microbes: Promising candidates for abiotic and biotic stress management for agricultural and environmental sustainability. *Microbial Ecology* 86(3), 1455–1486.

Ansabayeva, A., Makhambetov, M., Rebouh, N.Y., Abdelkader, M., Saudy, H.S. *et al.* (2025) Plant growth-promoting microbes for resilient farming systems: Mitigating environmental stressors and boosting crops productivity—a review. *Horticulturae* 11(3), 260. DOI: 10.3390/horticulturae11030260.

Bhantana, P., Rana, M.S., Sun, X., Moussa, M.G., Saleem, M.H. *et al.* (2021) Arbuscular mycorrhizal fungi and its major role in plant growth, zinc nutrition, phosphorous regulation and phytoremediation. *Symbiosis* 84(1), 19–37. DOI: 10.1007/s13199-021-00756-6.

Bhaskar, R., Xavier, L.S.E., Udayakumaran, G., Kumar, D.S., Venkatesh, R. *et al.* (2022) Biotic elicitors: A boon for the *in-vitro* production of plant secondary metabolites. *Plant Cell, Tissue and Organ Culture* 149(1–2), 7–24. DOI: 10.1007/s11240-021-02131-1.

Bhatla, S.C. and Lal, M.A. (2023) Secondary metabolites. In: *Plant Physiology, Development and Metabolism*. Springer Nature, Singapore, pp. 765–808. DOI: 10.1007/978-981-99-5736-1_33.

Cardillo, A.B., Perassolo, M., Giulietti, A.M. and Rodriguez Talou, J. (2021) Cyclodextrins: A tool in plant cell and organ culture bioprocesses for the production of secondary metabolites. *Plant Cell, Tissue and Organ Culture* 146(1), 1–19. DOI: 10.1007/s11240-021-02045-y.

Castronovo, L.M., Vassallo, A., Mengoni, A., Miceli, E., Bogani, P. *et al.* (2021) Medicinal plants and their bacterial microbiota: A review on antimicrobial compounds production for plant and human health. *Pathogens* 10(2), 106.

Chaudhary, P., Xu, M., Ahamad, L., Chaudhary, A., Kumar, G. *et al.* (2023) Application of synthetic consortia for improvement of soil fertility, pollution remediation, and agricultural productivity: A review. *Agronomy* 13(3), 643. DOI: 10.3390/agronomy13030643.

Chauhan, P., Sharma, N., Tapwal, A., Kumar, A., Verma, G.S. *et al.* (2023) Soil microbiome: Diversity, benefits and interactions with plants. *Sustainability* 15(19), 14643. DOI: 10.3390/su151914643.

Chen, Q., Song, Y., An, Y., Lu, Y. and Zhong, G. (2024) Mechanisms and impact of rhizosphere microbial metabolites on crop health, traits, functional components: A comprehensive review. *Molecules* 29(24), 5922. DOI: 10.3390/molecules29245922.

Daghino, S., Murat, C., Mita, S., Martino, E. and Perotto, S. (2025) Comparative genomics reveals substantial divergence in metal sensitive and metal tolerant isolates of the ericoid mycorrhizal fungus *Oidiodendron maius*. *Mycorrhiza* 35(2), 1–16.

Dai, Z., Ramesh, V. and Locasale, J.W. (2020) The evolving metabolic landscape of chromatin biology and epigenetics. *Nature Reviews Genetics* 21(12), 737–753. DOI: 10.1038/s41576-020-0270-8.

Davis, C.C. and Choisy, P. (2024) Medicinal plants meet modern biodiversity science. *Current Biology* 34(4), R158–R173. DOI: 10.1016/j.cub.2023.12.038.

Diacono, M., Persiani, A., Castellini, M., Giglio, L. and Montemurro, F. (2021) Intercropping and rotation with leguminous plants in organic vegetables: Crop performance, soil properties and sustainability assessment. *Biological Agriculture and Horticulture* 37(3), 141–167. DOI: 10.1080/01448765.2021.1891968.

Diagne, N., Ngom, M., Djighaly, P.I., Fall, D., Hocher, V. *et al.* (2020) Roles of arbuscular mycorrhizal fungi on plant growth and performance: Importance in biotic and abiotic stressed regulation. *Diversity* 12(10), 370. DOI: 10.3390/d12100370.

Díaz-Rodríguez, A.M., Parra Cota, F.I., Cira Chávez, L.A., García Ortega, L.F., Estrada Alvarado, M.I. *et al.* (2025) Microbial inoculants in sustainable agriculture: Advancements, challenges, and future directions. *Plants* 14(2), 191. DOI: 10.3390/plants14020191.

Ehlers, B.K., Berg, M.P., Staudt, M., Holmstrup, M., Glasius, M. *et al.* (2020) Plant secondary compounds in soil and their role in belowground species interactions. *Trends in Ecology and Evolution* 35(8), 716–730. DOI: 10.1016/j.tree.2020.04.001.

El-Saadony, M.T., Saad, A.M., Mohammed, D.M., Korma, S.A., Alshahrani, M.Y. *et al.* (2025) Medicinal plants: Bioactive compounds, biological activities, combating multidrug-resistant microorganisms, and human health benefits – a comprehensive review. *Frontiers in Immunology* 16, 1491777. DOI: 10.3389/fimmu.2025.1491777.

Enebe, M.C. and Babalola, O.O. (2018) The influence of plant growth-promoting rhizobacteria in plant tolerance to abiotic stress: A survival strategy. *Applied Microbiology and Biotechnology* 102(18), 7821–7835. DOI: 10.1007/s00253-018-9214-z.

Fei, S., Wu, R., Liu, H., Yang, F. and Wang, N. (2025) Technological innovations in urban and peri-urban agriculture: Pathways to sustainable food systems in metropolises. *Horticulturae* 11(2), 212. DOI: 10.3390/horticulturae11020212.

Folina, A., Kakabouki, I., Baginetas, K. and Bilalis, D. (2025) Integration of bioresources for sustainable development in organic farming: A comprehensive review. *Resources* 14(7), 102. DOI: 10.3390/resources14070102.

Fukami, J., Cerezini, P. and Hungria, M. (2018) *Azospirillum*: Benefits that go far beyond biological nitrogen fixation. *AMB Express* 8(1), 73.

George, A.S. and Brandl, M.T. (2021) Plant bioactive compounds as an intrinsic and sustainable tool to enhance the microbial safety of crops. *Microorganisms* 9(12), 2485. DOI: 10.3390/microorganisms9122485.

Goyal, M.R. and Chauhan, A. (2024) Holistic approach of nutrients and traditional natural medicines for human health: A review. *Future Integrative Medicine* 3, 197–208. DOI: 10.14218/FIM.2023.00089.

Greff, B., Sáhó, A., Lakatos, E. and Varga, L. (2023) Biocontrol activity of aromatic and medicinal plants and their bioactive components against soil-borne pathogens. *Plants* 12(4), 706. DOI: 10.3390/plants12040706.

Gupta, N., Wijenayake, W.P.T., Roy, D., Kumar, R., Rangot, M. *et al.* (2025) Smart recycling and sustainable lignocellulosic waste management. In: Mukherjee, G. and Dhiman, S. (eds) *Value Addition and Utilization of Lignocellulosic Biomass: Through Novel Technological Interventions*. Springer Nature, Singapore, pp. 221–250. DOI: 10.1007/978-981-96-2786-8_9.

Hasan, A., Tabassum, B., Hashim, M. and Khan, N. (2024) Role of plant growth promoting rhizobacteria (PGPR) as a plant growth enhancer for sustainable agriculture: A review. *Bacteria* 3(2), 59–75. DOI: 10.3390/bacteria3020005.

Heredia-Bátiz, J.M., Manjarrez-Quintero, J.P., Valdez-Baro, O., Rivera-Salas, M.M., Bayardo-Rosales, H. *et al.* (2025) Role of bioactive compounds in plant disease management. In: Devi, J. (ed.) *Biotechnological Intervention in Production of Bioactive Compounds: Biosynthesis, Characterization and Applications*. Springer, Cham, Switzerland, pp. 141–162. DOI: 10.1007/978-3-031-76859-0_9.

Huang, W., Wang, Y., Tian, W., Cui, X., Tu, P. *et al.* (2022) Biosynthesis investigations of terpenoid, alkaloid, and flavonoid antimicrobial agents derived from medicinal plants. *Antibiotics* 11(10), 1380. DOI: 10.3390/antibiotics11101380.

Jabborova, D., Mamarasulov, B., Davranov, K., Enakiev, Y., Bisht, N. *et al.* (2024) Diversity and plant growth properties of rhizospheric bacteria associated with medicinal plants. *Indian Journal of Microbiology* 64(2), 409–417. DOI: 10.1007/s12088-024-01275-w.

Jain, D., Bisht, S., Parvez, A., Singh, K., Bhaskar, P. *et al.* (2024) Effective biotic elicitors for augmentation of secondary metabolite production in medicinal plants. *Agriculture* 14(6), 796. DOI: 10.3390/agriculture14060796.

Jalmi, S.K. and Sinha, A.K. (2022) Ambiguities of PGPR-induced plant signaling and stress management. *Frontiers in Microbiology* 13, 899563. DOI: 10.3389/fmicb.2022.899563.

Jamiołkowska, A. (2020) Natural compounds as elicitors of plant resistance against diseases and new biocontrol strategies. *Agronomy* 10(2), 173. DOI: 10.3390/agronomy10020173.

Javed, Z., Tripathi, G.D., Mishra, M. and Dashora, K. (2021) Actinomycetes – the microbial machinery for the organic-cycling, plant growth, and sustainable soil health. *Biocatalysis and Agricultural Biotechnology* 31, 101893. DOI: 10.1016/j.bcab.2020.101893.

Ji, C., Liang, Z., Cao, H., Chen, Z., Kong, X. *et al.* (2023) Transcriptome-based analysis of the effects of compound microbial agents on gene expression in wheat roots and leaves under salt stress. *Frontiers in Plant Science* 14, 1109077. DOI: 10.3389/fpls.2023.1109077.

Johny, L., Cahill, D.M. and Adholeya, A. (2021) AMF enhance secondary metabolite production in ashwagandha, licorice, and marigold in a fungi-host specific manner. *Rhizosphere* 17, 100314. DOI: 10.1016/j.rhisph.2021.100314.

Kamboj, N., Chugh, P., Wijenayake, W.P.T., Mitra, D., Panneerselvam, P. *et al.* (2025) Multifunctionality and diversity of antagonistic potential fungi as biocontrol agent. In: Mitra, D., Villalobos, S., Rani, A., Sierra, B. and Andjelkovic, S. (eds) *Bio-Control Agents for Sustainable Agriculture: Diversity, Mechanisms and Applications*. Springer Nature, Singapore, pp. 167–208. DOI: 10.1007/978-981-96-3232-9_9.

Kaur, S. and Deb, A. (2025) The role of mycorrhizal fungi in orchids. *Recent Patents on Biotechnology* 19(3), 210–220.

Kaur, S. and Suseela, V. (2020) Unraveling arbuscular mycorrhiza-induced changes in plant primary and secondary metabolome. *Metabolites* 10(8), 335.

Kavusi, E., Shahi Khalaf Ansar, B., Dehghanian, Z., Asgari Lajayer, B., Nobaharan, K. *et al.* (2023) Delivery of beneficial microbes via seed coating for medicinal and aromatic plant production: A critical review. *Journal of Plant Growth Regulation* 42(2), 575–597. DOI: 10.1007/s00344-022-10597-2.

Khan, A., Ding, Z., Ishaq, M., Khan, I., Ali Ahmed, A. *et al.* (2020) Applications of beneficial plant growth promoting rhizobacteria and mycorrhizae in rhizosphere and plant growth: A review. *International Journal of Agricultural and Biological Engineering* 13(5), 199–208. DOI: 10.25165/j.ijabe.20201305.5762.

Khare, P. and Khare, N. (2025) Herbal medicine and chronic disease management. In: Sathibabu Uddandrao, V. and Saravanan, G. (eds) *Therapeutic Plants: Recent Advances in the Use of Herbs as Alternative Medications*. Bentham Science Publishers, Sharjah, UAE, pp. 24–49.

Khoza, T., Masenya, A., Khanyile, N. and Thosago, S. (2025) Alleviating plant density and salinity stress in *Moringa oleifera* using arbuscular mycorrhizal fungi: A review. *Journal of Fungi* 11(4), 328.

Khullar, S. and Reddy, M.S. (2019) Ectomycorrhizal diversity and tree sustainability. In: Satyanarayana, T., Johri, B. and Das, S. (eds) *Microbial Diversity in Ecosystem Sustainability and Biotechnological Applications*. Springer, Singapore, pp. 145–166. DOI: 10.1007/978-981-13-8487-5_6.

Korenblum, E., Massalha, H. and Aharoni, A. (2022) Plant–microbe interactions in the rhizosphere via a circular metabolic economy. *Plant Cell* 34(9), 3168–3182.

Koza, N.A., Adedayo, A.A., Babalola, O.O. and Kappo, A.P. (2022) Microorganisms in plant growth and development: Roles in abiotic stress tolerance and secondary metabolites secretion. *Microorganisms* 10(8), 1528.

Kumar, R., Saurabh, K., Kumawat, N., Sundaram, P.K., Mishra, J.S. *et al.* (2021) Sustaining productivity through integrated use of microbes in agriculture. In: Seneviratne, G. and Zavahir, J. (eds) *Role of Microbial Communities for Sustainability*. Springer, Singapore, pp. 109–145. DOI: 10.1007/978-981-15-9912-5_5.

Kumar, R., Swapnil, P., Meena, M., Selpair, S. and Yadav, B.G. (2022) Plant growth-promoting rhizobacteria (PGPR): Approaches to alleviate abiotic stresses for enhancement of growth and development of medicinal plants. *Sustainability* 14(23), 15514. DOI: 10.3390/su142315514.

Kumar, R., Kamboj, N., Mitra, D. and Ray, A. (2025a) Strengthening local communities through advanced recycling technology for sustainable growth. In: Mitra, D., Choudhury, T., Madan, A., Chattaraj, S. and Pellegrini, M. (eds) *AI Technologies for Enhancing Recycling Processes*. IGI Global Scientific Publishing, Palmdale, USA, pp. 393–420. DOI: 10.4018/979-8-3693-7282-1.ch017.

Kumar, R., Lakhera, A., Gulati, M., Rani, A., Thapliyal, A. *et al.* (2025b) Genome mining of biocontrol agents for bioprospecting action modes. In: Mitra, D., Villalobos, S., Rani, A., Sierra, B. and Andjelkovic, S. (eds) *Bio-Control Agents for Sustainable Agriculture: Diversity, Mechanisms and Applications*. Springer Nature, Singapore, pp. 41–62. DOI: 10.1007/978-981-96-3232-9_3.

Kumar, R., Farda, B., Mignini, A., Djebaili, R., Koolman, L. *et al.* (2025c) Microbial solutions in agriculture: Enhancing soil health and resilience through bio-inoculants and bioremediation. *Bacteria* 4(3), 28. DOI: 10.3390/bacteria4030028.

Kumar, R., Vasić, T.P., Živković, S.P., Panneerselvam, P., Santoyo, G. *et al.* (2025d) Mechanistic role of heavy metals in driving antimicrobial resistance: From rhizosphere to phyllosphere. *Applied Microbiology* 5(3), 79. DOI: 10.3390/applmicrobiol5030079.

Kumar, R., Pal, O.P., Kamboj, N. and Jaiswal, A. (2025e) Phytonutrients in neuroprotection and neurodegenerative disorders. *Current Research in Nutrition and Food Science Journal* 13(Special Issue July), 72–96. DOI: 10.12944/CRNFSJ.13.Special-Issue-July.05.

Lal, M.K., Tiwari, R.K., Adavi, S.B., Kumar, A., Bamatov, I. *et al.* (2024) Deciphering phenomics approaches for understanding plant–microbe interactions in nutrient absorption and utilization. *Plant Physiology Reports* 29(4), 769–785. DOI: 10.1007/s40502-024-00844-x.

Liu, H., Hao, D., Li, Y., Wang, X. and Chen, J. (2022) Approaches for the establishment of optimized co-culture system of multiple *Trichoderma* strains for culture metabolites highly effective in cucumber growth promotion. *Frontiers in Microbiology* 13, 1020077. DOI: 10.3389/fmicb.2022.1020077.

Marchev, A.S., Vasileva, L.V., Amirova, K.M., Savova, M.S., Balcheva-Sivenova, Z.P. *et al.* (2021) Metabolomics and health: From nutritional crops and plant-based pharmaceuticals to profiling of human biofluids. *Cellular and Molecular Life Sciences* 78(19–20), 6487–6503. DOI: 10.1007/s00018-021-03918-3.

Mashabela, M.D., Piater, L.A., Dubery, I.A., Tugizimana, F. and Mhlongo, M.I. (2022) Rhizosphere tripartite interactions and PGPR-mediated metabolic reprogramming towards ISR and plant priming: A metabolomics review. *Biology* 11(3), 346. DOI: 10.3390/biology11030346.

Meena, R., Kumar, S., Datta, R., Lal, R., Vijayakumar, V. *et al.* (2020) Impact of agrochemicals on soil microbiota and management: A review. *Land* 9(2), 34. DOI: 10.3390/land9020034.

Meraj, T.A., Fu, J., Raza, M.A., Zhu, C., Shen, Q. *et al.* (2020) Transcriptional factors regulate plant stress responses through mediating secondary metabolism. *Genes* 11(4), 346. DOI: 10.3390/genes11040346.

Mistry, V., Darji, S., Tiwari, P. and Sharma, A. (2022) Engineering *Catharanthus roseus* monoterpenoid indole alkaloid pathway in yeast. *Applied Microbiology and Biotechnology* 106(7), 2337–2347. DOI: 10.1007/s00253-022-11883-5.

Mitra, D., Kumar, R. and Kamboj, N. (2024) Microbial innovations in agriculture: Interdisciplinary approaches to leveraging microbes for food sustainability and security. *Indian Journal of Microbiology Research* 11(3), 129–139. DOI: 10.18231/j.ijmr.2024.026.

Modolon, F., Barno, A.R., Villela, H.D.M. and Peixoto, R.S. (2020) Ecological and biotechnological importance of secondary metabolites produced by coral-associated bacteria. *Journal of Applied Microbiology* 129(6), 1441–1457. DOI: 10.1111/jam.14766.

Mohanram, S. and Kumar, P. (2019) Rhizosphere microbiome: Revisiting the synergy of plant-microbe interactions. *Annals of Microbiology* 69(4), 307–320. DOI: 10.1007/s13213-019-01448-9.

Motolinía-Alcántara, E.A., Castillo-Araiza, C.O., Rodríguez-Monroy, M., Román-Guerrero, A. and Cruz-Sosa, F. (2021) Engineering considerations to produce bioactive compounds from plant cell suspension culture in bioreactors. *Plants* 10(12), 2762. DOI: 10.3390/plants10122762.

Nagrale, D.T., Chaurasia, A., Kumar, S., Gawande, S.P., Hiremani, N.S. *et al.* (2023) PGPR: The treasure of multifarious beneficial microorganisms for nutrient mobilization, pest biocontrol and plant growth promotion in field crops. *World Journal of Microbiology and Biotechnology* 39(4), 100.

Naz, M., Dai, Z., Hussain, S., Tariq, M., Danish, S. *et al.* (2022) The soil pH and heavy metals revealed their impact on soil microbial community. *Journal of Environmental Management* 321, 115770. DOI: 10.1016/j.jenvman.2022.115770.

Nosheen, S., Ajmal, I. and Song, Y. (2021) Microbes as biofertilizers, a potential approach for sustainable crop production. *Sustainability* 13(4), 1868. DOI: 10.3390/su13041868.

Pandey, P., Tripathi, A., Dwivedi, S., Lal, K. and Jhang, T. (2023) Deciphering the mechanisms, hormonal signaling, and potential applications of endophytic microbes to mediate stress tolerance in medicinal plants. *Frontiers in Plant Science* 14, 1250020. DOI: 10.3389/fpls.2023.1250020.

Pandit, A., Gunsola, D., Kumar, R., Panneerselvam, P. and Mitra, D. (2025) Advancing innovative techniques in arbuscular mycorrhizal fungi propagation: A key to sustainable agriculture and ecosystem management. *Journal of Microbiological Methods* 236, 107200.

Pinto, A.G., Kos, T., Puškarić, J., Vrandečić, K., Benković-Lačić, T. *et al.* (2024) Soil ecosystem functioning through interactions of nematodes and fungi *Trichoderma* sp. *Sustainability* 16(7), 2885. DOI: 10.3390/su16072885.

Qasim, M. (2023) Beneficial microbes in plant health: From biocontrol agent to plant growth promoting rhizobacteria. *International Journal of Research and Advances in Agricultural Sciences* 2(3), 39–49.

Raj, Y., Ali, N., Pati, A.M. and Kumar, R. (2022) Cleaner production technologies for the amelioration of soil health, biomass and secondary metabolites in *Ocimum basilicum* L.

under Indian Western Himalaya. *Frontiers in Plant Science* 13, 976295. DOI: 10.3389/fpls.2022.976295.

Ran, Z., Ding, W., Cao, S., Fang, L., Zhou, J. *et al.* (2022) Arbuscular mycorrhizal fungi: Effects on secondary metabolite accumulation of traditional Chinese medicines. *Plant Biology* 24(6), 932–938.

Ribeiro, V.P., Gomes, E.A., de Sousa, S.M., de Paula Lana, U.G., Coelho, A.M. *et al.* (2022) Co-inoculation with tropical strains of *Azospirillum* and *Bacillus* is more efficient than single inoculation for improving plant growth and nutrient uptake in maize. *Archives of Microbiology* 204(2), 143. DOI: 10.1007/s00203-022-02759-3.

Rizvi, A., Ahmed, B., Khan, M.S., El-Beltagi, H.S., Umar, S. *et al.* (2022) Bioprospecting plant growth promoting rhizobacteria for enhancing the biological properties and phytochemical composition of medicinally important crops. *Molecules* 27(4), 1407.

Sahu, P.K., Singh, S., Gupta, A.R., Gupta, A., Singh, U.B. *et al.* (2020) Endophytic bacilli from medicinal-aromatic perennial Holy basil (*Ocimum tenuiflorum* L.) modulate plant growth promotion and induced systemic resistance against *Rhizoctonia solani* in rice (*Oryza sativa* L.). *Biological Control* 150, 104353.

Samal, S., Kumar, R., Thapliyal, M., Gunsola, D., Apaarna *et al.* (2025) Insight into an effective development of biocontrol agent formulation for commercial production. In: Mitra, D., Villalobos, S., Rani, A., Sierra, B. and Andjelkovic, S. (eds) *Bio-Control Agents for Sustainable Agriculture: Diversity, Mechanisms and Applications*. Springer Nature, Singapore, pp. 385–407. DOI: 10.1007/978-981-96-3232-9_16.

Sammauria, R., Kumawat, S., Kumawat, P., Singh, J. and Jatwa, T.K. (2020) Microbial inoculants: Potential tool for sustainability of agricultural production systems. *Archives of Microbiology* 202(4), 677–693.

Sangwan, S. and Prasanna, R. (2022) Mycorrhizae helper bacteria: Unlocking their potential as bioenhancers of plant–arbuscular mycorrhizal fungal associations. *Microbial Ecology* 84(1), 1–10.

Santos, M.S., Rodrigues, T.F., Nogueira, M.A. and Hungria, M. (2021) The challenge of combining high yields with environmentally friendly bioproducts: A review on the compatibility of pesticides with microbial inoculants. *Agronomy* 11(5), 870. DOI: 10.3390/agronomy11050870.

Satrio, R.D., Fendiyanto, M.H. and Miftahudin, M. (2024) Tools and techniques used at global scale through genomics, transcriptomics, proteomics, and metabolomics to investigate plant stress responses at the molecular level. In: Shahid, M. and Gaur, R. (eds) *Molecular Dynamics of Plant Stress and Its Management*. Springer Nature, Singapore, pp. 555–607. DOI: 10.1007/978-981-97-1699-9_25.

Shen, S., Zhan, C., Yang, C., Fernie, A.R. and Luo, J. (2023) Metabolomics-centered mining of plant metabolic diversity and function: Past decade and future perspectives. *Molecular Plant* 16(1), 43–63. DOI: 10.1016/j.molp.2022.09.007.

Shokati, B. and Poudineh, Z. (2017) An overview of plant growth promoting rhizobacteria and their influence on essential oils of medicinal plants: A review article. *Iranian Journal of Plant Physiology* 7(3), 2051–2061.

Shukla, S., Patra, D., Sinha, A., Saha, R., Tripathi, Y. *et al.* (2025) Medicinal and aromatic plant cultivation, improvement and conservation for sustainable development. In: Kumar, N. (ed.) *Industrial Crops Improvement: Biotechnological Approaches for Sustainable Agricultural Development*. Springer, Cham, Switzerland, pp. 183–204. DOI: 10.1007/978-3-031-75937-6_11.

Singh, V.K. and Kumar, A. (2023) Secondary metabolites from endophytic fungi: Production, methods of analysis, and diverse pharmaceutical potential. *Symbiosis* 90(2), 111–125. DOI: 10.1007/s13199-023-00925-9.

Singh, A., Singh, D.K., Kharwar, R.N., White, J.F. and Gond, S.K. (2021) Fungal endophytes as efficient sources of plant-derived bioactive compounds and their prospective applications in natural product drug discovery: Insights, avenues, and challenges. *Microorganisms* 9(1), 197. DOI: 10.3390/microorganisms9010197.

Singh, R., Soni, P., Jain, R., Chauhan, A., Kumar, R. *et al.* (2026) Microbial metabolites: A sustainable solution for agricultural challenges. *Pedosphere* 36, 302–316. DOI: 10.1016/j.pedsph.2025.06.003.

Srour, A.Y., Ammar, H.A., Subedi, A., Pimentel, M., Cook, R.L. *et al.* (2020) Microbial communities associated with long-term tillage and fertility treatments in a corn-soybean cropping system. *Frontiers in Microbiology* 11, 1363. DOI: 10.3389/fmicb.2020.01363.

Suman, A., Govindasamy, V., Ramakrishnan, B., Aswini, K., SaiPrasad, J. *et al.* (2022) Microbial community and function-based synthetic bioinoculants: A perspective for sustainable agriculture. *Frontiers in Microbiology* 12, 805498. DOI: 10.3389/fmicb.2021.805498.

Tahat, M., Alananbeh, K., Othman, Y. and Leskovar, D. (2020) Soil health and sustainable agriculture. *Sustainability* 12(12), 4859. DOI: 10.3390/su12124859.

Tohge, T., Watanabe, M., Hoefgen, R. and Fernie, A.R. (2013) The evolution of phenylpropanoid metabolism in the green lineage. *Critical Reviews in Biochemistry and Molecular Biology* 48(2), 123–152. DOI: 10.3109/10409238.2012.758083.

Tripathi, A., Pandey, P., Tripathi, S.N. and Kalra, A. (2022) Perspectives and potential applications of endophytic microorganisms in cultivation of medicinal and aromatic plants. *Frontiers in Plant Science* 13, 985429. DOI: 10.3389/fpls.2022.985429.

Twaij, B.M. and Hasan, M.N. (2022) Bioactive secondary metabolites from plant sources: Types, synthesis, and their therapeutic uses. *International Journal of Plant Biology* 13(1), 4–14. DOI: 10.3390/ijpb13010003.

Vaghela, N. and Gohel, S. (2023) Medicinal plant-associated rhizobacteria enhance the production of pharmaceutically important bioactive compounds under abiotic stress conditions. *Journal of Basic Microbiology* 63(3–4), 308–325. DOI: 10.1002/jobm.202200361.

Verma, P., Yadav, A.N., Khannam, K.S., Saxena, A.K. and Suman, A. (2017) Potassium-solubilizing microbes: Diversity, distribution, and role in plant growth promotion. In: Panpatte, D., Jhala, Y., Vyas, R. and Shelat, H. (eds) *Microorganisms for Green Revolution: Volume 1: Microbes for Sustainable Crop Production*. Springer, Singapore, pp. 125–149. DOI: 10.1007/978-981-10-6241-4_7.

Vinale, F., Nicoletti, R., Borrelli, F., Mangoni, A., Parisi, O.A. *et al.* (2017) Co-culture of plant beneficial microbes as source of bioactive metabolites. *Scientific Reports* 7(1), 14330. DOI: 10.1038/s41598-017-14569-5.

Vishnoi, S. and Goel, R.K. (2024) Climate smart agriculture for sustainable productivity and healthy landscapes. *Environmental Science and Policy* 151, 103600. DOI: 10.1016/j.envsci.2023.103600.

Vishwakarma, K., Kumar, N., Shandilya, C., Mohapatra, S., Bhayana, S. *et al.* (2020) Revisiting plant-microbe interactions and microbial consortia application for enhancing sustainable agriculture: A review. *Frontiers in Microbiology* 11, 560406. DOI: 10.3389/fmicb.2020.560406.

Vurukonda, S.S.K.P., Fotopoulos, V. and Saeid, A. (2024) Production of a rich fertilizer base for plants from waste organic residues by microbial formulation technology. *Microorganisms* 12(3), 541. DOI: 10.3390/microorganisms12030541.

Wahab, A., Muhammad, M., Munir, A., Abdi, G., Zaman, W. *et al.* (2023) Role of arbuscular mycorrhizal fungi in regulating growth, enhancing productivity, and potentially influencing ecosystems under abiotic and biotic stresses. *Plants* 12(17), 3102. DOI: 10.3390/plants12173102.

Wang, G., Ren, Y., Bai, X., Su, Y. and Han, J. (2022) Contributions of beneficial microorganisms in soil remediation and quality improvement of medicinal plants. *Plants* 11(23), 3200. DOI: 10.3390/plants11233200.

Yu, L., Wang, Y., Wang, X., Han, S., Wang, L. *et al.* (2025) Transcriptomic, metabonomic and proteomic analyses reveal that terpenoids and flavonoids are required for *Pinus koraiensis* early defence against *Bursaphelenchus xylophilus* infection. *BMC Plant Biology* 25(1), 185. DOI: 10.1186/s12870-025-06192-8.

Yuan, M.-L., Zhang, M.-H., Shi, Z.-Y., Yang, S., Zhang, M.-G. *et al.* (2023) Arbuscular mycorrhizal fungi enhance active ingredients of medicinal plants: A quantitative analysis. *Frontiers in Plant Science* 14, 1276918. DOI: 10.3389/fpls.2023.1276918.

Zhang, Y., Chen, X., Geng, S. and Zhang, X. (2025) A review of soil waterlogging impacts, mechanisms, and adaptive strategies. *Frontiers in Plant Science* 16, 1545912. DOI: 10.3389/fpls.2025.1545912.

Zhao, Y., Cartabia, A., Lalaymia, I. and Declerck, S. (2022) Arbuscular mycorrhizal fungi and production of secondary metabolites in medicinal plants. *Mycorrhiza* 32(3), 221–256.

14 Mechanisms Underlying the Interactions Between Medicinal Plants and Rhizosphere Microorganisms: Reciprocal Influences, Ecological, and Evolutionary Contexts

SANIYA SHAKIR[1], SAMREEN ABBAS[1], MANOJ KUMAR MISHRA[2] AND NEHA VERMA[3*]

[1]*Department of Biotechnology, Institute of Engineering and Technology, Lucknow, India*
[2]*Department of Biotechnology, S.R. Institute of Management and Technology, Lucknow, India*
[3]*Department of Biotechnology, Meerut Institute of Engineering and Technology, Meerut, India*

14.1 Introduction

Medicinal plants are defined by their inherent capacity to produce a wide range of bioactive secondary metabolites (Yu *et al.*, 2024). These complex organic compounds are not essential for the plant's primary metabolic functions but serve crucial roles in defense, signaling, and adaptation. These secondary metabolites, including alkaloids, terpenoids, and flavonoids, are of immense value to human medicine, providing the basis for numerous pharmaceuticals and traditional remedies (Taghinasab and Jabaji, 2020). The concentration and profile of these metabolites are key determinants of a medicinal plant's quality and therapeutic efficacy (Yu *et al.*, 2024).

Beyond their direct utility to humans, medicinal plants are integral components of diverse ecosystems, where they contribute to soil health and overall biodiversity (Mohanram and Kumar, 2019). The cultivation of medicinal plants faces significant challenges, including a global decline in

*Corresponding author: neha.emo@gmail.com

 Medicinal Plants and the Rhizosphere (eds M.K. Mishra and S. Tiwari)
DOI: 10.1079/9781836990475.0014

soil fertility and a heavy reliance on synthetic chemical inputs such as fertilizers and pesticides. This dependency poses considerable environmental risks, including soil degradation, water pollution, and loss of biodiversity (Chen *et al.*, 2024). Consequently, there is a pressing need for the development of sustainable agricultural practices that can enhance the cultivation of these crops while minimizing their ecological footprint.

The rhizosphere is a dynamic and nutrient-rich zone of soil immediately surrounding plant roots. It is recognized as a hotspot of intense biological activity, where a dense and diverse consortium of microorganisms, including bacteria, fungi, and archaea, interacts directly with the plant host. This microbial community, often referred to as the rhizobiome, is fundamental to plant health and productivity. Rhizosphere microorganisms provide a suite of essential ecosystem services that are vital for plant survival and flourishing. These services include facilitating nutrient cycling, suppressing soilborne diseases, and enhancing plant tolerance to both biotic and abiotic stresses (Mohanram and Kumar, 2019). The profound relationship between plants and their associated microbes has led to the concept of the "holobiont," where the plant and its microbiome are viewed as a single, integrated biological entity (Kumar and Dubey, 2020). This synergistic partnership, forged over evolutionary timescales, significantly expands the genomic and metabolic capabilities of the plant, enabling it to adapt to its environment and resist various threats (Quiza *et al.*, 2015).

While the general benefits of rhizosphere microbes for plant health are well established (Fig. 14.1), their specific influence on the biosynthesis and accumulation of valuable secondary metabolites in medicinal plants

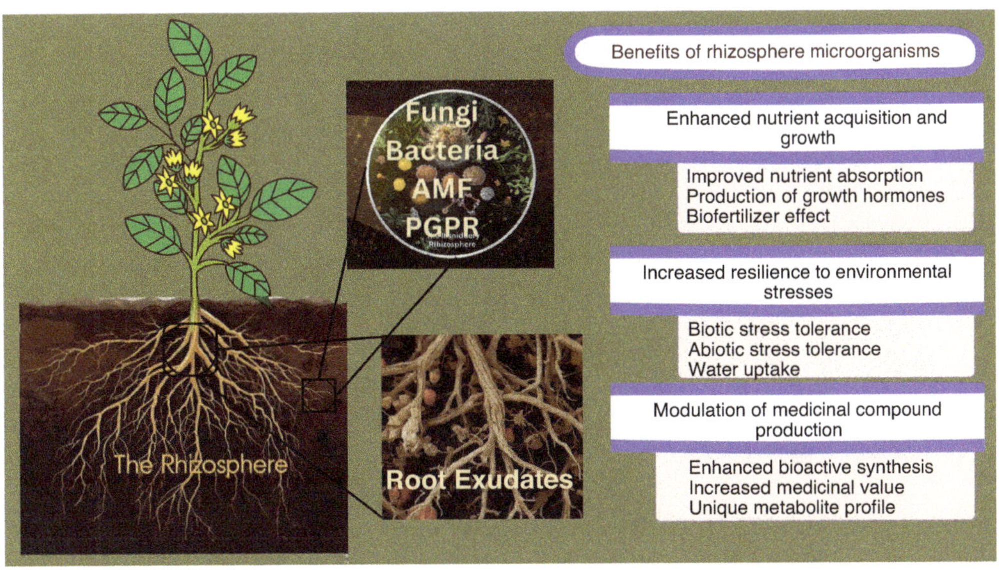

Fig. 14.1. Rhizosphere benefits for medicinal plants. This illustration details the root zone (rhizosphere), highlighting root exudates and diverse microbes that enhance plant health and medicinal compound production.

is a rapidly expanding field of research (Yu *et al.*, 2024). The chemical and biological dialogue that occurs in the rhizosphere has been found to directly impact the quality and quantity of the very compounds that define a medicinal plant's value. Understanding these intricate and mutually beneficial relationships is crucial for developing novel and sustainable agricultural strategies (Chen *et al.*, 2024). By harnessing these natural alliances, it is possible to cultivate more potent medicinal crops and reduce reliance on synthetic chemical inputs, thereby addressing the environmental challenges posed by conventional farming. This knowledge is essential for advancing global food security and environmental sustainability in an era of climate change and increasing demand for natural products (Mohanram and Kumar, 2019).

The primary aim of this chapter is to provide a comprehensive, expert-level analysis of the mechanisms, reciprocal influences, and broader ecological and evolutionary contexts of medicinal plant–rhizomicroorganism interactions. This review will synthesize recent scientific advancements, highlight key empirical case studies, and discuss the future trajectory of the field. It will also explore the immense potential of biotechnological applications and identify persistent knowledge gaps that must be addressed to unlock the full potential of these plant–microbe partnerships.

14.2 Mechanisms of Interaction Between Medicinal Plants and Rhizosphere Microorganisms

Understanding the complex interactions between medicinal plants and their environment is essential for optimizing their health and growth. These interactions influence the plants' ability to produce bioactive compounds, especially under stress conditions, which can enhance their therapeutic properties. When medicinal plants encounter environmental stresses such as drought, pests, or nutrient deficiencies, they often adapt by modifying their physiological and biochemical responses, leading to the synthesis of specific secondary metabolites that are valuable for medicinal purposes. Moreover, knowledge of how these interactions affect plant growth allows practitioners to develop standardized herbal preparations, ensuring consistency in efficacy and safety across different batches. Sustainable cultivation practices can also be guided by understanding plant responses to environmental stimuli, promoting stress-induced adaptations that strengthen plant resilience and medicinal quality. Additionally, recent research suggests that newly discovered compounds used in agriculture may influence the health of medicinal plants, potentially altering their bioactive profiles. Ultimately, by considering environmental factors and the plants' response mechanisms, it is possible to enhance both yield and therapeutic value, supporting the sustainable production of high-quality medicinal plants. Such insights are vital for advancing cultivation methods that maximize health benefits while preserving ecological balance (Sharma, 2024).

Microorganisms establish and maintain symbiotic relationships with medicinal plants through complex interactions that enhance plant health and resilience. These beneficial microbes, such as bacteria and fungi, produce a range of bioactive compounds like antibiotics, organic acids, and secondary metabolites that help plants suppress pathogenic organisms and reduce disease incidence. For example, *Pseudomonas frederiksbergensis* secretes chitinolytic enzymes and diketopiperazines, which inhibit the germination and growth of pathogenic fungi, thus directly protecting the plant. Additionally, microbes like actinomycetes and *Trichoderma harzianum* contribute to plant defense by producing antibiotics and enhancing stress tolerance respectively, leading to improved growth and photosynthetic efficiency under environmental stresses (Hou *et al.*, 2024). The in-depth mechanisms by which medicinal plants and rhizosphere microorganisms interact are described below.

14.2.1 Root exudates and microbial recruitment

The rhizosphere is a highly selective environment where plants actively shape the composition of their microbial communities (Mohanram and Kumar, 2019). This selective process is primarily driven by secretion by the plant of a diverse array of organic compounds, collectively known as rhizodeposits or root exudates, including sugars, amino acids, organic acids, vitamins, and secondary metabolites. Table 14.1 lists the major classes of root exudates, microbial responders, and their ecological functions in the rhizosphere. This intricate chemical signaling allows the plant to actively control the microbial communities in its immediate vicinity (Kumar and Dubey, 2020).

The composition and function of these root exudates are critical. They serve as a primary nutrient source, fueling the intense microbial activity that characterizes the root zone (Mohanram and Kumar, 2019). More importantly, the specific chemical profile of the exudates acts as a powerful chemoattractant, enabling the plant to selectively recruit beneficial microbial partners while simultaneously inhibiting the growth of pathogenic organisms. This is a crucial, plant-driven mechanism that establishes a favorable microbial environment from the very outset of the plant's life (Chen *et al.*, 2024).

14.2.2 Plant growth-promoting rhizobacteria and fungi

Plant growth-promoting rhizobacteria (PGPR) and beneficial fungi represent a broad and functionally diverse group of microorganisms that reside in the rhizosphere and promote plant health through an array of direct and indirect mechanisms (Díaz-Rodríguez *et al.*, 2025). Their metabolic outputs, particularly secondary metabolites, are key to these beneficial effects.

One of the most fundamental roles of PGPR is to enhance nutrient acquisition for the plant. They convert essential, but often unavailable, nutrients into forms that are accessible for plant uptake. This includes crucial processes such as nitrogen fixation, carried out by genera like

Table 14.1. Major classes of root exudates, microbial responders, and their ecological functions in the rhizosphere.

Compound type	Examples	Microbial responders	Regulation	Functional impact on plant	References
Sugars	Glucose, fructose, sucrose	*Pseudomonas*, *Bacillus*, *Azospirillum*	Serve as carbon/energy sources; chemoattractants enhance colonization	Promote root growth, enhance PGPR colonization	(Bais *et al*., 2006)
Organic acids	Citric acid, malic acid, oxalic acid	*Bacillus subtilis*, AMF (*Glomus*)	Acidify rhizosphere, solubilize P and Fe, mobilize micronutrients	Improved nutrient uptake, stress tolerance	(Nock *et al*., 2009)
Amino acids	Tryptophan, proline, glutamate	*Rhizobium*, *Bradyrhizobium*	Tryptophan is a precursor for IAA biosynthesis in microbes	Enhanced nodulation, nitrogen fixation	(Dakora and Phillips, 2002)
Flavonoids (phenolics)	Apigenin, quercetin, genistein	*Rhizobium*, antifungal bacteria	Induce nod genes in rhizobia, inhibit fungal pathogens	Symbiotic N fixation, protection from soil fungi	(Kruger and Ratcliffe, 2012)
Terpenoids	α-Pinene, β-caryophyllene	Antagonistic fungi, rhizobacteria	Antimicrobial activity; signaling molecules that shape the microbiome	Defense priming, microbial community balance	(Margalit and Kavanagh, 2015)

Coumarins	Scopoletin, umbelliferone	*Pseudomonas putida*, beneficial rhizobacteria	Released under Fe deficiency; recruit microbes with siderophore activity	Improved Fe uptake, pathogen suppression	(Zuzarte-Luís and Mota, 2018)
Fatty acids and lipids	Palmitic acid, linoleic acid	PGPR, AMF	Modify microbial membranes, act as carbon source	Stimulate colonization and stress resilience	(Jang *et al.*, 2009)
Alkaloids	Berberine, nicotine	Pathogenic fungi, rhizobacteria	Exhibit antimicrobial activity; regulate microbial density	Protect roots from pathogens	(Tiku, 2018)
Benzoxazinoids	DIMBOA, MBOA (maize)	*Pseudomonas*, *Burkholderia* spp.	Root-secreted defense metabolites, suppress pathogens, shape microbiome	Disease resistance, microbial selection	(Hu *et al.*, 2018)
Phenolic acids	Ferulic acid, caffeic acid	*Trichoderma*, rhizobacteria	Inhibit pathogenic fungi, act as carbon source for beneficials	Biocontrol, microbial recruitment	(Ahuja *et al.*, 2012)
Vitamins	Thiamine (B1), niacin (B3)	Symbiotic fungi, PGPR	Enhance microbial growth and symbiotic signaling	Increased plant vigor and defense	(Dakora and Phillips, 2002)
Exopolysaccharide precursors	Polysaccharide fragments	*Pseudomonas fluorescens*, biofilm-forming PGPR	Induce microbial biofilm formation and adhesion to roots	Stabilize root–microbe interaction, soil aggregation	(Carvalhais *et al.*, 2011)

AMF, arbuscular mycorrhizal fungi; DIMBOA, 2,4-dihydroxy-7-methoxy-1,4-benzoxazin-3-one; IAA, indole-3-acetic acid; MBOA, 6-methoxy-2-benzoxazolinone; PGPR, plant growth-promoting rhizobia.

Rhizobium and *Azotobacter*, and the solubilization of phosphate and iron, which are often limited in soil. Microbes achieve iron solubilization by secreting high-affinity iron-chelating compounds known as siderophores (Chen *et al.*, 2024).

Beyond nutrient provision, these beneficial microbes are prolific producers of phytohormones, a class of plant growth regulators. Phytohormones such as indole-3-acetic acid (IAA), gibberellins, and cytokinins, produced by various rhizosphere bacteria and fungi, have direct and profound influences on plant physiology. They promote root elongation and branching, which in turn increases the efficiency of nutrient and water uptake, and ultimately boosts overall plant growth and vigour (Chen *et al.*, 2024).

Furthermore, PGPR and fungi possess significant biocontrol potential against phytopathogens. They protect plants by producing a variety of antimicrobial compounds, including antibiotics and volatile organic compounds (VOCs). For example, some *Pseudomonas* species produce 2,4-diacetylphloroglucinol (DAPG), a tricyclic phenolic molecule with potent biocontrol properties. Beneficial microbes also compete with pathogens for nutrients and niche space, actively suppressing disease progression (Mohanram and Kumar, 2019). The diverse metabolic strategies employed by these microorganisms are fundamental to maintaining a healthy and resilient plant.

14.2.3 Mycorrhizal and endophytic symbioses

Plants also engage in more intimate and prolonged symbiotic relationships that extend beyond the rhizosphere. These partnerships involve mycorrhizal fungi and endophytic microorganisms that colonize the internal tissues of the plant (Wijesekara and Xu, 2023). Arbuscular mycorrhizal fungi (AMF) form a mutually beneficial relationship with an estimated 80% of all land plants, including many important medicinal species (Díaz-Rodríguez *et al.*, 2025). The fungal hyphae form an extensive network that extends far beyond the plant root system, significantly enhancing the uptake of immobile nutrients, particularly phosphorus and micronutrients like iron (Kumar and Dubey, 2020). In return, the plant provides the fungi with carbohydrates produced during photosynthesis. In addition to nutrient exchange, these symbioses confer enhanced resistance to a wide range of abiotic stresses, including drought, salinity, and heavy metal toxicity. For example, the medicinal plant *Salvia miltiorrhiza* has been shown to boost its stress resistance through signals transmitted via AMF hyphal networks after exposure to pathogen stress (Jain *et al.*, 2024).

Endophytic fungi and bacteria are microorganisms that live inside the living tissues of plants for a significant portion of their life cycle without causing any apparent disease symptoms (Wijesekara and Xu, 2023). Unlike

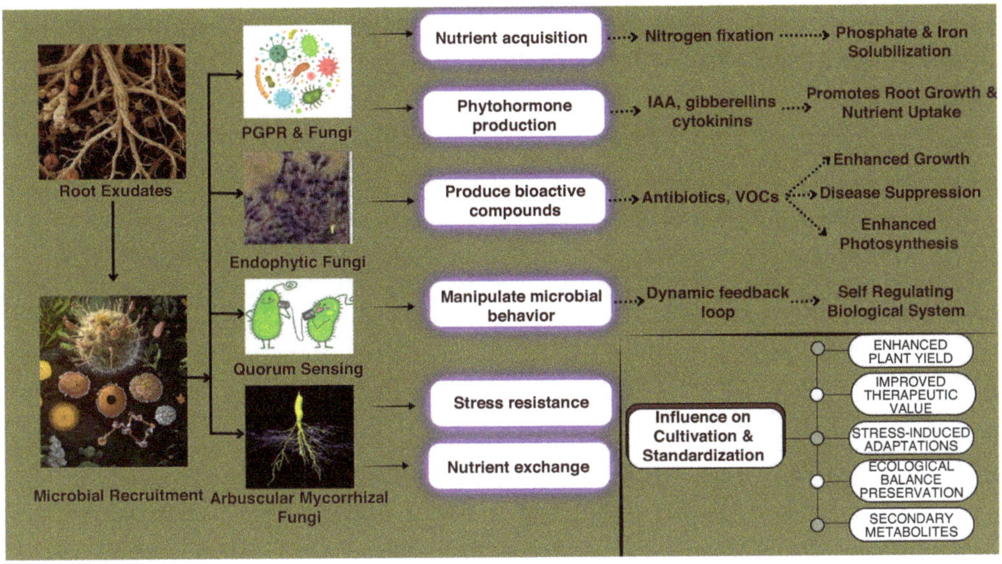

Fig. 14.2. Overview of interaction between medicinal plants and rhizosphere microorganisms.

mycorrhizal fungi, they can inhabit various plant parts, including roots, stems, leaves, and seeds (Wijesekara and Xu, 2023). Endophytes are of particular interest due to their remarkable ability to produce a wide range of bioactive compounds and secondary metabolites, many of which are chemically identical to those produced by their host plants (Yu *et al.*, 2024). This has led to the hypothesis of horizontal gene transfer or parallel evolution of biosynthetic pathways. These endophytic associations are a promising and largely untapped source of novel compounds for a variety of applications in medicine and industry (Toppo *et al.*, 2023).

The mechanisms of interaction between plants and their associated microorganisms are not merely a series of linear events but a complex, multilevel feedback loop of chemical communication, as shown in Fig. 14.2. A crucial, though often overlooked, aspect of this dialogue is quorum sensing (QS), a communication mechanism used by microorganisms to coordinate collective behaviors such as biofilm formation, virulence, and symbiosis in a cell density-dependent manner (Jamil *et al.*, 2022). These QS signaling molecules are not only internal to the microbial community but can also be perceived and responded to by the plant host (Shahni *et al.*, 2025). This signifies a more sophisticated, multikingdom dialogue where the plant can potentially manipulate microbial community behavior to its advantage. This intricate system of reciprocal communication, far from being a simple cause-and-effect relationship, represents a highly sophisticated, self-regulating biological system.

14.3 Reciprocal Influence: Microbial Impact on Medicinal Plant Growth and Metabolite Production

The concept of "reciprocal influence" captures the dynamic, bidirectional communication that takes place between plants and their associated microbial communities, particularly within the rhizosphere. Rather than a linear, one-sided interaction in which microbes merely support plant growth, this relationship represents a finely tuned feedback loop where both partners actively shape, respond, and adapt to one another. Over evolutionary time, this intimate exchange has given rise to the concept of the holobiont, wherein the plant and its microbiome are regarded not as separate entities but as a single, integrated ecological and functional unit.

Quantitative evidence of this reciprocity is exemplified in *Arabidopsis thaliana*, where distinct shifts in rhizosphere microbial composition accompany developmental transitions. During the seedling stage, the rhizosphere is enriched with Actinobacteria and Acidobacteria, phyla that are often associated with nutrient mobilization and early defences priming. As development progresses through the vegetative and bolting stages, there is a noticeable restructuring of the microbial community. By the time the plant reaches the flowering stage, Proteobacteria and Bacteroidetes emerge as dominant groups, reflecting their roles in nutrient cycling, stress tolerance, and signaling at later stages of growth (Chaparro *et al.*, 2014). These patterns of succession are visualized in Fig. 14.3, in which a stacked bar chart illustrates the dynamic assembly of the rhizosphere microbiome across developmental stages. The x-axis represents the plant's life cycle from seedling to senescence, capturing the temporal unfolding of plant–microbe interactions, while the y-axis shows the relative abundance of microbial taxa, with each color-coded segment highlighting the proportional contribution of different phyla. Together, these data underscore how plants and microbes act

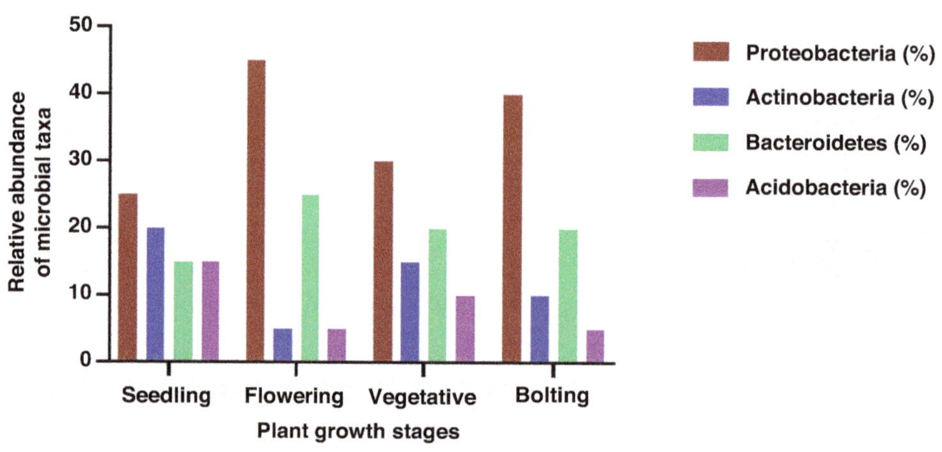

Fig. 14.3. Microbiome assembly across plant developmental stages.

in concert, constantly reshaping one another's physiology and ecology in a reciprocal dance that defines plant health and productivity.

14.3.1 Plant influence on the microbiome

Plants are not passive hosts; they actively shape the composition of their microbial communities. They do this primarily by secreting a variety of organic compounds, known as root exudates (or rhizodeposits), into the soil which serve as both a primary food source for microbes and as specific chemical signals. This intricate chemical signaling allows the plant to selectively recruit and nurture a community of beneficial microorganisms while inhibiting the growth of pathogenic ones. For example, when a plant is under stress from a pathogen or environmental condition like drought, it can alter the composition of its root exudates to attract specific microbial consortia that are known to aid in stress mitigation (Chen *et al.*, 2024).

14.3.2 Microbial influence on the plant

In return for the plant's resources, the microbial community provides a host of essential services. These beneficial microbes release a wide range of compounds and signals that directly affect the plant physiology and health. Key examples of this microbial influence include the following.

14.3.2.1 Enhancement of medicinal plant secondary metabolite synthesis

The presence of rhizosphere and endophytic microorganisms has a significant and well-documented impact on the biosynthesis and accumulation of secondary metabolites in medicinal plants, as shown in Fig. 14.4 (Yu *et al.*, 2024). Table 14.2 lists the different microbial influences on plant secondary metabolites and their mechanism of action. This beneficial influence can manifest through two primary pathways. First, certain microorganisms, particularly endophytes, have the genetic capacity to directly synthesize secondary metabolites themselves. For instance, the fungus *Streptomyces* SYP-7257, isolated from the rhizosphere of *Panax notoginseng*, can produce actinomycin X2 in its culture medium (Yu *et al.*, 2024). Second, and more commonly, microbes promote the biosynthesis of secondary metabolites within the plant itself through various indirect and integrated strategies.

Empirical studies provide compelling evidence for this phenomenon. A comprehensive analysis of *Cinnamomum migao* demonstrated a strong correlation between specific rhizobacterial genera and the accumulation of bioactive compounds in the plant fruit. The study found that genera such as *Acidothermus* and *Acidibacter* were positively correlated with the biosynthesis of important compounds like 1,8-cineole, cypressene, and limonene (Li *et al.*, 2023). Similarly, in the case of *Cannabis sativa*, microbial communities are known to modulate the production of valuable terpenes and cannabinoids.

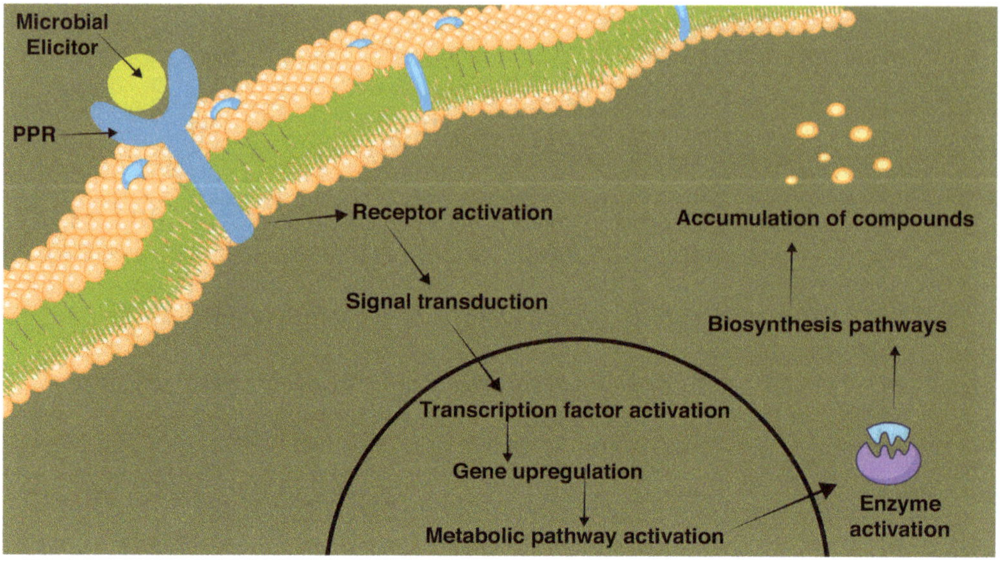

Fig. 14.4. Microbial elicitor–plant receptor interactions trigger gene cluster upregulation, leading to secondary metabolite accumulation.

They achieve this by influencing nutrient cycling, modulating plant hormone levels, and activating the plant's defense responses, such as induced systemic resistance (ISR) (FloraFlex, 2023). The presence of these microbial partners is therefore intricately linked to the complex biochemical pathways, such as the polyketide and methylerythritol 4-phosphate (MEP) pathways, that are responsible for the synthesis of these compounds (Ahsan *et al.*, 2024).

14.3.2.2 Microbial-mediated alterations in plant biochemical pathways

The influence of microorganisms on plant metabolism is not merely correlational; it is a direct result of the molecular crosstalk that occurs at the plant–microbe interface. Microbes can secrete plant hormones directly into the host plant to manipulate hormone-mediated biochemical pathways, thereby influencing growth and metabolism (Yu *et al.*, 2024). They can also activate the plant's immune signaling pathways, a process that is often linked to the promotion of secondary metabolite accumulation. A key example of this lies in the role of endophytic fungi, which have been shown to modulate the expression of genes responsible for the synthesis of key enzymes (e.g. HMGR, DXR) in the plant host. This targeted gene modulation directly affects the plant's capacity to produce important compounds like phenolics, alkaloids, and terpenoids (Toppo *et al.*, 2023). The relationship can be seen as an integrated part of the plant's stress response, where the microorganisms prime the host defense system, which is intrinsically linked to the synthesis of defensive secondary metabolites. The observed increase in metabolite yields is often a manifestation of this successful, microbe-supported stress resistance mechanism.

14.3.2.3 *Influence of rhizosphere microbiota on plant stress resistance*

Rhizosphere microorganisms are crucial for improving plant resistance to both biotic and abiotic stresses. In response to stressful conditions like drought, salinity, or pathogen attack, plants alter the chemical composition of their root exudates. This change in exudate profile enables the plant to actively attract specific microbial consortia that are known to aid in stress mitigation (Wankhade *et al.*, 2025). In a mutually beneficial feedback loop, the recruited microbes respond to these stress signals by producing a variety of protective compounds. These include osmoprotectants and other anti-stress molecules that help the plant cope with harsh environmental conditions, as well as antibiotics and other antimicrobial substances that help suppress pathogens (Chen *et al.*, 2024). This dynamic interaction highlights a symbiotic strategy where the plant leverages its microbial partners to enhance its own resilience and survival in a hostile environment.

14.3.2.4 *Nutrient provision and growth promotion*

Microbes convert essential, but often unavailable, nutrients in the soil into forms that the plant can absorb. This includes processes like nitrogen fixation and the solubilization of phosphate and iron. Many microbes produce phytohormones, such as auxins and cytokinins, which are absorbed by the plant and directly stimulate root development, nutrient uptake, and overall growth. This constant, intricate dialogue where the plant shapes its microbiome and the microbiome, in turn, influences plant growth and resilience is a fundamental aspect of the plant–microbe relationship and is integral to nutrient cycling, stress tolerance, and immune responses (Kumar and Dubey, 2020).

14.4 Ecological and Evolutionary Perspectives on Medicinal Plant–Microbe Interactions

14.4.1 Evolutionary adaptations and coevolutionary trajectories

The complex and synergistic relationships between plants and their microbiomes are the result of coevolutionary trajectories that have unfolded over millions of years. This long history has led to a highly selective relationship where plants have evolved to secrete specific compounds to recruit and maintain a particular microbial community in their rhizosphere. This coadaptation is a significant selective force that shapes the fitness of the entire holobiont, which functions as a cohesive biological unit (Quiza *et al.*, 2015). The genetic and biochemical traits of the plant, such as its root architecture, root exudate chemistry, and immune responses, have been shaped by this evolutionary pressure to foster and maintain these beneficial microbial symbioses. Conversely, the microorganisms have evolved to respond to

Table 14.2. Key rhizosphere microorganisms associated with medicinal plants and their roles in secondary metabolite modulation.

Microorganism / group	Plant host (s)	Functional role	Metabolites influenced	Mechanism of action	References
Pseudomonas fluorescens	*Curcuma longa*, general	Antagonistic activity, ISR, VOCs	Curcuminoids, antibiotics (2,4-DAPG, phenazines)	VOCs, ISR, pathogen suppression	(Haas and Défago, 2005)
Bacillus subtilis	*Withania somnifera*, general	PGPR, ISR, lipopeptide production	Withanolides, surfactins, iturins, fengycins	Induced systemic resistance, auxin modulation, antifungal lipopeptides	(Ongena and Jacques, 2008)
Glomus spp. (AMF)	*Ocimum sanctum*	Mycorrhizal symbiosis, nutrient uptake	Eugenol, flavonoids	Enhanced nutrient assimilation, altered carbon flux to secondary metabolism	(Zhao *et al*., 2022)
Trichoderma harzianum	*Artemisia annua*, general	Biocontrol, elicitor of SM pathways	Artemisinin, chitinase-inducible metabolites	Defense elicitation, JA/SA signaling, pathogen cell wall degradation	(Shoresh *et al*., 2010)
Azospirillum brasilense	*Aloe vera*, cereals	Nitrogen fixation, hormone modulation	Aloin, polysaccharides, phytohormones (IAA, gibberellins)	IAA production, root elongation, nutrient uptake	(Spaepen and Vanderleyden, 2011)
Rhizobium leguminosarum	*Glycyrrhiza uralensis*	Symbiotic nitrogen fixation, nodule formation	Glycyrrhizin, flavonoids	Symbiotic exchange of nutrients, induced nodulation	(Schwember *et al*., 2019)
Piriformospora indica	*Withania somnifera*	Endophytic fungus, stress tolerance	Withanolides	Induces stress signaling pathways, enhances metabolite biosynthesis	(Kumari *et al*., 2023)

Streptomyces spp.	*Panax ginseng*, *Zingiber officinale*	Secondary metabolite elicitation, antibiotic production	Actinomycin X2, gingerols, shogaols, taxol-like compounds	Direct biosynthesis of metabolites; elicitation of plant SM pathways	(Milojevic *et al.*, 2019)
Enterobacter cloacae	*Catharanthus roseus*	PGPR, phytohormone production (IAA)	Vincristine, vinblastine	IAA and ACC deaminase production; enhanced alkaloid biosynthesis	(Glick, 2014)
General PGPR and endophytes	Multiple	Elicitors, VOCs, siderophores, enzymes	Terpenes, alkaloids, flavonoids, phenolics	Chitin, LPS, β-glucans, siderophore-mediated Fe acquisition, VOC priming	(Richardson *et al.*, 2009)

AMF, arbuscular mycorrhizal fungi; DAPG, 2,4-diacetylphloroglucinol; IAA, indole-3-acetic acid; ISR, induced systemic resistance; JA, jasmonic acid; LPS, lipopolysaccharide; PGPR, plant growth-promoting rhizobia; SA, salicylic acid; SM, specialized metabolism; VOC, volatile organic compound.

these plant signals, providing a range of life support functions to their host (Sharma, 2024).

14.4.2 Ecological implications and soil microbiome diversity

The assembly and function of plant microbiomes are governed by a combination of external and internal factors. External, or allogenic, factors include prevailing environmental conditions such as soil type, climate, and the availability of nutrients. These factors provide the initial microbial inoculum and set the stage for community development. Internal, or autogenic, factors, such as the plant genotype, its developmental stage, and the chemistry of its root exudates, act as a powerful secondary filter that refines the microbial community, selecting for specific species and functions (Hakim *et al.*, 2021). Table 14.3 presents ecological and environmental factors which affect medicinal plant and rhizosphere microbial interactions.

The process of plant domestication and the widespread adoption of intensive agricultural practices have had a profound impact on these coevolved systems. Studies have shown that the microbiomes of modern crops are often simplified and less diverse compared to those of their wild relatives. This simplification, driven by practices such as intensive fertilization, may have inadvertently led to the loss of beneficial, coevolved microbial functions. This phenomenon directly relates to the modern challenge of low secondary metabolite production in cultivated medicinal plants (Taghinasab and Jabaji, 2020). By breeding for traits like high yield and uniformity, agricultural selection may have unintentionally discarded the symbiotic traits that recruit the very microbes responsible for enhancing secondary metabolites. The study of wild plant microbiomes and their native habitats, as proposed in the literature, represents a critical forward-looking strategy to identify and reintroduce these lost microbial functions, thereby improving the quality and resilience of cultivated medicinal crops (Sharma, 2024).

Furthermore, medicinal plants themselves can act as powerful regulators of soil microbial biodiversity. Through their specific rhizodeposition patterns, they create unique chemical environments that can enrich the microbial community, influencing the overall health and functionality of the soil ecosystem (Kumar and Dubey, 2020).

14.5 Biotechnological and Agricultural Applications

14.5.1 Use of microbial consortia for sustainable medicinal plant cultivation

The profound understanding of plant–microbe interactions offers a promising path towards developing sustainable agricultural systems, reducing the heavy reliance on synthetic chemical inputs that are detrimental to soil and water quality (Kumar and Dubey, 2020). A key application of this knowledge is the use of microbial inoculants, which are preparations containing live or dormant beneficial microorganisms (Díaz-Rodríguez *et al.*, 2025). These inoculants can be applied to the soil to enhance nutrient availability, suppress pests and diseases, and stimulate overall plant growth. Figure 14.5 represents a workflow for developing medicinal plant microbial inoculants, from selecting medicinal plants and sampling their associated microbes to using high-throughput profiling and functional assays to design beneficial synthetic microbial consortia for agricultural use.

A more advanced approach involves the use of microbial consortia or "synthetic communities" (SynComs). These are rationally designed mixtures of beneficial microbes that are intended to mimic the synergistic effects of a natural, healthy microbiome (Leggieri *et al.*, 2021). This approach moves beyond single-strain inoculants to create a more robust and functionally redundant system that is better equipped to handle diverse environmental stresses.

14.5.2 Biotechnological approaches for enhancing microbial symbiosis

Agricultural and biotechnological techniques are being developed to apply the principles of plant–microbe symbiosis to real-world cultivation. This includes the use of organic amendments like compost or well-decomposed manure, which can enrich microbial diversity and promote nutrient cycling. Companion planting, a traditional practice, can also be employed to attract specific beneficial microorganisms and create a more diverse and resilient microbial community.

A study on *Cinnamomum migao* provides a compelling example of how a scientific understanding of the links between soil properties, specific microbes, and plant-derived compounds can be used to inform targeted management strategies. By correlating rhizobacterial community composition with fruit bioactive compounds, growers can implement methods to increase the yield and quality of medicinal plant materials (Li *et al.*, 2023).

14.5.3 Microbial inoculants for improving secondary metabolite yields

The application of microbial inoculants can be specifically tailored to improve the yield and profile of secondary metabolites. Table 14.4 lists various case studies on the effect of microbial influence on secondary metabolite production in medicinal plants.In the cultivation of *Cannabis sativa*, for instance, targeted microbial inoculants can be used to augment the microbial community, thereby enhancing the production of valuable terpene and cannabinoid profiles (FloraFlex, 2023). The immense potential of this approach is further demonstrated by *in vitro* studies in which endophytic fungi have been used to successfully enhance production of host secondary metabolites,

Table 14.3. Ecological and environmental factors shaping medicinal plant and rhizosphere microbial interactions.

Factor	Impact on rhizobiome	Mechanism	Microbial examples	Plant examples	References
Soil pH	Alters microbial community composition	Changes ion solubility, affects enzyme activity	Acidic soils → Actinobacteria Neutral soils → *Rhizobium* spp.	*Glycyrrhiza glabra*, *Camellia sinensis*	(Labouyrie *et al.*, 2023)
Soil texture and organic matter	Determine microbial diversity and nutrient availability	High clay content retains organic carbon; sandy soils leach nutrients	*Pseudomonas fluorescens*, *Glomus* spp.	*Curcuma longa*, *Ocimum sanctum*	(Fierer and Jackson, 2006)
Climate (temperature and humidity)	Influences microbial metabolic activity	High humidity → fungal colonization; heat → stress-tolerant bacteria	*Trichoderma harzianum*, thermophilic *Bacillus* spp.	*Aloe vera*, *Withania somnifera*	(Bahram *et al.*, 2018)
Plant genotype	Determines root exudation profile	Wild vs domesticated plants differ in metabolites	*Bradyrhizobium* (legumes), AMF communities	*Glycine max* (soybean) vs wild *Glycine soja*	(Saleem *et al.*, 2018)
Agronomic practices	Alter microbial community stability	Fertilizers, pesticides, tillage disrupt natural microbiome	Reduced AMF under heavy fertilizer; decline in PGPR	*Ocimum sanctum*, *Artemisia annua*	(Bano *et al.*, 2021)
Biotic interactions	Shape rhizobiome through competition or facilitation	Soil fauna (nematodes, insects) regulate microbial balance	Nematode predation promotes *Pseudomonas*	*Mentha piperita*	(Islam *et al.*, 2020)

Secondary metabolites of plants	Select for specialized microbial guilds	Flavonoids and alkaloids act as selective filters	Flavonoids recruit rhizobia; alkaloids suppress fungi	*Catharanthus roseus*, *Withania somnifera*	(Badri and Vivanco, 2009)
Soil nutrient availability	Modulates microbial recruitment	Low nitrogen favors diazotrophs; low phosphorus → AMF recruitment	*Rhizobium*, *Glomus* spp.	*Cicer arietinum*, *Ocimum sanctum*	(Schilling *et al*., 1998)
Seasonality	Temporal shifts in microbial composition	Root exudates vary across growth stages	Early growth → PGPR; late stage → saprophytes	*Curcuma longa*, *Zingiber officinale*	(Cui *et al*., 2021)
Geographical location	Determines biogeographic microbial patterns	Soil origin influences core microbiome	Distinct AMF in Asia vs Europe	*Panax ginseng*	(Nemergut *et al*., 2011)

AMF, arbuscular mycorrhizal fungi; PGPR, plant growth-promoting rhizobia.

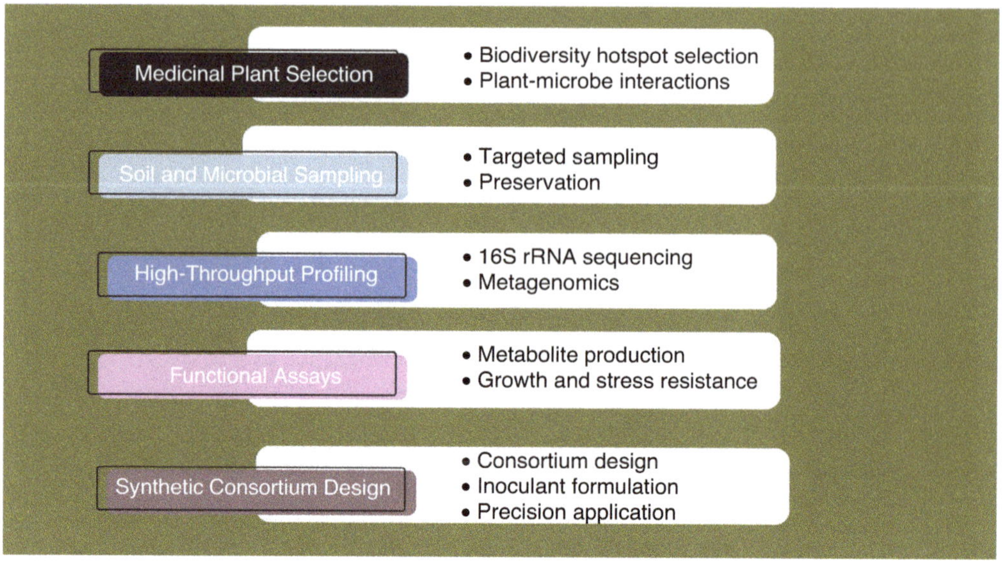

Fig. 14.5. Workflow for developing medicinal plant microbial inoculants.

as exemplified by the case of *Centella asiatica* (Shubhpriya Gupta, 2019). By isolating and culturing these beneficial microorganisms, it is possible to create a sustainable and reliable source for the production of pharmaceutically relevant compounds.

14.6 Challenges and Future Perspectives

14.6.1 Knowledge gaps and methodological constraints

Despite significant progress in understanding plant–microbe interactions, critical knowledge gaps and methodological constraints persist (Chiquito-Contreras *et al.*, 2024). A fundamental challenge is elucidating the molecular mechanisms by which a plant immune system distinguishes between beneficial symbionts and antagonistic pathogens. The precise metabolic pathways and signal transduction cascades that underpin these decisions are not yet fully understood.

Furthermore, there are substantial methodological limitations. It remains difficult to accurately replicate the complexity of natural environmental conditions in a controlled laboratory setting (Chen *et al.*, 2024; Yang *et al.*, 2024). The challenge of scaling up from simplified, binary lab interactions to the dynamic complexities of a whole ecosystem is a major obstacle. The analysis and interpretation of the massive and ever-changing datasets generated by modern research also present a significant technical and intellectual burden.

These issues highlight the need for a shift towards more sophisticated, integrated research methodologies.

14.6.2 Potential of omics and synthetic biology

The future of the field lies in the integration of advanced technologies that can overcome these inherent complexities. Omics approaches, including metagenomics, transcriptomics, and metabolomics, provide invaluable tools to move beyond single-species studies and characterize the entire microbial community's composition and functional potential at once (Table 14.5). For example, metagenomics can provide insights into a microbiome's phylogenetic composition and functional potential, while transcriptomics can reveal changes in gene expression during symbiosis (Leggieri *et al.*, 2021). Metabolomics can be used to identify the key signaling molecules that mediate the interactions between plants and microbes (Jain *et al.*, 2024).

Synthetic biology offers the ultimate potential to engineer microbiomes from the bottom up. By leveraging the vast genetic and biochemical diversity of the biosphere, researchers can design and assemble microbial consortia with predictable, robust, and a priori defined functions (Levin and Budisa, 2023). This approach is poised to create novel bio-based technologies with applications in agriculture, medicine, and environmental remediation. These next-generation tools are not merely incremental improvements; they represent a fundamental paradigm shift that is required to surmount the inherent complexity of plant–microbiome systems.

14.7 Conclusion

The relationship between medicinal plants and their rhizomicrobiome is a sophisticated, reciprocal feedback loop governed by complex molecular communication and evolutionary coadaptation. This synergistic partnership enhances plant growth, resilience, and, critically, the production of valuable secondary metabolites. The microbial community functions as an extension of the plant itself, providing essential services that are vital for its survival and productivity. By harnessing these natural alliances, we can revolutionize agricultural practices through the development of microbial inoculants and consortia. This approach offers a sustainable, eco-friendly alternative to chemical inputs, ensuring both the quality of medicinal products and the long-term health of our ecosystems.

Future research should focus on a multipronged approach that integrates omics data with ecological theory to understand the interplay of selection, dispersal, and diversification in microbiome assembly (Cordovez *et al.*, 2019). A critical need is for more robust, long-term field studies to validate the findings from laboratory and greenhouse experiments (Díaz-Rodríguez *et al.*, 2025). The goal is to develop tailored, strain-specific microbial inoculants and

Table 14.4. Case studies of microbial influence on secondary metabolite production in medicinal plants.

Medicinal plant	Associated microorganism	Enhanced secondary metabolites	Mechanism type	Mechanism	Study type	References
Cinnamomum migao	*Acidothermus*, *Acidibacter*, *Bryobacter*, *Candidatus solibacter*, *Acidimicrobiales*	1,8-cineole, cypressene, limonene, α-terpineol	Microbiome correlation	Positive correlation between specific rhizobacteria and accumulation of fruit bioactive compounds	Field microbiome-metabolome correlation	(Li *et al*., 2023)
Cannabis sativa	Rhizospheric and endophytic microbial consortia	Terpenes (limonene, myrcene), cannabinoids (CBD, THC)	Nutrient cycling, ISR, hormone modulation	Microbial communities influence metabolite pathways via improved nutrient availability and stress signaling	Pot and greenhouse trials	(Borovsky *et al*., 2020; Taghinasab and Jabaji, 2020)
Panax notoginseng	*Streptomyces* SYP-7257	Actinomycin X2, novel compounds	Direct microbial biosynthesis	Strain produces secondary metabolites in culture medium independent of plant	*In vitro* fermentation study	(Peng *et al*., 2020)
Centella asiatica	*Colletotrichum gloeosporioides*	Asiaticoside	Elicitation	Endophytic fungal elicitors trigger secondary metabolite accumulation in coculture	Tissue culture and elicitor addition	(Mandal *et al*., 2023)
Withania somnifera (ashwagandha)	*Bacillus subtilis*, *Pseudomonas fluorescens*	Withanolides	PGPR-induced signaling	PGPR enhances SM production through ISR and auxin modulation	Greenhouse and metabolomics profiling	(Ahmad *et al*., 2024)

Curcuma longa (turmeric)	*Trichoderma harzianum*, *Pseudomonas putida*	Curcuminoids	Biocontrol + VOC elicitation	VOCs and ISR enhance curcuminoid synthesis and disease resistance	Pot culture and metabolomics	(Khan *et al*., 2023)
Artemisia annua	Endophytic fungi (*Piriformospora indica*)	Artemisinin	Symbiosis and signaling	Symbiotic association improves artemisinin yield via jasmonate pathway stimulation	Greenhouse and transcriptomics	(Zheng *et al*., 2021)
Ocimum sanctum (holy basil)	AMF (*Glomus* spp.), *Trichoderma viride*	Eugenol, ursolic acid	Symbiosis and nutrient uptake	AMF improves nutrient assimilation and secondary metabolite flux	Field trial	(Şirin *et al*., 2022)
Glycyrrhiza uralensis (licorice)	*Rhizobium* spp., *Bacillus amyloliquefaciens*	Glycyrrhizin	PGPR-mediated nutrient and hormone modulation	PGPR influences saponin biosynthesis via IAA and cytokinin regulation	Field and greenhouse trials	(Chen *et al*., 2022)

AMF, arbuscular mycorrhizal fungi: CBD, cannabinoids; IAA, indole-3-acetic acid; ISR, induced systemic resistance; SM, specialized metabolism; THC, tetrahydrocannabinol; VOC, volatile organic compound.

Table 14.5. Omics technologies for studying plant–microbe interactions.

Technology	Measures	Application to plant–microbe interactions	Key output	Advantages	Limitations	References
Metagenomics	Collective genetic material of microbial communities	Characterizes taxonomic composition and functional gene potential; compares healthy and diseased plant microbiomes	Taxonomic profiles, functional gene catalogues	Culture-independent, captures unculturable microbes	Limited resolution on active functions	(Fadiji and Babalola, 2020)
Metatranscriptomics	RNA transcripts from microbial communities	Identifies active microbial genes during interaction phases (e.g. nodulation, defense)	Functional gene expression data	Reveals “who is doing what” in real time	RNA is unstable; requires high sequencing depth	(Zhang *et al.*, 2021)
Metaproteomics	Proteins produced by plants and microbes	Identifies signaling proteins, enzymes, and defense molecules	Protein catalogues, posttranslational modifications	Functional confirmation beyond RNA	Protein extraction from soil is challenging	(Pan *et al.*, 2024)
Metabolomics	Small-molecule metabolites (plant + microbe)	Identifies signaling molecules and secondary metabolites (e.g. flavonoids, alkaloids, VOCs)	Metabolic fingerprints, pathway mapping	Directly links chemical signaling to phenotype	Complex data interpretation; requires spectral libraries	(Gupta *et al.*, 2022)
Phenomics (high-throughput phenotyping)	Plant growth, physiology, and traits under microbial influence	Links plant phenotype to microbial function (e.g. growth promotion, stress tolerance)	Quantitative phenotypic data	Scalable, automated	Indirect; does not reveal molecular mechanisms	(Araus and Cairns, 2014)

Single-cell genomics	Genetic material of individual microbial cells	Identifies rare and unculturable microbes with key functional roles	Single-cell genomes	Resolves microdiversity; rare biosphere mapping	Technically demanding, costly	(de Jager and Siezen, 2011)
Synthetic biology	Engineering microbial systems	Assembles microbial consortia for plant growth, biocontrol, or metabolite production	Engineered microbial strains	Predictable functions, tailored consortia	Ethical/regulatory issues; ecological risks	(Kumar *et al.*, 2016)
CRISPR-based functional genomics	Gene editing of microbes or plants	Identifies genes essential for plant–microbe symbiosis or antagonism	Knock-out/knock-in mutants	Precise functional validation	Off-target effects; regulatory issues	(Prabhukarthikeyan *et al.*, 2020)
Multi-omics integration	Combined genomics, transcriptomics, proteomics, and metabolomics	Builds holistic models of plant–microbe interactions across scales	Systems-level interaction networks	Comprehensive; crossvalidates datasets	Requires advanced computational tools	(Jain *et al.*, 2024)

CRISPR, clustered regularly interspaced short palindromic repeats; VOC, volatile organic compound.

synthetic communities that are not only effective under controlled conditions but are also robust and efficacious in a variety of real-world agricultural systems (Zhuang *et al.*, 2023). This will require a deeper understanding of the molecular mechanisms that govern these interactions under different environmental stresses, paving the way for a new era of sustainable cultivation.

References

Ahmad, Z., Shareen, Ganie, I.B., Firdaus, F., Ramakrishnan, M., Shahzad, A. *et al.* (2024) Enhancing withanolide production in the *Withania* species: Advances in *in vitro* culture and synthetic biology approaches. *Plants* 13(15), 2171. DOI: 10.3390/PLANTS13152171.

Ahsan, S., Hoque, M., Kumar, A., Shaffique, S., Rahman, M. *et al.* (2024) Plant improvement and metabolite production in *Cannabis sativa*: Recent biotechnological advances. *Plant Trends* 2(3), 74–91.

Ahuja, I., Kissen, R. and Bones, A.M. (2012) Phytoalexins in defense against pathogens. *Trends in Plant Science* 17(2), 73–90. DOI: 10.1016/J.TPLANTS.2011.11.002.

Araus, J.L. and Cairns, J.E. (2014) Field high-throughput phenotyping: The new crop breeding frontier. *Trends in Plant Science* 19(1), 52–61. DOI: 10.1016/j.tplants.2013.09.008.

Badri, D.V. and Vivanco, J.M. (2009) Regulation and function of root exudates. *Plant, Cell and Environment* 32(6), 666–681. DOI: 10.1111/J.1365-3040.2009.01926.X.

Bahram, M., Hildebrand, F., Forslund, S.K., Anderson, J.L., Soudzilovskaia, N.A. *et al.* (2018) Structure and function of the global topsoil microbiome. *Nature* 560(7717), 233–237. DOI: 10.1038/S41586-018-0386-6.

Bais, H.P., Weir, T.L., Perry, L.G., Gilroy, S. and Vivanco, J.M. (2006) The role of root exudates in rhizosphere interactions with plants and other organisms. *Annual Review of Plant Biology* 57, 233–266. DOI: 10.1146/ANNUREV.ARPLANT.57.032905.105159.

Bano, S., Wu, X. and Zhang, X. (2021) Towards sustainable agriculture: Rhizosphere microbiome engineering. *Applied Microbiology and Biotechnology* 105(19), 7141–7160. DOI: 10.1007/S00253-021-11555-W.

Borovsky, Y., Mohan, V., Shabtai, S. and Paran, I. (2020) CaFT-LIKE is a flowering promoter in pepper and functions as florigen in tomato. *Plant Science* 301, 110678. DOI: 10.1016/J.PLANTSCI.2020.110678.

Carvalhais, L.C., Dennis, P.G., Fedoseyenko, D., Hajirezaei, M.R., Borriss, R. *et al.* (2011) Root exudation of sugars, amino acids, and organic acids by maize as affected by nitrogen, phosphorus, potassium, and iron deficiency. *Journal of Plant Nutrition and Soil Science* 174(1), 3–11. DOI: 10.1002/JPLN.201000085.

Chaparro, J.M., Badri, D.V. and Vivanco, J.M. (2014) Rhizosphere microbiome assemblage is affected by plant development. *ISME Journal* 8(4), 790–803. DOI: 10.1038/ISMEJ.2013.196.

Chen, C., Zhong, C., Gao, X., Tan, C., Bai, H. *et al.* (2022) *Glycyrrhiza uralensis* Fisch. Root-associated microbiota: The multifaceted hubs associated with environmental factors, growth status and accumulation of secondary metabolites. *Environmental Microbiomes* 17(1), 1–14. DOI: 10.1186/S40793-022-00418-0.

Chen, Q., Song, Y., An, Y., Lu, Y. and Zhong, G. (2024) Mechanisms and impact of rhizosphere microbial metabolites on crop health, traits, functional components: A comprehensive review. *Molecules* 29(24). DOI: 10.3390/MOLECULES29245922.

Chiquito-Contreras, C.J., Meza-Menchaca, T., Guzmán-López, O., Vásquez, E.C. and Ricaño-Rodríguez, J. (2024) Molecular insights into plant–microbe interactions: Comprehensive review of key mechanisms. *Frontiers in Bioscience - Elite* 16(1), 9. DOI: 10.31083/J.FBE1601009.

Cordovez, V., Dini-Andreote, F., Carrión, V.J. and Raaijmakers, J.M. (2019) Ecology and evolution of plant microbiomes. *Annual Review of Microbiology* 73, 69–88. DOI: 10.1146/ANNUREV-MICRO-090817-062524.

Cui, Z., Huntley, R.B., Zeng, Q. and Steven, B. (2021) Temporal and spatial dynamics in the apple flower microbiome in the presence of the phytopathogen *Erwinia amylovora*. *ISME Journal* 15(1), 318–329. DOI: 10.1038/S41396-020-00784-Y.

Dakora, F.D. and Phillips, D.A. (2002) Root exudates as mediators of mineral acquisition in low-nutrient environments. *Plant and Soil* 245(1), 35–47. DOI: 10.1023/A:1020809400075.

de Jager, V. and Siezen, R.J. (2011) Single-cell genomics: Unravelling the genomes of unculturable microorganisms. *Microbial Biotechnology* 4(4), 431. DOI: 10.1111/J.1751-7915.2011.00271.X.

Díaz-Rodríguez, A.M., Parra Cota, F.I., Cira Chávez, L.A., García Ortega, L.F., Estrada Alvarado, M.I. *et al.* (2025) Microbial inoculants in sustainable agriculture: Advancements, challenges, and future directions. *Plants* 14(2). DOI: 10.3390/PLANTS14020191.

Fadiji, A.E. and Babalola, O.O. (2020) Metagenomics methods for the study of plant-associated microbial communities: A review. *Journal of Microbiological Methods* 170, 105860. DOI: 10.1016/J.MIMET.2020.105860.

Fierer, N. and Jackson, R.B. (2006) The diversity and biogeography of soil bacterial communities. *Proceedings of the National Academy of Sciences USA* 103(3), 626–631. DOI: 10.1073/PNAS.0507535103.

FloraFlex (2023) The Impact of Microbial Communities on Terpene Production in Cannabis. Available at: www.floraflex.com/blogs/floraflex-media/the-impact-of-microbial-communities-on-terpene-production-in-cannabis (accessed 6 September 2025).

Glick, B.R. (2014) Bacteria with ACC deaminase can promote plant growth and help to feed the world. *Microbiological Research* 169(1), 30–39. DOI: 10.1016/J.MICRES.2013.09.009.

Gupta, S., Schillaci, M. and Roessner, U. (2022) Metabolomics as an emerging tool to study plant–microbe interactions. *Emerging Topics in Life Sciences* 6(2), 175–183. DOI: 10.1042/ETLS20210262.

Haas, D. and Défago, G. (2005) Biological control of soil-borne pathogens by fluorescent pseudomonads. *Nature Reviews Microbiology* 3(4), 307–319. DOI: 10.1038/NRMICRO1129.

Hakim, S., Naqqash, T., Nawaz, M.S., Laraib, I., Siddique, M.J. *et al.* (2021) Rhizosphere engineering with plant growth-promoting microorganisms for agriculture and ecological sustainability. *Frontiers in Sustainable Food Systems* 5, 617157. DOI: 10.3389/FSUFS.2021.617157.

Hou, M., Leng, C., Zhu, J., Yang, M., Yin, Y. *et al.* (2024) Alpine and subalpine plant microbiome mediated plants adapt to the cold environment: A systematic review. *Environmental Microbiome* 19(1), 1–18. DOI: 10.1186/S40793-024-00614-0.

Hu, L., Robert, C.A., Cadot, S., Zhang, X.I., Ye, M. *et al.* (2018) Root exudate metabolites drive plant-soil feedbacks on growth and defense by shaping the rhizosphere microbiota. *Nature Communications* 9(1), 1–13. DOI: 10.1038/S41467-018-05122-7.

Islam, W., Noman, A., Naveed, H., Huang, Z. and Chen, H.Y. (2020) Role of environmental factors in shaping the soil microbiome. *Environmental Science and Pollution Research* 27(33), 41225–41247. DOI: 10.1007/S11356-020-10471-2.

Jain, A., Sarsaiya, S., Singh, R., Gong, Q., Wu, Q. *et al.* (2024) Omics approaches in understanding the benefits of plant-microbe interactions. *Frontiers in Microbiology* 15, 1391059. DOI: 10.3389/FMICB.2024.1391059.

Jamil, F., Mukhtar, H., Fouillaud, M. and Dufossé, L. (2022) Rhizosphere signaling: Insights into plant–rhizomicrobiome interactions for sustainable agronomy. *Microorganisms* 10(5), 899. DOI: 10.3390/MICROORGANISMS10050899.

Jang, E.K., Min, K.H., Kim, S.H., Nam, S.H., Zhang, S. *et al.* (2009) Mitogen-activated protein kinase cascade in the signaling for polyamine biosynthesis in tobacco. *Plant and Cell Physiology* 50(3), 658–664. DOI: 10.1093/PCP/PCP009.

Khan, S., Ambika, A., Rani, K., Sharma, S., Kumar, A. *et al.* (2023) Rhizobacterial mediated interactions in *Curcuma longa* for plant growth and enhanced crop productivity: A systematic review. *Frontiers in Plant Science* 14, 1231676. DOI: 10.3389/FPLS.2023.1231676.

Kruger, N.J. and Ratcliffe, R.G. (2012) Pathways and fluxes: Exploring the plant metabolic network. *Journal of Experimental Botany* 63(6), 2243–2246. DOI: 10.1093/JXB/ERS073.

Kumar, A. and Dubey, A. (2020) Rhizosphere microbiome: Engineering bacterial competitiveness for enhancing crop production. *Journal of Advanced Research* 24, 337. DOI: 10.1016/J.JARE.2020.04.014.

Kumar, V., Baweja, M., Singh, P.K. and Shukla, P. (2016) Recent developments in systems biology and metabolic engineering of plant–microbe interactions. *Frontiers in Plant Science* 7, 221182. DOI: 10.3389/FPLS.2016.01421.

Kumari, P., Deepa, N., Trivedi, P.K., Singh, B.K., Srivastava, V. *et al.* (2023) Plants and endophytes interaction: A 'secret wedlock' for sustainable biosynthesis of pharmaceutically important secondary metabolites. *Microbial Cell Factories* 22(1), 1–19. DOI: 10.1186/S12934-023-02234-8.

Labouyrie, M., Ballabio, C., Romero, F., Panagos, P., Jones, A. *et al.* (2023) Patterns in soil microbial diversity across Europe. *Nature Communications* 14(1), 1–21. DOI: 10.1038/S41467-023-37937-4.

Leggieri, P.A., Liu, Y., Hayes, M., Connors, B., Seppälä, S. *et al.* (2021) Integrating systems and synthetic biology to understand and engineer microbiomes. *Annual Review of Biomedical Engineering* 23, 169. DOI: 10.1146/ANNUREV-BIOENG-082120-022836.

Levin, D.B. and Budisa, N. (2023) Synthetic biology encompasses metagenomics, ecosystems, and biodiversity sustainability within its scope. *Frontiers in Synthetic Biology* 1, 1255472. DOI: 10.3389/FSYBI.2023.1255472.

Li, L., Yang, X., Tong, B., Wang, D., Tian, X. *et al.* (2023) Rhizobacterial compositions and their relationships with soil properties and medicinal bioactive ingredients in *Cinnamomum migao*. *Frontiers in Microbiology* 14, 1078886. DOI: 10.3389/FMICB.2023.1078886.

Mandal, S., Das, T., Nandy, S., Ghorai, M., Saha, S.C. *et al.* (2023) Biotechnological and endophytic-mediated production of centellosides in *Centella asiatica*. *Applied Microbiology and Biotechnology* 107(2–3), 473–489. DOI: 10.1007/S00253-022-12316-Z.

Margalit, A. and Kavanagh, K. (2015) The innate immune response to *Aspergillus fumigatus* at the alveolar surface. *FEMS Microbiology Reviews* 39(5), 670–687. DOI: 10.1093/FEMSRE/FUV018.

Milojevic, T., Albu, M., Blazevic, A., Gumerova, N., Konrad, L. *et al.* (2019) Nanoscale tungsten-microbial interface of the metal immobilizing thermoacidophilic archaeon *Metallosphaera sedula* cultivated with tungsten polyoxometalate. *Frontiers in Microbiology* 10, 415638. DOI: 10.3389/FMICB.2019.01267.

Mohanram, S. and Kumar, P. (2019) Rhizosphere microbiome: Revisiting the synergy of plant-microbe interactions. *Annals of Microbiology* 69(4), 307–320. DOI: 10.1007/S13213-019-01448-9.

Nemergut, D.R., Costello, E.K., Hamady, M., Lozupone, C., Jiang, L. *et al.* (2011) Global patterns in the biogeography of bacterial taxa. *Environmental Microbiology* 13(1), 135–144. DOI: 10.1111/J.1462-2920.2010.02315.X.

Nock, C.A., Geihofer, D., Grabner, M., Baker, P.J., Bunyavejchewin, S. *et al.* (2009) Wood density and its radial variation in six canopy tree species differing in shade-tolerance in western Thailand. *Annals of Botany* 104(2), 297–306. DOI: 10.1093/AOB/MCP118.

Ongena, M. and Jacques, P. (2008) *Bacillus lipopeptides*: Versatile weapons for plant disease biocontrol. *Trends in Microbiology* 16(3), 115–125. DOI: 10.1016/j.tim.2007.12.009.

Pan, H., Wattiez, R. and Gillan, D. (2024) Soil metaproteomics for microbial community profiling: Methodologies and challenges. *Current Microbiology* 81(8), 1–10. DOI: 10.1007/S00284-024-03781-Y.

Peng, F., Zhang, M.Y., Hou, S.Y., Chen, J., Wu, Y.Y. *et al.* (2020) Insights into *Streptomyces* spp. isolated from the rhizospheric soil of *Panax notoginseng*: Isolation, antimicrobial activity and biosynthetic potential for polyketides and non-ribosomal peptides. *BMC Microbiology* 20(1), 1–16. DOI: 10.1186/S12866-020-01832-5.

Prabhukarthikeyan, S.R., Parameswaran, C., Keerthana, U., Teli, B., Jagannadham, P.T. *et al.* (2020) Understanding the plant-microbe interactions in CRISPR/Cas9 era: Indeed a sprinting start in marathon. *Current Genomics* 21(6), 429–443. DOI: 10.2174/1389202921999200716110853.

Quiza, L., St-Arnaud, M. and Yergeau, E. (2015) Harnessing phytomicrobiome signaling for rhizosphere microbiome engineering. *Frontiers in Plant Science* 6, 507. DOI: 10.3389/FPLS.2015.00507.

Richardson, A.E., Barea, J.M., McNeill, A.M. and Prigent-Combaret, C. (2009) Acquisition of phosphorus and nitrogen in the rhizosphere and plant growth promotion by microorganisms. *Plant and Soil* 321(1–2), 305–339. DOI: 10.1007/S11104-009-9895-2.

Saleem, M., Law, A.D., Sahib, M.R., Pervaiz, Z.H. and Zhang, Q. (2018) Impact of root system architecture on rhizosphere and root microbiome. *Rhizosphere* 6, 47–51. DOI: 10.1016/J.RHISPH.2018.02.003.

Schilling, G., Gransee, A., Deuhel, A., Ležoviž, G. and Ruppel, S. (1998) Phosphorus availability, root exudates, and microbial activity in the rhizosphere. *Zeitschrift Für Pflanzenernährung Und Bodenkunde* 161(4), 465–478. DOI: 10.1002/JPLN.1998.3581610413.

Schwember, A.R., Schulze, J., Del Pozo, A. and Cabeza, R.A. (2019) Regulation of symbiotic nitrogen fixation in legume root nodules. *Plants* 8(9), 333. DOI: 10.3390/PLANTS8090333.

Shahni, Y.S., Singh, K.P., Pandey, S., Panigrahi, C.K., Das, N. *et al.* (2025) Role of quorum sensing molecules in plant-microbe interaction for sustainable agriculture: A review. *Journal of Advances in Biology and Biotechnology* 28(5), 889–903. DOI: 10.9734/JABB/2025/V28I52352.

Sharma, M.K. (2024) Exploring the biochemical profiles of medicinal plants cultivated under stressful environmental conditions. *Current Agriculture Research Journal* 12(1), 81–103. DOI: 10.12944/CARJ.12.1.07.

Shoresh, M., Harman, G.E. and Mastouri, F. (2010) Induced systemic resistance and plant responses to fungal biocontrol agents. *Annual Review of Phytopathology* 48, 21–43. DOI: 10.1146/ANNUREV-PHYTO-073009-114450.

Shubhpriya Gupta, P.C. (2019) Enhancing secondary metabolite production in medicinal plants using endophytic elicitors: A case study of *Centella asiatica* (Apiaceae) and asiaticoside. In: Hodkinson, T., Doohan, F., Saunders, M. and Murphy, B. (eds) *Endophytes for a Growing World*. Cambridge University Press, Cambridge, UK, pp. 310–327.

Şirin, E., Ertürk, Y. and Kazankaya, A. (2022) Effects of PGPR, AMF and Trichoderma applications on adaptation abilities to different biotic and abiotic conditions in medicinal and aromatic plants. *Turkish Journal of Agriculture - Food Science and Technology* 10(2), 166–173. DOI: 10.24925/TURJAF.V10I2.166-173.4615.

Song, C., Saloner, A., Fait, A., Bernstein, N. (2023) Nitrogen deficiency stimulates cannabinoid biosynthesis in medical cannabis plants by inducing a metabolic shift towards production of low-N metabolites. *Industrial Crops and Products* 202, 116969. DOI: 10.1016/j.indcrop.2023.116969.

Spaepen, S. and Vanderleyden, J. (2011) Auxin and plant-microbe interactions. *Cold Spring Harbor Perspectives in Biology* 3(4), a001438. DOI: 10.1101/CSHPERSPECT.A001438.

Taghinasab, M. and Jabaji, S. (2020) Cannabis microbiome and the role of endophytes in modulating the production of secondary metabolites: An overview. *Microorganisms* 8(3), 355. DOI: 10.3390/MICROORGANISMS8030355.

Tiku, A.R. (2018) Antimicrobial compounds and their role in plant defense. *Molecular Aspects of Plant-Pathogen Interaction* 283–307. DOI: 10.1007/978-981-10-7371-7_13.

Toppo, P., Kagatay, L.L., Gurung, A., Singla, P., Chakraborty, R. *et al.* (2023) Endophytic fungi mediates production of bioactive secondary metabolites via modulation of genes involved in key metabolic pathways and their contribution in different biotechnological sector. *3 Biotech* 13(6). DOI: 10.1007/S13205-023-03605-Z.

Wankhade, A., Wilkinson, E., Britt, D.W. and Kaundal, A. (2025) A review of plant–microbe interactions in the rhizosphere and the role of root exudates in microbiome engineering. *Applied Sciences* 15(13), 7127. DOI: 10.3390/APP15137127.

Weinert, C.H., Sonntag, F., Egert, B., Pawelzik, E., Kulling, S.E. *et al.* (2021) The effect of potassium fertilization on the metabolite profile of tomato fruit (*Solanum lycopersicum* L.). *Plant Physiology and Biochemistry* 159, 89-99. DOI: 10.1016/j.plaphy.2020.12.010.

Wijesekara, T. and Xu, B. (2023) Health-promoting effects of bioactive compounds from plant endophytic fungi. *Journal of Fungi* 9(10), 997. DOI: 10.3390/JOF9100997.

Yang, L., Qian, X., Zhao, Z., Wang, Y., Ding, G. *et al.* (2024) Mechanisms of rhizosphere plant-microbe interactions: Molecular insights into microbial colonization. *Frontiers in Plant Science* 15, 1491495. DOI: 10.3389/FPLS.2024.1491495.

Yu, J., Bai, M., Wang, C., Wu, H. and Liang, X. (2024) Regulation of secondary metabolites accumulation in medicinal plants by rhizospheric and endophytic microorganisms. *Medicinal Plant Biology* 3, 1–10.

Zhang, Y., Thompson, K.N., Branck, T., Yan, Y., Nguyen, L.H. *et al.* (2021) Metatranscriptomics for the human microbiome and microbial community functional profiling. *Annual Review of Biomedical Data Science* 4, 279–311. DOI: 10.1146/ANNUREV-BIODATASCI-031121-103035.

Zhao, Y.Y., Cartabia, A., Lalaymia, I. and Declerck, S. (2022) Arbuscular mycorrhizal fungi and production of secondary metabolites in medicinal plants. *Mycorrhiza* 32(3–4), 221–256. DOI: 10.1007/S00572-022-01079-0.

Zheng, L.P., Li, X.P., Zhou, L.L. and Wang, J.W. (2021) Endophytes in *Artemisia annua* L.: New potential regulators for plant growth and artemisinin biosynthesis. *Plant Growth Regulation* 95(3), 293–313. DOI: 10.1007/S10725-021-00751-3.

Zhuang, X., Liu, Y., Fang, N., Bai, Z. and Gao, J. (2023) Quorum sensing improves the plant growth-promoting ability of *Stenotrophomonas rhizophila* under saline-alkaline stress by enhancing its environmental adaptability. *Frontiers in Microbiology* 14, 1155081. DOI: 10.3389/FMICB.2023.1155081.

Zuzarte-Luís, V. and Mota, M.M. (2018) Parasite sensing of host nutrients and environmental cues. *Cell Host and Microbe* 23(6), 749–758. DOI: 10.1016/j.chom.2018.05.018.

Index

www.ingramcontent.com/pod-product-compliance
Lightning Source LLC
LaVergne TN
LVHW072028250926
848956LV00010B/34
* 9 7 8 1 8 3 6 9 9 0 4 5 1 *